Development in E-waste Management

This book concerns the developments in the field of e-waste management with a particular focus on urban mining, sustainability, and circular economy aspects. It explains e-waste recycling technologies, supply chain aspects, and e-waste disposal in IT industries, including health and environmental effects of e-waste recycling processes, and associated issues, challenges, and solutions. Further, it describes the economic potential of resource recovery from e-waste.

Features:

- Covers recent developments in e-waste management.
- Explores technological advances, such as nanotech from e-waste, MREW, fungal biotech, and so forth.
- Reviews electronic component recycling aspects.
- Discusses the implementation of circular economy in the e-waste sector.
- Includes urban mining and sustainability aspects of e-waste.

This book is aimed at graduate students and researchers in environmental engineering, waste management, urban mining, circular economy, waste processing, electronics, and telecommunication engineering, electrical and electronics engineering, and chemical engineering.

Development in E-waste Management

Sustainability and Circular Economy Aspects

Edited by
Biswajit Debnath, Abhijit Das,
Potluri Anil Chowdary, and Siddhartha Bhattacharyya

CRC Press
Taylor & Francis Group
Boca Raton London New York

CRC Press is an imprint of the
Taylor & Francis Group, an **informa** business

Designed cover image: © Shutterstock

First edition published 2023
by CRC Press
6000 Broken Sound Parkway NW, Suite 300, Boca Raton, FL 33487-2742

and by CRC Press
4 Park Square, Milton Park, Abingdon, Oxon, OX14 4RN

CRC Press is an imprint of Taylor & Francis Group, LLC

ISBN: 978-1-032-29507-7 (hbk)
ISBN: 978-1-032-29508-4 (pbk)
ISBN: 978-1-003-30189-9 (ebk)

DOI: 10.1201/9781003301899

Typeset in Times
by Apex CoVantage, LLC

Biswajit would like to dedicate this book to Prof. Amar Chandra Das (Bodda) and Late Dr. Sudhangsu Ranjan Dutta.

Abhijit would like to dedicate this book to his beloved wife Amrita.

Anil would like to dedicate this book to Dr. Divis Murali.

Siddhartha would like to dedicate this volume to his beloved wife Rashni.

Contents

PART I *Policies and Case Studies for Sustainable E-waste Management: Features Policies and Case Studies Adopted in Different Countries for Sustainable E-waste Management*

PART II *Technological Advancement towards Sustainable E-waste Management: Features Progression in the Technological Aspect of E-waste Management towards Sustainability*

PART III   Sustainability Aspects of E-waste Management: Features Three Pillars of Sustainability, i.e., Environment, Economic, and Social with Respect to E-waste Management

PART IV   Critical Aspects of Sustainable E-waste Management: Focuses on Relatively Unexplored but Critical Area of E-waste Management

PART V Toward Sustainable Circular Economy in E-waste Sector: Features Approaches that Dictate Sustainable Circular Economy in the E-waste Sector

About the Editors

Biswajit Debnath is Senior Research Fellow at the Chemical Engineering Department, Jadavpur University. He was Commonwealth Split-Site Scholar at Aston University, Birmingham, UK (2019 to 2020). He received his B.Tech. and M.E. in Chemical Engineering in 2013 and 2015, respectively. His area of specialization is waste valorization and sustainability with special focus on e-waste and plastic waste. His other research interests include Circular Economy, Climate Change, SDGs, Supply Chain Management, Sustainable Smart City, Environmental Chemical Engineering, etc. He has worked in UKIERI projects and published nearly 60 articles including conference proceedings, peer- reviewed journals and contributed to nearly 30 book chapters in his research area. His h-index is 12 with 572 citations. He is a CPCB certified trainer for the six waste management rules and provided training on e-waste and plastic waste rules on invitation. He has won best paper award several times in international conferences, which offered him ten invited lectures including webinars. Since December 2018, he has been one of the most read authors from his department in ResearchGate. He has completed five collaborative (unfunded) projects with colleagues in United States, Finland, Saudi Arabia, and India. He is a reviewer of reputed journals. He is the co-creator of 'La Sustaina', which is the Kolkata's first upcycled café-cum-store. He also works as a sustainability consultant for national and international clients.

Abhijit Das has received his B.Tech. (IT) from the University of Kalyani, M.Tech. (IT) from the University of Calcutta, and Ph.D. (Engineering) from the Department of CSE Jadavpur University, India.

Dr. Das has over 17 years of teaching and research experience and more than 40 publications and 3 edited books of international repute to his credit. Presently, he is serving as Associate Professor in the Department of IT, RCC Institute of Information Technology, Kolkata, India. He had been Head of the Department of IT (January 2018 to January 2020) and has convened various committees at an institutional level.

Dr. Das has organized and chaired various international conferences and seminars. He has served as a resource person in various institutes and universities and television channels at state and national levels. Currently, six scholars are working with him on different research topics like IoT, E-Waste Management, Data Science, Quantum Computing, and Object-Oriented Categorization.

Abhijit has published four patents and eight copyrights to date. He serves as a reviewer for many reputed journals and is a professional member of IEEE, IETE (Fellow), and ACM. Abhijit is a professional singer as well and frequently performs in All India Radio Kolkata, Doordarshan, and various private television channels and has more than 10 music albums on his credit.

Potluri Anil Chowdary is Managing Director of Green Waves Environmental Solutions, the only First Authorized e-waste collection and handling unit in Andhra Pradesh (unit in Visakhapatnam) since April 2015. Mr. Potluri graduated with Diploma in Environment Resource Management from Waiariki Institute of Technology (New Zealand). He completed his Master's in Science (M.Sc.) in Environmental Science branch from GITAM Institute of Science, GITAM University (India). He did his Engineering in Chemical Technology from Chaitanya Bharathi Institute of Technology, Osmania University (India). He has worked in several projects in New Zealand including plantation of native flora and Pest Management (2014); 'Love Your Water' program organized by Sustainable Coastline on cleaning of fresh and marine water bodies (2014); and Composting and Vermi Composting for the Linton Park Community Centre under the guidance of Mr. Rick Mansell (Centre Coordinator, Linton Park Community Centre, Rotorua, New Zealand) in the year 2014. Under his reign, Green Waves Environmental Solutions has won multiple national and international awards such as National Awards for its excellence in E-Waste Recycling at Indian Industry Session (at 8th Regional 3R Forum in Asia and the Pacific); Seva Puraskar award by Andhra Pradesh Pollution Control Board for its great contribution toward sensitizing the people on e-waste management and for effective recycling of e-waste on World Environmental Day 2018; IconSWM Award for its excellence in e-waste recycling at 8th International Conference on Sustainable Waste Management on 22 November 2018 in Acharya Nagarjuna University, Guntur, Andhra Pradesh. He has delivered invited lectures and participated in workshops on waste management, e-waste, upcycling, and nature conservation.

Siddhartha Bhattacharyya did his Bachelor's in Physics, Bachelor's in Optics and Optoelectronics, and Master's in Optics and Optoelectronics from University of Calcutta, India, in 1995, 1998, and 2000, respectively. He completed his Ph.D. in Computer Science and Engineering from Jadavpur University, India, in 2008. He is the recipient of the University Gold Medal from the University of Calcutta for his Master's. He is the recipient of several coveted awards including the Distinguished HoD Award and Distinguished Professor Award conferred by Computer Society of India, Mumbai Chapter, India, in 2017; the Honorary Doctorate Award (D. Litt.) from the University of South America and the South East Asian Regional Computing Confederation (SEARCC) International Digital Award ICT Educator of the Year in 2017. He has been appointed the ACM Distinguished Speaker for the tenure 2018–2020. He has been inducted into the People of ACM hall of fame by ACM, the United States, in 2020. He has been appointed as the IEEE Computer Society Distinguished Visitor for the tenure 2021–2023. He has been elected as Full Foreign Member of the Russian Academy of Natural Sciences. He has been elected as Full Fellow of The Royal Society for Arts, Manufacturers and Commerce (RSA), London, UK.

He is currently serving as Principal of Rajnagar Mahavidyalaya, Rajnagar, Birbhum and also a Scientific Advisor to Algebra University College, Zagreb, Croatia. He served as Professor in the Department of Computer Science and Engineering of Christ University, Bangalore. He served as Principal of RCC Institute of Information Technology, Kolkata, India, during 2017–2019. He has also served as Senior Research Scientist in the Faculty of Electrical Engineering and Computer Science of VSB Technical University of Ostrava, Czech Republic (2018–2019). Prior to this, he was Professor of Information Technology of RCC Institute of Information Technology, Kolkata, India. He served as Head of the Department from March 2014 to December 2016. Prior to this, he was Associate Professor of Information Technology of RCC Institute of Information Technology, Kolkata, India, from 2011 to 2014. Before that, he served as Assistant Professor in Computer Science and Information Technology of University Institute of Technology, the University of Burdwan, India, from 2005 to 2011. He was Lecturer in Information Technology of Kalyani Government Engineering College, India, during 2001–2005. He is a co-author of 6 books and the co-editor of 94 books and has more than 400 research publications in international journals and conference proceedings to his credit. He has got 2 PCTs and 19 patents to his credit. He has been the member of the organizing and technical program committees of several national and international conferences.

He is founding Chair of ICCICN 2014, ICRCICN (2015, 2016, 2017, 2018), and ISSIP (2017, 2018) (Kolkata, India). He was General Chair of several international conferences like WCNSSP 2016 (Chiang Mai, Thailand), ICACCP (2017, 2019) (Sikkim, India), ICICC 2018 (New Delhi, India), and ICICC 2019 (Ostrava, Czech Republic).

He is Associate Editor of several reputed journals including *Applied Soft Computing, IEEE Access, Evolutionary Intelligence,* and *IET Quantum Communications.* He is Editor of *International Journal of Pattern Recognition Research* and founding Editor in Chief of *International Journal of Hybrid Intelligence, Inderscience and Applied Quantum Computing, Bon View Publishing.* He has guest edited several issues with several international journals. He is serving as Series Editor of IGI Global Book Series Advances in Information Quality and Management, De Gruyter Book Series Frontiers in Computational Intelligence and Intelligent Biomedical Data Analysis, CRC Press Book Series Computational Intelligence and Applications, Quantum Machine Intelligence, and Intelligent Data Driven Technology for Sustainable Environment, Elsevier Book Series Hybrid Computational Intelligence for Pattern Analysis and Understanding, and Springer Tracts on Human Centered Computing.

His research interests include Hybrid Intelligence, Pattern Recognition, Multimedia Data Processing, Social Networks, and Quantum Computing. He is Life Fellow of Optical Society of India (OSI), India; Life Fellow of International Society of Research and Development (ISRD), UK; Fellow of Institution of Engineering and Technology (IET), UK Fellow of Institute of Electronics and Telecommunication Engineers (IETE), India; and Fellow of Institution of Engineers (IEI), India. He is also Senior Member of Institute of Electrical and Electronics Engineers (IEEE), USA; Asia-Pacific Artificial Intelligence Association (AAIA), Hong Kong; International Institute of Engineering and Technology (IETI), Hong Kong; and Association for Computing Machinery (ACM), USA. He is Life Member of Cryptology Research Society of India (CRSI), Computer Society of India (CSI),

Indian Society for Technical Education (ISTE), Indian Unit for Pattern Recognition and Artificial Intelligence (IUPRAI), Center for Education Growth and Research (CEGR), Integrated Chambers of Commerce and Industry (ICCI), and Association of Leaders and Industries (ALI). He is Member of Institution of Engineering and Technology (IET), UK; International Rough Set Society, International Association for Engineers (IAENG), Hong Kong; Computer Science Teachers Association (CSTA), USA; International Association of Academicians, Scholars, Scientists and Engineers (IAASSE), USA; Institute of Doctors Engineers and Scientists (IDES), India; The International Society of Service Innovation Professionals (ISSIP); and The Society of Digital Information and Wireless Communications (SDIWC). He is also Certified Chartered Engineer of Institution of Engineers (IEI), India. He is on the Board of Directors of International Institute of Engineering and Technology (IETI), Hong Kong.

Contributors

Chayan Acharya
Department of Film Studies
Jadavpur University
Kolkata, West Bengal, India

Rachna Arora
Deutsche Gesellschaft für Internationale
  Zusammenarbeit (GIZ) GmbH
New Delhi, India

Abhijit Banerjee
School of Liberal Arts & Humanities
O. P. Jindal Global University
Sonipat, Haryana, India

Pousali Bhattacharjyya
Gouridevi Institute of Medical
  College & Hospital
Durgapur, West Bengal, India

Suparna Bhattacharyya
Department of Chemical
  Engineering
Jadavpur University
Kolkata, West Bengal, India

Yuriy Danko
Professor, Vice-Rector for Scientific
  Work
Sumy National Agrarian University
Sumy, Ukraine
Laboratoire Genie Industriel
Université Paris-Saclay
CentraleSupélec, 91190 Gif-sur-Yvette,
  France

Sanchari Das
University of Denver
Ritchie School of Engineering and
  Computer Science
Denver, Colorado

Biswajit Debnath
Chemical Engineering Department
Jadavpur University
Kolkata, West Bengal, India
Department of Mathematics, ASTUTE
Aston University
Birmingham, West Midlands, UK

Hema Diwan
Sustainability Management
National Institute of Industrial
  Engineering (NITIE)
Mumbai, Maharashtra, India

Saswati Gharami
Chemistry Department
Jadavpur University
Kolkata, West Bengal, India

Subrata Hait
Department of Civil and Environmental
  Engineering
Indian Institute of Technology
Patna, Bihar, India

Sunil Herat
School of Engineering and Built
  Environment
Griffith University
Queensland, Australia

A.K.M. Salman Hosain
Islamic University of Technology
Electrical and Electronic Engineering
  Department
Dhaka, Bangladesh

Kavita Kanaujia
Department of Civil and Environmental
  Engineering
Indian Institute of Technology, Patna,
  Bihar, India

Prasad Kaparaju
School of Engineering and Built
    Environment
Griffith University
Queensland, Australia

Harveen Kaur
University of Delhi
Delhi, India

Rayan Kundu
Consortium of Researchers for
    Sustainable Development
Kolkata, West Bengal, India

Jeremy Lin
Director, New Grid Solution
Washington, Pennsylvania

Gautam Mehra
Deutsche Gesellschaft für
    Internationale Zusammenarbeit
    (GIZ) GmbH
New Delhi, India

Shweta Mitra
School of Engineering and Built
    Environment
Griffith University
Queensland, Australia

Dr. Anisha Modak
Gouridevi Institute Of Medical College
    & Hospital
Durgapur, India

Tathagata Mukherjee
Department of Mechanical
    Engineering
BIT Mesra
Ranchi, Jharkhand, India

Tirthankar Mukherjee
School of Engineering and Built
    Environment
Griffith University
Queensland, Australia

Rohit Panchal
Sustainability Management
National Institute of Industrial
    Engineering (NITIE)
Mumbai, Maharashtra, India

Reva Prakash
Deutsche Gesellschaft für Internationale
    Zusammenarbeit (GIZ) GmbH
New Delhi, India

Indrashis Saha
Chemical Engineering Department
Calcutta University
Kolkata, West Bengal, India

Lorenzo Sasso
School of Governance and Politics
Moscow State Institute of International
    Relations (MGIMO)
Moscow, Russia

Tetiana Shevchenko
Université Paris-Saclay
Gif-sur-Yvette, France
Marketing and Logistics Department
Sumy National Agrarian University
Sumy, Ukraine

Satoka Shimizu
Director, Japan India Women's Forum
Japan

Anju Singh
Sustainability Management
National Institute of Industrial
    Engineering (NITIE)
Mumbai, Maharashtra, India

Amber Trivedi
Department of Civil and Environmental
    Engineering
Indian Institute of Technology
Patna, Bihar, India

Kumar Upvan
Department of Civil and Environmental
    Engineering
Indian Institute of Technology
Patna, Bihar, India

Preface

With express development in technology, the advancement and demand for electronic and electrical devices have radically amplified. The exponential escalation in the use of technology to fulfill the requirements of the fast-paced information era has led to electronic and electrical devices being rejected or disposed of at a quicker rate than in the past. This has in turn resulted in e-waste becoming the fastest-growing form of solid waste. E-waste is classified as unsafe waste and can be harmful to the environment and health if not disposed of properly. Therefore, alternatives, such as exporting, landfilling, and recycling (from different economic, social, technical, and environmental criteria), are of interest to minimize the disposal of this toxic waste form into landfill sites. The low economic and environmental awareness of stakeholders regarding e-waste shows that e-waste management system initiatives must be expedited to adhere to e-waste best practices implemented nationally and internationally.

Hence, the changing global environment, exhaustion of natural resources, and increasing pressure on managing waste generated in the complex eco-system forced us to think of sustainable solutions for e-waste management ensuring resource efficiency and achieving the sustainable development goals. Circulating various types of metals and materials has become an essential task, transforming the economy from linear to circular so that the waste sector and manufacturing industries take advantage of minimized environmental impacts and better productivity, with an increased job creation and overall well-being of human beings. On the other hand, e-waste is a valuable source of secondary materials that can be used to manufacture a new product. Universities and research organizations worldwide are actively involved in researching the issue of sustainable solutions to the waste problem. However, substantial knowledge gaps exist in the scientific landscape in the e-waste management and recycling research field.

Standing in 2022, when the whole world is recovering from the COVID-19 pandemic, the implementation of a sustainable circular economy in the e-waste management sector is imperative. This book aims to disseminate the knowledge of this field to a broader audience regardless of the country's status. The development in e-waste management toward a sustainable future and proper circular economical aspect needs to be understood while keeping in mind the targets of the SDGs. In this line, the book is divided into 5 major parts containing total of 13 chapters with authors from 8 countries.

Part I describes the e-waste management policies and regulations in the EU and countries in the Asia-Pacific region covering policy topics such as EPR, ARF, and transboundary movement issues. Case studies at the country level are presented with a focus on the COVID-19 pandemic.

Part II focuses on technological advancement toward sustainable e-waste management. This section includes emerging technologies such as fungal biotechnology and nanotechnology discussing their contributions to e-waste valorization and resource efficiency.

Part III offers detailed information about the sustainability aspects of e-waste management. This section focuses on three pillars of sustainability, i.e., environment, economic, and social and offers three separate chapters addressing each pillar. Social issues such as e-waste awareness have also been addressed.

Part IV highlights two critical aspects of e-waste management, namely, Electronic Component Recycling and Security Threats from e-waste. These two are usually not discussed and are unique features of this book.

Part V contributes to the vision of industry and academia toward sustainable circular economy in the e-waste management sector with a focus on informal sector amalgamation, consumer behaviour, and chemical engineering perspective on circular economy in e-waste systems.

The book is intended for researchers, academicians, policymakers, and practitioners. This will serve as a 'holy grail' for the sustainability seekers as well as researchers as it covers multiple aspects of e-waste management with a focus on sustainability and circular economy. The editors would feel rewarded if the concepts presented in the book are used in the social cause.

May 2022

Biswajit Debnath
(Kolkata, India)

Abhijit Das
(Kolkata, India)

Potluri Anil Chowdary
(Vizag, India)

Siddhartha Bhattacharyya
(Birbhum, India)

Acknowledgments

The editors express their gratitude to CRC Press for publishing the book. We are humbled and grateful to all those who have helped to materialize all the ideas. We sincerely thank the contributing authors and the following persons who helped in bringing the volume to life.

Dr. Gagandeep Singh
Prof. Amar Chandra Das
Ms. Ankita Das
Mrs. Munmun Modak
Mr. Abhijit Debnath
Ms. Abhipsa Das
Mr. Amritesh Das
Dr. Divis Murali
Dr. Kameswar Rao
Mr. Pranesh Varma
Dr. Abhijit Banerjee
Mr. Sushanto Sen Sharma
Late Manindra Debnath
Late Guru Ramkrishna Goswami
Team Green Waves

Part I

Policies and Case Studies for Sustainable E-waste Management

Features Policies and Case Studies
Adopted in Different Countries for
Sustainable E-waste Management

1 Economic Instruments for Electronic Waste Management in India

Harveen Kaur and Jeremy Lin

CONTENTS

1.1 INTRODUCTION

With the global proliferation of innovative electronic products and accelerated product obsolescence, the generation and management of electronic waste (e-waste) have become a more serious and difficult issue (Fan, 2018). Several studies have identified e-waste as a major and potential environmental problem around the globe (Nnorom & Osibanjo, 2008; Bouvier & Wagner, 2011; Rahmani et al., 2014; Garlapati, 2016; Borthakur, 2016) because of its hazardous content and poor recycling rates (Nixon & Saphores, 2007).

DOI: 10.1201/9781003301899-2

Most developed countries find it commercially beneficial to give their e-waste to developing countries for recycling or reuse. That is because the cost of recycling a single device in developed countries like United States is $20, while the same computer can be recycled in India for just $2, resulting in a net savings of US $18 if the computer is shipped to India (Chatterjee & Kumar, 2009). It is quite evident that initiatives to manage e-waste in both developed and developing countries are ubiquitous, and the allocation of responsibility to sponsor management programmes is lacking. At present, two financing models are being adopted globally to manage e-waste, i.e. consumers' funding and producers'/manufacturers' funding. e.g. in Canada, legislations for e-waste management put the financial burden for waste management primarily on consumers and not on producers, whereas in India, it is on producers. Nevertheless, these two models can be combined as they have the potential to overlap with each other (Borthakur & Govind, 2017).

The economics of e-waste in India varies significantly from those of developed countries. In India, the economics of the e-waste stream include a complex and massive system of traders, recyclers, and collectors who make a monetary profit through repair, reuse, and recycling. And consumers or users of electronic products are not liable to pay recycling fee, rather waste collectors reimburse a positive cost to them for their discarded items. In comparison to developed nations such as the European Union, consumers or users pay a recycling expense to the waste collector. The e-waste collected by small waste collectors is sold to traders who cumulate and sort various types of waste and afterwards offer it to recyclers who then recover the precious metals. The recovered metals and constituents segregated attract exorbitant cost in the market (European Union Report, 2011).

Over the last couple of years, many developing countries, including China and India, are gradually increasing their focus on EPR to handle their rising quantity of e-waste in an eco-friendly manner (Borthakur, 2016). It is also found that Extended Producer Responsibility (EPR) is being widely used in other countries, but it has yet to reach its maximum potential in India. The Batteries (Management and Handling) Rules, 2001, is the only environmental regulation with EPR elements. However, its implementation is not yet deemed successful. Another example of EPR proposed in Indian legislation concerns plastics. As part of EPR policy, the government is accountable for legislating regulations, fixing particular collection targets, recycling rates, identifying e-waste categories, recommending data collection, and monitoring to track compliance in the implementation of EPR (Toxics Link, 2006; European Union Report, 2011).

Given the severity of the crisis, there is a pressing necessity for appropriate policy and funding systems that can enable greater accountability of stakeholders for e-waste management, as shown in Figure 1.1.

1.1.1 CHALLENGES TO E-WASTE FINANCING

While EPR-based e-waste legislation has been in place in India since 2016, the industry is only now reluctantly accepting the requirement for a comprehensive and securely funded e-waste management system. The mandatory take-back goals have necessitated the need for the creation of an evidence-based and trackable scheme. Central Pollution Control Board and State Pollution Control Boards will in any case

Regulatory Instruments	Economic Instruments	Information Instruments
• Minimum recycled content standards • Mandatory take back • Energy efficient standards • Secondary utilization rates • Product/material restrictions • Disposal restrictions	• Advance disposal fee • Material taxes • Removing subsidy on virgin material • Deposit refund scheme • Tax relief on green products.	• Environmental information labelling • Product profile for life cycle of materials • Product hazard warnings • Product durability labelling

FIGURE 1.1   Policy and funding instruments.

Source: Adapted from Toxics Link Report (2006)

be needed to screen and uphold consistency with the guidelines indicated for dismantlers, collection centers, recyclers, and Producer Responsibility Organizations (PROs).

The enactment, critically, doesn't indicate how such a framework ought to be financed, rather provides flexibility to them to plan the framework that accomplishes the overall sustainability of the environment (Sinha-Khetriwal, 2020). The major financial difficulties are outlined here.

a. **E-waste recycling of all fractions is rare:** Although regulations for e-waste cover both IT waste (laptops, tablets and PCs, etc.) and consumer electronic waste (TVs, refrigerators, washing machines, air conditioners, and lamps), there is a widespread misconception that e-waste is just about IT waste, especially PCs and cell phones, which are a gold mine. Thus, electronics such as PCs and laptops are considered for recycling whereas TVs, CRTs, and lamps are found less appealing to recyclers and consequently not accepted.

b. **Complex composition and technological development of e-products:** With technical advancements, the material structure of electronic devices is also evolving. Miniaturization, changed material configuration, and technical advancements all affect inherent material composition and the economics of reprocessing. In the course of time, non-ferrous and precious metal content in PCBs has also declined as manufacturers aim to make more economical products by substituting or reducing pricy materials. Modern circuits, for example, have a thin layer of contact between 300 and 600 nm as opposed to the dense layer of 1–2.5 μm in the 80s. There is also a big push towards plastics replacing metal components to reduce weight. Digitalization and increased hardware intelligence have been major development in the past decade. This has resulted in programmed lamps with WIFI, self-working vacuum cleaning robots, and even umbrellas with advanced circuits that empower them to associate with cell phones. In terms of reprocessing, this has made items not only more complicated to dismantle, but also more problematic to assemble as items are now more broadly distributed in urban and rural markets.

c. **Nominal compliance costs adoption by producers:** Producers are adopting minimal compliance costs. Major global Original Equipment Manufacturers (OEMs) to small distributers of electronic products firms are concentrating on keeping compliance budgets as low as possible for e-waste management and are able to cut corners wherever probable.

d. **Lack of funding for surveillance and control:** As both recyclers and producers believe that surveillance and control are vital—recyclers want surveillance to guarantee that more producers fund authorized recycling so that there is curtained informal recycling; producers expect surveillance to ensure that recyclers comply standards and are not paper-trading. However, there is no clear funding scheme available to guarantee a trusted and well-monitored process, leaving regulatory authorities liable for its absence.

e. **Poor logistics add to e-waste transportation cost:** The costs of collection and transportation are increased by poor logistics network. These factors, along with the country's recycling capability being concentrated mostly in a few metropolitan areas, make e-waste transportation costly. Despite the fact that the single tax regime and e-way bills have reduced the regulatory burden of e-waste transportation, there are still inefficiencies in the system that make processing and transportation costs exorbitant.

f. **Lower revenues due to insufficient recycling materials:** Poor recycle and recovery processes result in reduced material revenues, resulting in wider funding disparities. Many vital raw materials are currently either not recovered because they are destroyed in conventional treatment and recovery systems or they are not commercially feasible. An efficient system with a decent recovery rate captures a larger portion of the inherent value than a low recovery, high collection rate system, e.g. the informal sector receives a huge amount of e-waste but uses inadequate disposal and recovery procedures, resulting in a large proportion of the inherent value being lost.

1.1.2 E-WASTE STAKEHOLDERS AND DEGREE OF THEIR INFLUENCE IN INDIA

To understand the economies of various instruments concerning e-waste, it is important to recognize the key stakeholders of e-waste supply chain management. The key stakeholders in e-waste production and management are the consumers, manufacturers, importers, refurbishers, recyclers, and regulators (Table 1.1). Along these lines, for preparing an effective regulation, it is critical to know the interests of different stakeholders (European Union Report, 2011).

1.2 RESEARCH APPROACH

Wath et al. (2010) suggests various systems, approaches, and practices for the management of e-waste. The basic idea behind all these approaches in managing e-waste is to collect certain fees from some sources which can be used to do recycling of e-waste. The sources for these fees or funds can be the producers, consumers,

TABLE 1.1

E-waste Stakeholders and Degree of Their Influence in India

Stakeholders	Industry	Interest	Degree of influence
Producers • IT • Household appliances • Office appliances	Industry associations such as MAIT, CEAMA, TEMA, ELCINA, Large producers	To comply with legal obligations in the country at minimum cost	High
Government: **Environment sector** • Policymakers • Enforcement agencies	Ministry of Environment and Forests and Central Pollution Control Board State Pollution Control Board	Protect the environment and health of the population. Set standards for environmental norms for processing e-waste and monitor compliance	High but limited in capacity High but limited in capacity High but limited in capacity
Government: **Other sectors** • IT • Industry	Ministry of Information Technology and Ministry of Industry & Commerce	Protect the interest of the IT industry for compliance with the regulations to reduce threats caused by e-waste	Limited
Collectors	Individual scrap collectors and traders (*kabadiwallahs*)	Maintain gainful employment	Low
Local government	State Level Pollution Control Boards and Municipalities	Enforcement of the regulation and minimize interstate dumping	Low
NGOs	Toxics Link, Greenpeace, Chintan and TERI	Protection of environment and health. Safeguard interests of public. Protect livelihood for the informal sector	High
Consumers • Households • Corporate • Public	Ministry of Consumer Affairs Food and Public Distribution	Safeguard consumer rights	Low
Recyclers: **Informal sector**	Unorganized and hence not represented	Maintain gainful employment Healthy work environment	Low
Recyclers: **Formal sector**	Authorized recyclers, ERA (E-waste Recyclers Association)	Interested in business and increasing scale of operation	Low

Source: Adapted from European Union Report (2011)

government subsidies, or some types of taxes. The well-known model which collects fees from the producers is the EPR model, while Advance Recycling Fee (ARF) model collects fees directly from the consumers. These two models are most well-known while the difference between the two models is about the fundamental philosophical question on whether the producers or consumers or even both should pay for managing that e-waste. From the societal point of view, both the producers and consumers should be responsible for managing the end-of-life (EoL) electronic products which simply becomes e-waste.

The participation by both the producers and the consumers makes up the so-called 'market'. To solve this type of e-waste recycling problem, it is sensible to rely on the invisible hand of the market. However, market can fail or do not achieve the goals it intends to achieve. In this case, such as a market failure, it will fall back into the hands of the government. So, the government has to step in to resolve that societal problem.

1.3   VARIOUS APPROACHES AND ECONOMIC INSTRUMENTS FOR THE MANAGEMENT OF E-WASTE

1.3.1   EXTENDED PRODUCER RESPONSIBILITY (EPR)

The strategy of EPR has become one of the worldwide principles for the management of e-waste. It is also an application of the polluter pays principle. In nutshell, it is a statutory obligation that entails a manufacturer or producer to be responsible for the entire life pattern of its product—from product designing to last disposal (OECD, 2001; Rau et al., 2020). *Features of EPR strategy are as follows*:

- the shifting of accountability (physically and/or monetarily, completely or partly) upstream towards the producer and away from municipalities and
- to provide the impetuses to producers to include eco-friendly contemplations in the design of their products.

It also includes 'Producer Responsibility Model' (compulsory reclaim) in which it is obligatory for the retailers and manufacturers to reclaim their own items and a fractional percentage of 'orphaned electronic items' for reprocessing. In India, there exist numerous non-recognizable small- and medium-scale ventures, which produce low-cost electronic products and trade them in the white as well as the grey market.

Thus, this alternative may not be viable for certain producers as the majority of them do not have the infrastructural and financial capacities for embracing this framework. Table 1.2 sums up the different methodologies, frameworks, and practices embraced by different nations for the execution of EPR. It is clear from the table that most of the legislatures have chosen to make producers liable for reclaiming and eliminating their e-products, considering monetarily self-supporting method of framework (Wath et al., 2010).

TABLE 1.2
Various EPR Approaches

Types of EPR Approach	
Types of Tools	**Examples of EPR Applied**
Product take-back programmes	
• Mandatory take back	• Packaging (Germany)
• Voluntary or negotiated take back programmes	• Packaging (Norway, The Netherlands)
Regulatory approaches	
• Minimum product standards	• EEE, batteries
• Prohibitions of certain hazardous materials or products	• Cadmium in batteries (Sweden)
• Disposal bans	• EEE in landfills (Switzerland)
• Mandated recycling	• Packaging (Austria, Germany, Sweden)
Voluntary industry practices	
• Voluntary codes of practice	• Transport packaging (Denmark)
• Public/private servicizing partnerships	• Photocopiers, vehicles
• Leasing, servicizing, labelling	
Financial instruments	
• Deposit refund schemes	• Beverage packaging (Canada, Korea)
• Advance recycling fees	• EEE (Sweden, Switzerland)
• Fees on disposal, material taxes/subsidies	• EEE (Japan)
Some other approaches	
• Tax credits, advance recovery fee, producer responsibility model, deposit refund system	• EEE, batteries (EU countries)

Source: Adapted from OECD (2001)

1.3.2 OBJECTIVES OF EXTENDED PRODUCER RESPONSIBILITY (EPR)

The Report of the European Union (2011) outlines the following three EPR objectives (Figure 1.2).

1.4 WASTE DISPOSAL POLICIES

According to Shinkuma (2007), waste disposal policies are classified into two types: Disposal Fee (DF) Policies and Advance Disposal Fee (ADF) Policies. They vary in terms of who pays for the disposal and when. The DF policy allows to pay the charge at the time of disposal of item while in the ADF policy, consumers must pay the fee in advance, i.e. while buying the item. In several developing countries, an ADF strategy has been implemented specifically for the recycling and disposal of e-waste materials at the end of their lives.

Environmental objectives Focus on reducing natural resources, utilization through high product and material recovery. The EPR system should offer incentives to producers to design and manufacture their products in a way that improves their environmental performance.	**Economic objectives** Focus on to shifting all or a portion of the physical and economic burden of post-consumer product management from local governments to producers and other commercial operators in a product's supply chain management.	**Social objectives** An EPR-based system should lead to greater recycling under favourable environmental, health, and societal conditions and create meaningful labour-intensive jobs in the recycling sector industry.

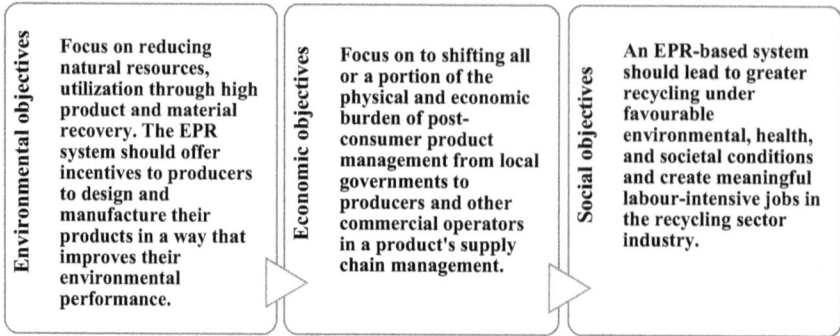

FIGURE 1.2 Objectives of Extended Producer Responsibility (EPR).

Source: European Union (2011)

Furthermore, the author has contrasted the efficacy of DF and ADF policies and concluded that implementing a DF policy could entice consumers to illegally dispose of outdated or faulty used items. Such consumer behaviour may be followed by the economically disadvantaged sections of the society, making it difficult to implement a DF strategy in developed countries such as India. Furthermore, tracking such illegal e-waste disposal could be cost-intensive. On the plus side, DF can postpone the entry of electronic items into the e-waste stream and can ensure that electronic items are used optimally by consumers (Borthakur, 2016).

1.4.1 ADVANCE RECYCLING FEE (ARF)

In contrast to the idea of Extended Producer Responsibility which puts financial responsibility on the producers of electronic products, the burden of this financial responsibility is put on the consumers in the Advance Recycling Fee (ARF) model. To facilitate the funding of a system, there is often a charge designated as an 'ecological expense', 'eco-expense', 'eco-levy', or 'advance recycling charge', etc., subject to the guidelines. It is an extra charge paid by a consumer when he purchases an electronic item. Once the product is used up and no longer required, the consumer can return the item to the retailer or manufacturer who at that point utilizes the ARF to discard the electronic item. EPR funding mechanism of advanced recycling charge as utilized in Switzerland (Africa Clean Energy Technical Assistance Facility, 2019) is shown in Figure 1.3.

Consumers in Switzerland pay remuneration for e-waste disposal, but they get no incentive for the recycling of e-waste. In India, however, *kabadiwallahs* have an efficient and well-built network in each area for door-to-door e-waste collection. Furthermore, Indian consumers don't hesitate to give *kabadiwallahs* their e-waste as they take benefit from their e-waste. For efficient e-waste management in India, ARF can significantly minimize the infrastructural and operating costs associated with the processing and transportation of e-waste through recyclers. Therefore, ARF should be explicitly disclosed on the product price as an extra cost for the expense of recycling, as remunerated by the Swiss system.

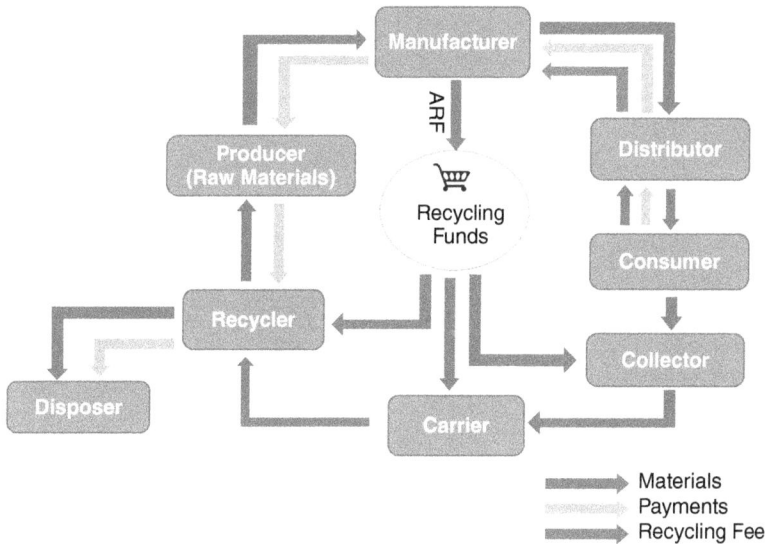

FIGURE 1.3   System financing through advance recycling fees (ARF).

Source: Africa Clean Energy Technical Assistance Facility (2019)

Wath et al. (2010) suggest that ARF should be reviewed from time to time on the basis of the category of electronic items in consultation with industry experts and in accordance to various electronics organizations such as the Electronic Industries Association of India (ELCINA), Consumer Electronics and Appliance Manufacturers Association (CEAMA), Telecom Equipment Manufacturers Association of India (TEMA), and the Manufacturers Association of Information Technology (MAIT).

The State of California, in the United States, is another State under which a funding mechanism for the collection and disposal of specific electronic waste is well implemented under Electronic Waste Recycling Act 2003 i.e., it also includes a preliminary recovery fee (ARF).

The Covered Electronic Waste (CEW) Recycling Programme is designed to make it easy for California residents to dispose of discarded products. The main feature of the legislation is the imposition of a waste disposal tax charged on retail selling of specific electronic products that have been listed as covered electronic devices (CEDs) by the Department of Toxic Substances Control (DTSC). This concept is established on the basis of the polluter pay principle.

The retailer of CEDs register with the California Department of Tax and Fee Administration (CDTFA) in order to remit the tax and keep 3% of the e-waste charge they receive as payment for price linked with tax collection. The tax was imposed to aid payment for the safe recycling of toxic materials in products such as PCs and televisions. The tax amount is determined by the diagonal scale of the video monitor. Figure 1.4 illustrates various fees/taxes imposed on video displays of varying sizes in different years (California Department of Resources Recycling and Recovery, 2020 and California Department of Tax and Fee Administration, 2021).

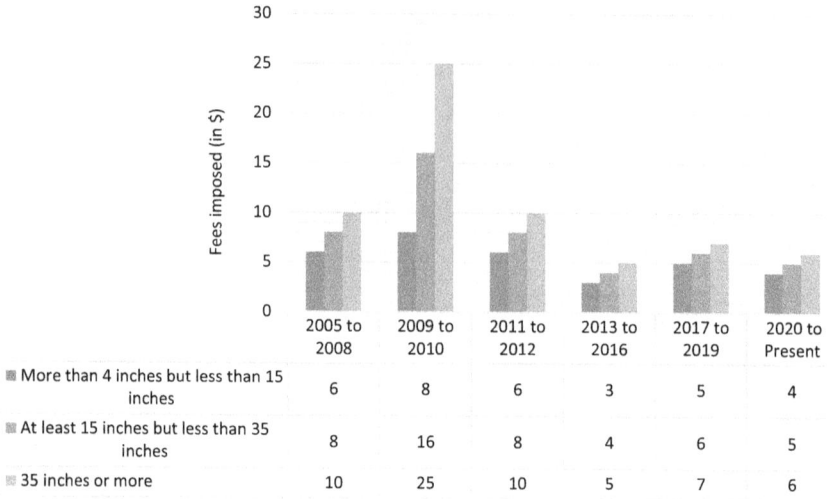

	2005 to 2008	2009 to 2010	2011 to 2012	2013 to 2016	2017 to 2019	2020 to Present
▦ More than 4 inches but less than 15 inches	6	8	6	3	5	4
▦ At least 15 inches but less than 35 inches	8	16	8	4	6	5
▦ 35 inches or more	10	25	10	5	7	6

FIGURE 1.4   Fees/taxes imposed on video displays of varying sizes in different years.

Source: Adapted from California Department of Tax and Fee Administration website

1.4.2   Deposit Refund Scheme

In this arrangement, the manufacturer charges an extra fee as a deposit at the time of sale of the product and refunds it to the consumer, along with an interest, when electronics and electrical products are returned at the end of their life (MoEFCC, 2016).

A deposit refund scheme is compliant with the EPR concept, and implementation of the deposit refund system has been introduced in many countries since EPR has gained traction globally.

In reality, the German mechanism functions in this manner. Most producers do not take physical responsibility for collection of e-waste themselves; instead, they associate themselves with producer responsibility organizations (PROs) that take care of collection and disposal while charging member companies for the operation.

These payments are calculated per pound of material and vary depending on the material type. Unfortunately, the scheme does not have a rebate on each pound of material recycled. The most well-known model of a deposit refund scheme is found in ten states in the United States for reimbursements of beverage containers. This so-called reimbursement was first enacted to address litter issues, but the strategy has since been expanded to include much more than bottles and soft drink cans. All types of cans, lead-acid batteries, tyres, motor oil, toxic chemicals, electronics, and other items are eligible for deposit returns.

Furthermore, the deposit refund technique can be used to resolve a wide range of environmental issues other than waste management, such as air and water contamination. Forty-four states in the United States have a lead-acid battery deposit refund scheme. Many of the states have employed Battery Council International (BCI) model law, which proposes that retailers can charge a $10 fee on all sold batteries, with the fee waived or returned if the customer returns the battery used within

30–45 days of buying it. The reprocessing percentage for lead-acid batteries rose to 97% due to the extensive implementation of the deposit refund solution. In addition, plastic is now being recycled in batteries at very high rates, between 60% and 80% (Walls, 2011).

In view of the increasing population size of India, short life span of the electronic products (varies from 1 to 10 years in relation to an electronic product type), and increased dynamics of the electronic industry, it is virtually impossible to maintain records of each and every product. In addition, if such a scheme is implemented, the manufacturer, recycler, and government will face increased administrative responsibility.

Thus, economic tools, such as Advance Recycling Fees (ARFs) and Advance Disposal Fees (ADFs), would alleviate the physical liability of the manufacturers for sold products in the marketplace, and revenues produced could be used to develop market for products that reached the end of their lives and may be used in many different forms as explained here (Mohana et al., 2020; Turaga, 2020).

- Incentivize people to return their e-waste at the designated sites,
- Directly finance recyclers or Producer Responsibility Organizations (PROs), and
- Support training or skill development of workers in informal sector.

1.4.3 PRODUCT TAXES

The products which cause pollution when produced, consumed, or disposed of are subject to product taxes. These taxes are intended to alter relative values and to fund collection and treatment mechanism. These taxes may take up several forms in practice, such as tax differentiation which results in lower rates for environmentally friendly goods and higher prices for 'unfriendly' products and tax differentiation between leaded and unleaded gasoline (European Union Report, 2011).

1.4.4 RECYCLING SUBSIDIES/GRANTS

Recycling subsidies cover a wide range of financial aids provided by manufacturers to municipalities and organizations, many of which are social enterprises engaged in collection and reprocessing of products. These subsidies can be granted either through payment per unit or kilogram of recycled content or through lump sum subsidies (European Union Report, 2011).

1.4.5 MATERIAL TAXES

Manufacturers must pay material taxes when they utilize materials that have a substantial environmental effect, such as virgin materials, materials that are impossible to recover, or materials that contain hazardous chemicals. Material taxes would not be needed if manufacturers use environment-friendly products. Material taxes are intended to promote the use of environment-friendly resources. Such a tax has an obvious impact on product design. The tax can be used to collect, sort, and reclaim post-consumer products (European Union Report, 2011).

1.4.6 Tax Credit Scheme

The tax credit scheme also promotes the wider use of non-toxic or non-hazardous materials by producers and manufacturers of electronic products. According to Wath et al. (2010), the tax credit subsidy should be extended to producers, manufacturers, and traders who enrich products for easy recycling or recovery and limit the use of hazardous materials in electronic products. This scheme can be beneficial for e-waste recyclers too.

Although not as popular as the EPR model, the Advance Recycling Fee (ARF) model is a good tool and has high potential if applied in the context of India. It is our fundamental belief that it is the consumers who demand for technologically advanced electronic products, so they should be responsible for managing or recycling these products when they reach their end of life. It is obvious that the consumers of those electronic products reside in India, while those products can be manufactured, produced, and transported from another country or countries. While it is difficult to levy an extra fee on those foreign companies, the retailers who reside in India should be held responsible for covering some level of fees that will be used to recycle that e-waste. Also, the final EoL products typically end up in the land of India so the focus should be on managing that e-waste by using the funds that can be collected from the consumers or the retailers. Hence, it is recommended that ARF model should be piloted and implemented at a jurisdiction in India to gauge the success of adopting that model, with the goal of reducing and recycling the e-waste.

1.5 SUGGESTED FUNDING OPTIONS FOR INDIA

In the Indian context, it is vital that the charge is small and ideally does not disturb market vitality and consumer's buying power. Otherwise, the success of framework may fail because of the absence of acceptance and rebelliousness. The following are some of the funding options for India.

a. **Visible expense:** consumers can be charged on this by the producers for handling the expense of products put in the market. (In the EU, it is utilized for orphaned products.)

b. **Recycling expense:** another tool to charge consumers at the time of acquisition of new items and identified with the future expense of reusing of items.

c. **Reimbursed expense:** it refers to an expense deposited by the consumer for the management of e-waste, which is refunded after the receipt of the obsolete product.

d. **Primary materials taxes:** duties on essential materials may act to decrease the utilization of materials, support re-use, and proliferate the utilization of reused materials. This could be an alluring alternative for improving the sustainability of materials management (European Commision, 2012).

1.5.1 COMPARISON OF FINANCIAL RESPONSIBILITY OF PRODUCER, CONSUMER, AND RECYCLER IN INDIA

Table 1.3 illustrates the financial responsibility framework for different categories of electronic products. It is evident that for all categories of products, the manufacturer has the foremost responsibility followed by the responsibility of the recycler. The consumer can be charged ARF; however, it is important that the retailer or manufacturer issues a certification card for payment of ARF along with a warranty card (as it is done in China) to the consumer. The certification card can later be utilized to get a return on the deposited amount from the formal collection system. In case, the consumer wishes to sell the product in second-hand market, the certification card can be given along with the product so that it can be handed over to the new consumer of the product (Yu et al., 2010).

Manufacturing of environment-friendly or green products by manufacturers not only benefits them (promotional benefit) but also consumers and recyclers as the consumers get an efficient product with an increased useable life, and recyclers can easily dismantle them without impacting the environment. The most challenging task is the collection and recycling of orphaned products for both the manufacturer as well as the recycler due to the responsibility to take care of them as per the regulations. ARF model, if piloted, can be helpful here for an easy discard of such products.

Information and Communication Technologies Report (2017) suggests various models for financing used e-waste, such as producers' financing (EPR), consumers' financing (polluter pays), and financing by the public and private sectors, as well as international organizations. In developing countries, a combination of the three alternatives can help in achieving the sustainability of the system.

TABLE 1.3
Comparison of Financial Responsibility of Producer, Consumer, and Recycler for Recycling in India

	Manufacturer Issue/Concerns	Consumer Issue/Concerns	Recycler Issue/Concerns
Orphan products	Should Pay	Can pay the expense in advance (invisible or visible charge	Tie up with manufacturer and producer
Environment-friendly products/green products	Initial cost of product can be kept higher to compensate for recycling expense	Encouraged to buy. If affordable, may prefer such products.	Easy to recycle
New products	Plan take back before selling	Check for buy-back policies and easy discard	Tie up with manufacturer and producer

1.5.2   Recommendations

For strengthening the economic instruments to support the recycling of e-waste in developing countries, the model implemented by the EU can be suggested which includes all aspects of e-waste management, for example, collection, transportation, and treatment expenses of e-waste (European Union Report, 2011). Key elements of the model are listed here.

- Producers are responsible for financing supply chain management of e-waste. We believe that producers are at least partially responsible for instigating that amount for e-waste which requires recycling when the electronic products reach their end of life. Also, the producers and retailers have better knowledge about the supply chain process so they are in a better position to do e-waste recycling.
- Some e-waste management frameworks utilize an advance disposal fee (either visible or invisible) which is collected by the vendor when a first-hand item is purchased by the customer. This scheme based on consumer-pay-principle was successfully implemented in both Switzerland and the State of California. So we believe that this type of scheme can be implemented successfully in India. It is also our belief that the consumers are also partially responsible for creating that e-waste.
- Intermittent charges are imposed on the producers to pay for the actual costs incurred to keep the system functional as the actual cost of recycling e-waste can be different or higher than those fees collected by the EPR or ARF model. The reason that charges are imposed only to the producers is that it is difficult to ask consumers for additional charges or fees once they bought the electronic products.
- The directive obligates that the visible fee can only be levied for historical waste.
- A hybrid system can be used for funding both historical as well as existing products.

1.6   CONCLUSION

As financial resources are generally limited in India, affordable solutions regarding e-waste disposal must be adopted whenever possible. Additionally, as Indian consumers are exceptionally cost cognizant, if the upfront payment is too high, they will refuse to buy the product. In such circumstances, consumer fees and deposit reimbursement plans can be publicized to integrate additional expenses (if any); yet, a careful consideration must be given to the Indian consumer mentality, enforceability, and price elasticity.

Other instruments such as material charges, grants (e.g. for pilot projects), landfill and incineration bans, and green procurement could be potential choices for deliberation while deciding an effective blend of mechanisms for commercializing and executing enhanced management of e-waste in developing countries.

Material bans, consistent with global guidelines such as Restriction of Hazardous Substances (RoHS) Directive, could likewise be considered in India. Finally, the usage of minimum recyclable substance principles is exceptionally significant for India, as it will encourage more recycling of components and materials used in those electronic products (in comparison to the EU, it is extremely labour-intensive).

From the perspective of businesses, consistent frameworks ought to be shaped, and the government can welcome the business to recommend their ideal model for e-waste management in the country. Collective schemes can be favoured from the administrative perspective, although mechanisms set up by producers themselves as part of their Individual Producer Responsibility (IPR) can likewise be acceptable. Tie up with third-party vendors or adoption of outsourced model for numerous services, Public-Private Partnerships (PPP) model for the execution of shared income, and equal accountability of producers and the third party should be considered as part of the overarching of compliance framework.

REFERENCES

Africa Clean Energy Technical Assistance Facility. (2019). *E-Waste Policy Handbook.* Coffey International Development Ltd..

Borthakur, A. (2016). Policy implications of e-waste in India: A review. *International Journal of Environment and Waste Management, 17*(3/4), 301. doi:10.1504/IJEWM.2016.078600

Borthakur, A., & Govind, M. (2017). Emerging trends in consumer e-waste disposal behaviour and awareness: A worldwide overview with special focus on India. *Resources, Conservation and Recycling, 117,* 102–113. doi:10.1016/j.resconrec.2016.11.011

Bouvier, R., & Wagner, T. (2011). The influence of collection facility attributes on household collection rates of electronic waste: The case of televisions and computer monitors. *Resources, Conservation and Recycling, 55*(11), 1051–1059. doi:10.1016/j.resconrec.2011.05.019

California Department of Resources Recycling and Recovery. (2020). *Electronic Waste Recycling Act of 2003.* Retrieved January 28, 2021, from www.calrecycle.ca.gov/electronics/act2003

California Department of Tax and Fee Administration. (2021). *Electronic Waste Recycling Fee.* Retrieved January 28, 2021, from www.cdtfa.ca.gov/taxes-and-fees/tax-rates-stfd.htm#elecrecyclefee

Chatterjee, S., & Kumar, K. (2009). Effective electronic waste management and recycling process involving formal and non-formal sectors. *International Journal of Physical Sciences, 4*(13), 893–905.

European Commision. (2012). *Use of Economic Instruments and Waste Management Performances.* Bio Intelligence Services.

European Union Report. (2011). *Waste Electrical and Electronic Equipment: The EU and India Sharing Best Practices.* Eu-India Action Plan Support Facility—Environment.

Fan, C., Fan, S., & Wang, C. S. et al. (2018). Modelling computer recycling in Taiwan using system dynamics. *Resources, Conservation and Recycling, 128,* 167–175.

Garlapati, V. K. (2016). E-waste in India and developed countries: Management, recycling, business and biotechnological initiatives. *Renewable and Sustainable Energy Reviews, 54,* 874–881. doi:10.1016/j.rser.2015.10.106

Information and Communication Technologies Report. (2017). *Guidelines for the Definition of a Legal Framework on Electronic Waste.* International Telecommunication Union (ITU). Geneva.

MoEFCC. (2016). *E-Waste (Management) Rules, 2016.* Ministry of Environment, Forest and Climate Change.

Mohana, R., Turaga, R., Bhaskar, K., Sinha, S., Hinchliffe, D., Hemkhaus, M., & Sharma, H. (2020). E-waste management in India: Issues and strategies. *VIKALPA: The Journal for Decision Makers, 44.* doi:10.1177/0256090919880655

Nixon, H., & Saphores, J. D. M. (2007). Financing electronic waste recycling Californian households' willingness to pay advanced recycling fees. *Journal of Environmental Management, 84*(4), 547–559. doi:10.1016/j.jenvman.2006.07.003

Nnorom, I. C., & Osibanjo, O. (2008). Overview of electronic waste (e-waste) management practices and legislations, and their poor applications in the developing countries. *Resources, Conservation and Recycling, 52*(6), 843–858. doi:10.1016/j.resconrec.2008.01.004

OECD. (2001). *Extended Producer Responsibility : A Guidance Manual for Governments.* Organisation for Economic Co-Operation and Development, p. 159. doi:10.1787/978926 4189867

Rahmani, M., Nabizadeh, R., Yaghmaeian, K., Mahvi, A. H., & Yunesian, M. (2014). Estimation of waste from computers and mobile phones in Iran. *Resources, Conservation and Recycling, 87,* 21–29. doi:10.1016/j.resconrec.2014.03.009

Rau, H., Bisnar, A. R., & Velasco, J. P. (2020). Physical responsibility versus financial responsibility of producers for e-wastes. *Sustainability (Switzerland), 12*(10). doi:10.3390/SU12104037

Shinkuma, T. (2007). Reconsideration of an advance disposal fee policy for end-of-life durable goods. *Journal of Environmental Economics and Management, 53*(1), 110–121. doi:10.1016/j.jeem.2006.07.003

Sinha-Khetriwal, D. (2020). *E-Waste Roadmap 2023 for India A Compilation of Thought Pieces.* International Finance Corporation, World Bank Group.

Toxics Link. (2006). *EPR : Sustainable Solution to Electronic Waste.* Toxics Link.

Turaga, R. M. R. (2020). *E-Waste Roadmap 2023 for India: A Compilation of Thought Pieces by Sector Experts.* International Finance Corporation, World Bank Group.

Walls, M. (2011). *Deposit-Refund Systems in Practice and Theory.* Resources for the Future.

Wath, S. B., Vaidya, A. N., Dutt, P. S., & Chakrabarti, T. (2010). A roadmap for development of sustainable e-waste management system in India. *Science of the Total Environment, 409*(1), 19–32. doi:10.1016/j.scitotenv.2010.09.030

Yu, J., Williams, E., Ju M., et al. (2010). Managing e-waste in China: Policies, pilot projects and alternative approaches. *Resources, Conservation and Recycling, 54*(11), 991–999.

2 The EU E-waste Policy and Regulation

Lorenzo Sasso

CONTENTS

2.1 INTRODUCTION

In the last decades of significant economic development and advances in technology and communication, there has been an enormous increase in the consumption of electronic and electrical equipment (EEE). Nowadays, a major percentage of the world's population connects over the Internet through electronic devices such as computers, smartphones, and tablets. Even a more significant part uses electrical equipment such as TV sets, washing machines, and vacuum cleaners. The problem is that all these EEE sooner or later—at the end of their life cycle—become electronic waste (WEEE or e-waste). Because e-waste contains materials and components of hazardous content, it can cause significant environmental and health problems if not disposed of properly. The phenomenon of e-waste has grown very fast. The Global E-waste Monitor 2020 reports that the world generated 53.6 million metric tons (Mt) of e-waste in 2019, up by 9.2 million Mt since 2014 and may surpass 74.7 million Mt by 2030 (Forti et al., 2020). Nevertheless, only 17.4 percent of the world's e-waste was collected and delivered to formal recyclers. This percentage is substantially

DOI: 10.1201/9781003301899-3

19

higher in the European Union and correspond to less than 40% (EU Parliament on 23.12.2020), with recycling practices varying from 95–81% of Switzerland and Croatia to 21% of Malta.

On the flip side, e-waste also includes traces of precious metals such as gold, silver, and copper, which can be recovered and reused into the production cycle. These precious minerals, found primarily in circuit boards, are more concentrated in e-waste than in the most productive mines. In 2016, the amount of gold recovered by e-waste equalled more than a tenth of the total gold mined. However, most methods used to extract value from e-waste have been costly, inefficient, and hazardous. However, these factors have not discouraged many operators in less developed countries, with loose regulations and controls, from "mining" e-waste for decades. As a result, places like India, Thailand, and Indonesia have recovered gold at the expense of poisoning the environment and their communities. Agbogbloshie—an impoverished commercial district in Accra (Ghana)—in 2014 reflected a concentration of hazardous elements in the environment hundred times higher than safe levels. E-waste is generally shipped from higher-income countries to lower-income countries, often under the justification of "reuse and repair" of the e-waste, to hide its illegal export (Basel Action Network. Project of Earth Economics, 2005). The cross-border transport of hazardous wastes won the media attention since the late 80s, following a series of catastrophes caused by the offload of these toxic cargoes (Greenhouse, 1988; see also incidents such as Khian Sea waste disposal case in 1986 at Haiti beach and the Koko case in 1988 in Nigeria).

Arguably, strict environmental regulations of industrialized countries pushed those in charge of waste disposal to ship hazardous waste to other more impoverished regions with more flexible, if not absent, regulation. However, the short-term financial benefits of e-waste importing from developing countries, which rely on recycling e-waste as a source of revenue, unaware of the impact these substances have on their emerging nations, should not outweigh the long-term cost to society and the environment. In the attempt to find those responsible, public opinion has pointed fingers at multinational enterprises and their business models, including "single-use" products and "planned obsolescence" of products. In the years, device life spans have been intentionally shortened as companies design their products for obsolescence by updating the design or software and discontinuing support for older models, making it cheaper and simpler to buy a new device than to fix an old one. In their defence, multinational enterprises argue that this approach stimulates profits and innovation that drive our global markets. On the other side, the problem would not exist if consumers resist the temptation to buy a new device every couple of years, even when unnecessary. Indeed, the problem of e-waste involves the responsibility of a large community across the world. The Internet generation increases worldwide every year. Therefore, the proper treatment and prevention of dangerous e-waste require an active engagement of a diverse set of actors across the borders. Those who are not part of the solution may be part of the problem.

This chapter discusses the legislative efforts of the international community in the introduction of a regulatory framework to enforce the proper environmental treatment of e-waste both at the national and international levels. The chapter focuses on the European Union's approach and regulation implemented to shape EU e-waste policy and manage e-waste at the operational level. Indeed, the European Union imposes some of the highest environmental standards in the world and, for this reason, provides an excellent case of study. The chapter continues by introducing the legal framework on e-waste existing at the macroeconomic level with a historical evolution before examining the EU regulation and directives dealing with it. Improvements made as well as criticalities and future challenges are discussed. Some final considerations conclude.

2.2 THE LEGAL FRAMEWORK AT THE MACROECONOMIC LEVEL: A HISTORICAL EVOLUTION

2.2.1 THE BASEL CONVENTION OF 1989

The first efforts to tackle the ecological problem linked to the harmfulness of hazardous waste at an international level started as early as the 70s with the 1972 UN Conference on the Environment in Stockholm, which resulted in adopting a series of principles for sound management of the environment (Stockholm Declaration). One of its significant results was the creation of the United Nations Environment Programme (UNEP). After treaty negotiations initiated by 1980 under the auspices of the UNEP, the international community approved in March 1989 the Basel Convention on the "Control of Transboundary Movements of Hazardous Wastes and their Disposal" (Basel Convention). The treaty entered into force in 1992. As of today, 187 countries have ratified it, including the EU countries.

The Basel Convention has ambitious objectives, namely, regulating transnational movements and the disposal of hazardous waste with appropriate rules and legal procedures, providing for liability for those who ignore them. The Convention also seeks to promote self-sufficiency in waste management. The restriction of the movement of toxic waste is necessary to avoid any environmental degradation and protect human health. *Waste* is defined according to Art. 2 section 1 of the Basel Convention by reference to the words *disposal* and *recovery*:

[W]aste are substances or objects which are disposed of or are intended to be disposed of or are required to be disposed of by the provisions of national law;

By linking the definition of waste to disposal operations, the Basel Convention applies an operational approach. In its Art. 2, section 4 refers to Annex IV (Sections A and B), where the disposal operations are specified and divided into operations leading or not to the possibility of resource recovery, recycling reclamation, direct reuse or alternative uses. Consequently, the Basel Convention not only tackles waste for disposal, as could be deducted from the wording of Art. 2 section 1 but also encompasses waste destined for recycling and recovery under the title "disposal operations" (Grosz, 2011).

The Basel Convention classifies hazardous waste in Article 1 and its Annexes by category and by characteristic. It applies to wastes belonging to any category contained in Annex I ("Categories of wastes to be controlled") of the Convention unless they do not possess any of the characteristics contained in Annex III named "List of hazardous characteristics". In 1998, in its fourth meeting, the Conference of the Parties (COP)—an organism established by the Convention to promote the harmonization of relevant policies and create the subsidiary body needed for the implementation of the Convention—added Annexes VIII and IX to provide further elaboration as to the wastes listed in Annexes I and III. Annex VIII contains a list of hazardous waste under Article 1(a), while Annex IX contains the list of those wastes not considered hazardous under the same article unless they do not show characteristics of elements included in Annex III. Annex III contains elements explicitly categorized as dangerous by scientists for their potential harm to the environment (Sands and Peel, 2018). Both these Annexes are regularly updated. The last changes integrated recently became effective on 1 January 2021. Besides, the Basel Convention also applies to wastes defined as dangerous by the domestic legislation of the respective Contracting States as expressed in Article 1 section 1(b).

The same Article 1 in section 3 excludes radioactive wastes from the Convention's scope because those are subject to other international control systems, as for instance the IAEA or the MARPOL 73/78 Convention, which covers wastes deriving from the regular operation of a ship. Article 3 of the Basel Convention allows each country party to the Convention to notify the other parties of any other waste considered or defined as hazardous under its national legislation that is not listed in Annexes I and II and any requirements concerning transboundary movement procedures applicable to such wastes by informing the Secretariat of the Basel Convention. To harmonize the definition of waste, the Secretariat has produced a model applicable to the control of transboundary movements of hazardous wastes and other wastes and their disposal based on different national legislations (Model National Legislation). However, this model remains a "guide" with no binding force for signatory states. Finally, the expression "transboundary movement" implicates the participation of at least two states. As a consequence, the Basel Convention does not apply, for instance, to transports of hazardous wastes to the high seas for incineration or dumping as formulated in Article 2 section 3 (Grosz, 2011).

2.2.2   THE BAN AMENDMENT

In 1995, the COP-3 formalized the total ban to the trade of hazardous waste from/to the member countries. They proposed developing a protocol on liability and compensation to assign accountability for damages occurring in a transboundary movement of hazardous waste. The Ban Amendment prohibited all countries included in Annex VII, mainly OECD members along with Liechtenstein, from exporting hazardous wastes intended for recovery, recycling, or final disposal to countries not listed in Annex VII. The reason was that countries lacking regulation, facilities, finance, and technology usually do not have the institutional capacity for managing these wastes in an environmentally sound way. Nevertheless, since its adoption, the ban amendment has been criticized by the industrialized countries perceiving it as a too harsh interference in free trade and, thus, a barrier to economic development. The effectiveness of the ban amendment is weakened by the vague and unclear conceptual division of wastes into hazardous and *other* wastes, which are household wastes and residues from their incineration enumerated in Annex II that the Basel Convention draws in Article 1 section 2. The confusion arises from the fact that "other wastes" do not mean "non hazardous" wastes, as the wording could imply (Nemeth, 2015).

The ambiguity has given exporters the flexibility to bypass the prohibition and continue their export of waste as usual under the pretence of it being "commodities". E-waste has a double classification according to the elements that compose it. In Annex VIII of the Convention, it can be found more precisely in the A1180 list, as hazardous waste, and in Annex IX, in the list B1110 non-hazardous waste. E-waste is dangerous when it contains accumulators and other batteries, mercury switches, capacitors, cathode ray tubes or when it is contaminated by lead, mercury, or cadmium. Simultaneously, the Basel Convention provides an exemption for the equipment destined for reuse to prevent waste generation, one of its prime environmental goals. Reuse prolongs the lifecycle of EEE and, therefore, decreases the generation of hazardous waste. However, the distinction of whether something is waste or not and intended for reuse has been quite controversial under the Basel Convention (Albers, 2015; Widmer et al., 2005). The difficulty in defining what is considered hazardous is further complicated because while the Convention is a static list of dangerous substances, its definition is in continuous evolution. This reason limits the efficacy of the Convention's ban since a trade-in of new hazardous substances could be considered legal only because the Convention list of dangerous substances does not cover it (Widawsky, 2008; Kummer, 1998; De La Fayette, 1995). Maybe also for those reasons, the Ban Amendment has remained ineffective for many years, lacking the minimum number of ratifying countries, to finally enter into force on 5 December 2019, becoming the new Article 4a of the Basel Convention.

2.2.3 THE POLICY APPROACH OF THE BASEL CONVENTION

The Basel Convention has a threefold target to reduce production and international circulation of dangerous waste by treating it as near to its place of origin as possible. Accordingly, the Basel Convention sets out a three-step strategy. The first step aims to develop environmentally sound management of hazardous wastes. For this reason, the Basel Convention promotes a social responsibility for treating wastes along the supply chain that implies a strict control of the storage, transport, treatment, reuse, recycle, recovery, and final disposal of wastes. Governments and policymakers have called for an "integrated lifecycle approach" for multinational manufacturers of EEE. This strategy could provide companies with the right incentives to control every step in their production processes, thereby obtaining a more practical knowledge of the actual expenses of creating hazardous wastes. Optimistically, the industry should share responsibility for the generated wastes because only the industry has the tools, technologies, and financial resources for reducing these wastes, managing them better, and helping to destroy old stocks. Consumers, of course, play a vital role. It is essential to spread public awareness of the methods adopted by manufacturers to produce their devices. Knowing these processes, a consumer must consider himself or herself as part of the problem and, therefore, by changing attitude and mentality, as a vital part of the solution.

The second step is to incentivize e-waste disposal locally. E-waste often contains dangerous elements that should not be traded freely, like ordinary commercial goods. In this way, it is possible to reduce the accidents that sometimes happen during transport and ensure these waste generators bear hazardous waste disposal costs.

Several prohibitions are designed under Article 4 of the Basel Convention to attain this aim. It is prohibited to export hazardous waste to Antarctica to a State that is not a party to the Basel Convention or a party having banned the import of hazardous wastes. Notwithstanding the prohibitions, Article 11 allows countries party to the Convention to enter bilateral or multilateral agreements on hazardous waste management with other countries party or non to the Convention, assuming that such agreements are "environmentally sound" at least as much as the standards provided by the Basel Convention (Pasquali, 2005). This particular provision has been criticized for its vagueness since the words "no less environmentally friendly" leave too much room for interpretation. A non-signatory country interested in commercial waste transactions could use this article to avoid the Convention's provisions without being refused contracts with signatory states (Vu, 1994; Sundram, 1997). Finally, the Basel Convention provides its signatory countries with a list of recommendations included in Technical Guidelines detailing the technical requirements, legislation, and infrastructure needed for optimal waste management. These Technical Guidelines support governments by improving the local treatment of hazardous waste for safety and effectiveness and reducing pressure for moving these wastes elsewhere.

The third step aims to reduce the international movements of hazardous wastes (Kummer, 2017). The regime proposed by the Basel Convention starts by enumerating the types of wastes considered hazardous and thus submitted to the regulation on transboundary movement. Then, the Convention establishes for any company or broker willing to export hazardous wastes a specific Prior Informed Consent (PIC) procedure with stringent obligations for transboundary movements of hazardous wastes and other wastes. The procedure is at the core of the Basel Convention control system and articulates in four key stages: notification, consent and issuance of movement document, transboundary movement, and confirmation of disposal. Accordingly, by written notice, the exporting State must inform, the State of import and any transit States to move such wastes. Following Articles 6, 7, and 9 of the Basel Convention, the importing and transit States must first consent in writing to the cross-border movement in writing. The manufacturer must also conclude a contract in the case of waste disposal with the person performing the transport operations. The contract must guarantee that disposal will take place in an environmentally friendly manner, and it must comply with the national laws of the countries to which the contractual parties belong (Langlet, 2009). All hazardous waste shipments made in contravention to these rules are illegal. Suppose an illegal transboundary movement of wastes violates the provisions of Articles 6 and 7 or cannot be completed as foreseen. In that case, the Convention imposes the duty to one or more states involved to ensure safe disposal by re-importing into the State of generation or otherwise following Articles 8 and 9. The Basel Convention's Articles 10 and 13 also provide collaboration and assistance between parties, ranging from exchanging information on issues relevant to the Convention's implementation to technical support, especially to developing countries.

The Convention also helps the country parties with guidance on drafting and implementing national legislation to prevent and punish illegal waste traffic. For instance, Article 14 promotes the formation of training centres and technology transfers at a regional or sub-regional level to improve the management of dangerous wastes and minimize their production. Moreover, the Secretariat cooperates with national authorities to develop legislation at the country level, set up inventories of hazardous waste, strengthen local institutions, prepare plans and policy tools for dangerous waste management, and strengthen enforcement efforts. It is always ready to coordinate with governments and international organizations to support with expertise and equipment for any emergency or hazardous waste spill. The first decade of the 21st century has witnessed several partnership programmes as the Mobile Phone Partnership Initiative (MPPI) and the Partnership for Action on Computing Equipment (PACE). The MPPI is a public–private partnership set up to develop guidelines for the environmentally sound management of end-of-life (EoL) mobile phones. Relevant pilot projects at the country level, including in companies, have been initiated. The PACE, recently re-launched in 2018 by the COP-14, aims to improve the environmentally sound management of used and EoL computing equipment.

2.2.4 THE PROTOCOL ON LIABILITY AND COMPENSATION

An outstanding issue of the Basel Convention was the lack of a harmonized liability and compensation for damages resulting from transboundary movements. At the time of the negotiations, the parties did not reach an agreement on this essential question. The parties were left free to establish their liability regime and sanctioning practices. However, this solution could give rise to divergent national rights and policies. At the same time, it was said that without coercive sanctions for the offenders, the Convention would be a meaningless piece of paper (Kummer, 1998).

An attempt to fill this vacuum was effected 10 years later, when the work around the Protocol on Liability and Compensation carried out by the COP-5 was completed and adopted on 10 December 1999 (1999 Protocol). Under Article 3, paragraph 1, the 1999 Protocol provides the first mechanism in international environmental law to assign liability and provide adequate and prompt compensation for damages resulting from the transboundary movement of hazardous wastes and other wastes, including incidents occurring because of illegal traffic in those wastes (Ajibo, 2016). However, the burden of proof for this type of liability, placed on the victim, has complicated the 1999 Protocol's efficacy. The victim must prove the damage and establish a causal link between the perpetrator's act and the damage suffered and the defendant's existence of fault or negligence (Choski, 2001).

Besides an unlimited fault-based liability for damages arising out of the carriage of waste in Article 5, the 1999 Protocol also established a strict liability regime in Article 4. Strict liability does not need to prove fault or negligence of the defendant. However, it remains to prove the damage and establish causation between the defendant's conduct and end result. The advantage of strict liability is that as long as a deliberate or negligent act caused the damage, compensation is always available in the event of an accident or misfortune. Strict liability shall not be attached to the Carrier, but to all the entities acting as a Notifier, Exporter, Importer, or waste Disposer subject to either the exporting or the importing State jurisdiction. Although the term natural or legal persons also includes States acting in a private capacity, the 1999 Protocol chooses not to impose any civil liability on States. According to Article 4 sections 1 and 2, the State notifying the transport is free from liability. At the same time, there is no explicit rule imposing subsidiary liability on the State if the liable person cannot provide sufficient compensation. That approach seems justified. The State normally fails to control and supervise this kind of illegal activities due to a lack of sufficient information. Therefore, imposing a subsidiary civil liability to the State would appear to be an excessive financial burden.

Moreover, the damage resulting under circumstances beyond the control of the liable person usually excludes liability. Unlike fault-based liability, strict liability is limited in amount and time. Article 12 refers to the financial limits indicated in Annex B to the Protocol, which exclude any interest and costs granted by the court, while Article 13 enumerates the limitation periods relating to the actions for damages. For these reasons, strict liability has been the dominant form of liability in international law not to hamper insurance availability. However, although necessary instruments to the liability regime for transnational waste movements, they appear not the most appropriate (Louka, 1993). Not surprisingly, one foremost critic of the 1999 Protocol lamented "its failure to create a fund to minimize the damage from international hazardous waste accidents under Article 14 of the Convention" (Choski, 2001). The availability of such funds is critical for developing nations that may not always be able to afford the costs of clean-up in the case of environmental damages due to the dumping of hazardous waste. This possibility was provided by Articles 14 and 15 of the 1999 Protocol, where it is expressed that supplementary financial mechanisms like compulsory insurance or the establishment of a compensation fund can support strict liability.

Another critical aspect of the 1999 Protocol concerns its overlaps with other liability and limitation regimes. Article 3(7) governs the Protocol's relationships to other regimes regulating the transboundary movement of hazardous waste, while Article 11 governs its relationships to other general regimes of liability and compensation. According to Article 3(7),

> [T]he Protocol shall not apply to damage due to an incident occurring during a transboundary movement of hazardous wastes and other wastes and their disposal pursuant to a bilateral, multilateral or regional agreement or arrangement concluded and notified in accordance with Article 11 of the Convention,

if certain conditions happen.

These conditions are: the damage occurred in an area under the national jurisdiction of any of the parties to the agreement; there exists an applicable liability and compensation regime providing at least the same standards of protection offered by the 1999 Protocol; the parties have excluded the application of the Protocol. Requiring the alternative agreement or arrangement between the parties to meet the same standards of the Basel Protocol ensures complete protection. However, the 1999 Protocol does not indicate how to determine and compare the level of protection of any alternative agreement (Albers, 2015).

The Articles 11 of the Protocol stipulates that

> [W]henever the provisions of the Protocol and the provisions of a bilateral, multilateral or regional agreement apply to liability and compensation for damage caused by an

incident arising during the same portion of a transboundary movement, the Protocol
shall not apply,

assuming that the other agreement is in force between the parties and had been
opened for signature before the Basel Protocol, even if amended afterwards.

The provision has been criticized for several reasons. It does not take into consideration the approach of Article 3(7) to avoid the application of two different legal
regimes in respect of one movement of dangerous wastes. Article 11 allows the exclusion of the 1999 Protocol only for certain stages of the transport, causing uncertainty
regarding what regime is applicable and facilitating the protraction of litigation in
these situations. Another ambiguity arises from the use of the unclear words "portion
of a transboundary movement", where extent or duration is not defined anywhere in
the Protocol or Convention. Last but not least is the adoption of a formal, temporal
rather than a qualitative condition for the application of the Protocol. The temporal
requirement practically rejects any possible future ratification of the Basel Protocol
even if it provides a more specialized and sophisticated regime of liability, such as
damages caused by PCBs or POPs. Using a qualitative requirement would have rendered the legal distinction between the two scenarios regulated by Article 3(7) and
Article 11 trivial, and one single provision could have been adopted for both cases
(Albers, 2015).

Maybe that is why the Protocol has yet to enter into force, pending ratification of
a minimum number of countries. Among the States that have not ratified the Basel
Convention on hazardous waste and, therefore, the 1999 Protocol, there is the United
States of America. This lack of support from one of the largest producers of hazardous wastes has undermined the impact of the Basel Convention and the 1999
Protocol. The effectiveness of the 1999 Protocol is also limited by the persistence of
national legal traditions (i.e. common law and civil law) regulating the many existing liability regimes worldwide and hampering a proper harmonization of the rules
and a homogeneous application of themselves. This problem would suggest whether
it would not be better to develop a global administrative law for global governance.

2.2.5   THE BAMAKO CONVENTION

In the wake of the Basel Convention, the African countries, fearing the dumping
of dangerous Western waste on African lands and the harmful consequences that
this market could provoke on nature and human health, adopted the Multilateral
Convention of Bamako on the ban on the import to Africa and the control of transboundary movement and management of hazardous wastes within Africa in 1991,
which came into force only in 1998 (Bamako Convention). The Bamako Convention
limits the importation and movement of hazardous waste into and within Africa.
Like the Basel Convention, the Bamako Convention does not explicitly define the
concept of hazardous waste. Therefore, it is up to each member country to define
what turns out to be a hazardous waste on their territory. However, unlike the Basel
Convention, the Bamako Convention extends prohibited waste and illegal activities
concerning hazardous waste disposal. Similar to the Basel Convention, two bodies have been established under Articles 15 and 16 of the Bamako Convention to

ensure the proper implementation of the Convention. These are "The Conference of the Parties" (COP) and "The Secretariat". The COP facilitates the dialogue bringing together all the signatory countries' ministers in their attributions to the environment (UNEP, 2018a). Notwithstanding the initial impetus, the COP has only met three times so far, and the first time was in 2013, 15 years after its formation (Decision of June 2013). Like the Basel Protocol, African countries party to the Convention shall introduce a civil liability regime against damages caused by the illegal import of dangerous or toxic waste into African territory or its movement to strengthen environmental protection. This legal tool should assist the States in prosecuting those engaged in such illegal activities. However, the African countries encountered the same struggles experienced in continental Europe to prove someone guilty, and the lack of an enforcement arm of the Convention has largely restrained its capacity to achieve its targets. Likewise, the definition of hazardous waste remains vague and imprecise (Assemboni-Ogunjimi, 2015; Eze, 2007). The success of the Convention also depends on the skills and technical tools of the signatory states. However, most parties to the Bamako Convention are impoverished countries, needing more effective means to apply the Convention properly. African states are mostly aid-seeking states and are currently hardly concerned with development (Ilankoon et al., 2018; Olowu, 2006). The dumping incidents of Probo Koala in 2006 in Ivory Coast and the net inward flow of e-waste to countries such as Ghana and Nigeria despite their being signatories to the Convention are two good examples of it (Cox, 2010). Although the importation of dangerous waste can cause damage to the environment, some developing countries still believe that the activity is a source of income and an aid to economic development. Therefore, the Bamako Convention's advisory role, which follows the example of the Basel Convention, is strongly diminished, and a more coercive force would be needed (UNEP, 2018b).

2.2.6 THE STOCKHOLM AND MINAMATA CONVENTIONS

Another major UNEP convention tackles critical aspects of the life cycle of the chemicals: the Stockholm Convention on "Persistent Organic Pollutants (POPs)" (Stockholm Convention). Indeed, several POPs manufacture specific electronic devices by industries such as the polychlorinated biphenyls in electrical transformers. The Stockholm Convention aims to avoid all the potential and harmful consequences of persistent organic pollutants with toxic characteristics for the sake of the environment and human health. These POPs have the particularity of resisting biological degradation, accumulate in living beings, and spread quickly over long distances by air and water. Biodiversity and the inhabitants of the Arctic are the main ones affected by these POPs. When elements have chemicals identified by the Convention, these must be disposed of in an environmentally sound and rational manner when they become waste. Twelve pollutants are considered "POPs". The Stockholm Convention commits the parties to eliminate or limit the production of these polluting substances. The Convention also applies to two unintentionally generated substances during the incineration of e-waste: furan and dioxin. By May 2017, 181 parties ratified the Convention signed in Stockholm on May 23, 2001. The

European Union integrated this Convention through Regulation No. 850/2004. Then, Regulation No. 1021/2019 recently repealed it and replaced it.

Finally, the international community signed at Kumamoto (Japan) on 10 October 2013, the Minamata Convention on Mercury, which entered into force on 16 August 2017. The Convention aims to reduce the widespread mercury pollution from certain human activities. Mercury is used to extracting materials and manufacture goods such as electrical and electronic products for its unique characteristics. The UN treaty addresses the direct mining of mercury, its import and export, its safe storage, and its disposal once as waste (Minamata Convention).

2.3   THE EU LEGAL FRAMEWORK FOR E-WASTE

2.3.1   The Waste Shipment Regulation

The European Union is equipped with one of the strictest enforcement laws on e-waste, banning exports to developing countries and compelling manufacturers to help fund recycling. In Europe, a legal framework for managing and transporting various kinds of waste has been in place since the 70s and then revised and supplemented at different times. The Framework Directive on Waste (FDW), first adopted in 1975 (Directive 75/442/EEC), was amended in 1991 (Directive 91/156/EEC) to recast into a new directive on waste in 2006 (Directive 2006/12/EC). The FDW was complemented by the Hazardous Waste Directive (HWD), enacted for the first time in 1978 (Directive 78/319/EEC) to promote the environmentally sound management of hazardous waste and later repealed in 1991 (Directive 91/689/EEC). Two years later, a new FDW Directive on waste (Directive 2008/98/EC), based on Article 192(1) TFEU, replaced both the Directives. Finally, the FDW was recently amended by Directive 2018/851/EC.

The FDW directive clarifies several critical issues, such as the definitions of waste, recovery, and disposal. It introduces criteria for distinguishing by-products from waste and for determining the condition when waste ceases to be waste ("end-of-waste status"). It struggles to increase awareness of growing waste and strengthens waste prevention by encouraging waste recovery and reusing recycled materials. As discussed later in the chapter, the FDW and the related directives are based on several principles, one of which is the waste management hierarchy. Accordingly, waste management has to primarily prevent waste generation and reduce its harmfulness by re-using, recycling, or recovering waste material as an energy source. Waste disposal in incineration or landfill sites should be a last resort. Furthermore, Member States must cooperate to create an integrated network of waste disposal facilities, adopting all the "Best Available Techniques (to achieve this goal) Not Entailing Excessive Cost" (so-called BATNEEC principle). Accordingly, emissions from installation to the environment should be reduced as much as possible and in the most cost-effective ways. The third and fourth are the principles of self-sufficiency and proximity. The EU and its Member States should remain self-sufficient concerning waste disposal and dispose of it near their source (Grosz, 2011).

The European Union has been a party to the Basel Convention since 1994. However, since 1984, supervision and control of the trans-frontier shipment of hazardous waste have been guaranteed within the European Community (Directive 84/ 631/EEC). In 1993, Regulation (EEC) No 259/93 on the supervision and control of waste shipment within and out of the European Community replaced the Directive. In 2006, to address the uncontrolled flow of waste across borders, the European Parliament and the Council adopted the Waste Shipment Regulation (WSR) by repealing and replacing the Regulation of 1993 (Regulation (EC) No 1013/2006). WSR incorporated the ban issued by the Basel Convention into European legislation. However, the EU regulation applies to all waste, defined as Directive 2008/98/EC. Article 1(a) of the FDW defines *waste* as "any substance or object in the categories set out in Annex I which the holder discards or intends or is required to discard". Accordingly, the concept of waste is broader than that of the Basel Convention, regardless of whether the waste is dangerous or harmless (Article 3). The Regulation advocates the free movement of waste within the European Union because it considers goods to include waste. However, this freedom of movement is restricted if environmental protection becomes more critical than trade-in waste. Moreover, this freedom of movement cannot take place outside the borders of the EU member countries (Articles 34 and 40).

This Regulation aims to protect the environment in the European Union and in third countries to which waste is exported by prohibiting the export of hazardous e-waste for disposal within the European community and most non-OECD States. The 2006 Regulation establishes a supervisory regime for the movement of waste. According to the origin, destination, route of shipment, type of waste shipped, and treatment applied to the waste at its destination, a particular procedure and regime apply. For instance, this regulation does not cover shipments of radioactive waste or spent fuels because they are subject to the Euratom Directive (Directive 2006/117/ Euratom). Unfortunately, several e-waste components did not fulfil the WSR's definition of a hazardous substance and continued to be shipped to non-OECD countries under other provisions (Brandt and Holberg, 2015; Scotford, 2009; Maxianova, 2008).

In 2002, the EU legislator put in place two pieces of legislation to complement the legal framework. These are Directive 2002/95/EC on "the Restriction of the use of certain Hazardous Substances in electrical and electronic equipment" (RoHS Directive) and Directive 2002/96/EC on "Waste Electrical and Electronic Equipment" (WEEE Directive). The ROHS and WEEE Directives are complementary measures: while the former operates at the level of product manufacturing, the latter focuses on collecting and treating e-waste by the official schemes. The collection includes storage, transport, and regrouping of e-waste before treatment. Three main steps characterize the treatment:

1. The pre-treatment includes clean up, dismantling, shredding, and sorting of e-waste
2. The non-recyclable fractions' disposal
3. The recovery and production of secondary raw materials

The WEEE Directive's focus, scope, and monitoring are instead the first two steps because the fractions, once they achieved the end-of-waste status, are seen as products and not as waste anymore. Therefore, the supervisor applies a lower monitoring level for the compliance schemes than in the previous steps (Langlet and Mahmoudi, 2016).

2.3.2   THE ROHS DIRECTIVE

The Directive on the "Restriction of the use of certain Hazardous Substances" (RoHS Directive), which entered into force in February 2003, concerns reducing the content of certain hazardous substances in EEE (Directive 2002/95/EC). The RoHS Directive emphasizes the need to modify the product design and packaging to reduce dangerous substances and replace them with environment-friendly materials. It also aims to increase the recycling rate for domestically generated e-waste. Thus, the RoHS Directive is concerned with the beginning of the EEE life cycle and tries to eliminate (or replace with safer alternatives) dangerous substances like lead, cadmium, mercury, and hexavalent chromium and fire retardants such as polybrominated biphenyls (PBB) or polybrominated diphenyl ethers (PBDEs) in domestically produced or imported electrical and electronic products (Article 4). These banned substances could be found in several components of EEE. For example, lead could be detected in solder on printed circuit boards, mercury in thermostats, and cadmium in connectors, switches, and rechargeable computer batteries. The Directive aims not only to prevent those substances which are harmful to the environment from being components of electronic devices but also to avoid all the health problems that the employees may suffer during the product's manufacturing at the beginning and during the treatment at the end of the product's life. For this reason, in 2006, the EU Commission established maximum concentration values of these substances (Commission Decision 2005/618/EC). Besides, it required that all EEE placed on the market must not contain substances scientifically classified as dangerous, as included in the list contained in Annex II of the RoHS Directive (Commission Decision 2005/717/EC). The list is not exhaustive, and the EU Commission agreed to adapt it from time to time to scientific and technical progress by adding new substances every time the tests show dangerousness (Article 5). For instance, in December 2008, four new substances were included: bis phthalate, benzyl butyl phthalate, dibutyl phthalate, and finally, hexabromocyclododecane (Directive 2011/65/EU). At the same time, the RoHS directive imposes a declaration of conformity on manufacturers, importers, and distributors of a product, which involves the assumption of responsibility for the compliance of the EEE with the Directive's requirements (Article 13). Devices marked with a "CE" will be presumed to comply with the Directive. However, certain manufacturers have independently created their labelling system, such as "PB free" and "RoHS compliant". The market surveillance is left to the single-Member States by Articles 15 to 29 of Regulation (EC) No 765/2008. It is left to the Member States to decide what penalties apply to breaches of the national provisions adopted under this Directive and their implementation to have practical, proportionate, and dissuasive penalties (Article 23).

The RoHS Directive established several (nine) exemption cases from the ban on using lead, mercury, cadmium, and hexavalent chromium included in the Directive Annexes. In particular, the use of certain hazardous substances is authorized when, scientifically and technically, it is not possible to replace them. When their replacement does not improve the environment, health, and consumer safety or does not ensure the reliability of substitutes—the substitution of these hazardous substances in EEE must not harm the health and safety of their users. In practice, the new

Directive offers more exemptions than the old one. These exemption requests cannot relate to a new dangerous substance but must relate to a specific use of this substance and must meet the conditions mentioned in the Directive (Annex III). In addition, the new Directive widens the scope compared to the old one to all EEE, including medical devices, standard and industrial monitoring and control instruments, and in vitro diagnostic medical devices. According to RoHS, the definition of EEE is extended to all products using an electrical source serving to perform one of these functions that fall under the scope of EEE (Article 3).

2.3.3 THE WEEE DIRECTIVE

In 2002, the EU Commission also adopted the first version of Directive 2002/96/EC concerning producer responsibility for waste electrical and electronic equipment (WEEE Directive) that entered into force in February 2003. The WEEE Directive was then recast in 2012 (Directive 2012/19/EU) and recently amended in 2018 (Directive 2018/849/EC). The Directive is a comprehensive law established to enable environmentally sound management of WEEE or e-waste. Its goal is twofold. To prevent or reduce the adverse impacts of the generation and management of e-waste on the one hand, and to reduce the overall impacts of resource use and improve the efficiency of such use, thereby contributing to sustainable development (Article 1).

It applies to all electrical and electronic equipment (EEE), with a few exceptions. These include equipment necessary for protecting the vital interests of the security of the Member States, extensive-scale stationary industrial tools, and large-scale fixed installations. Since the 2002 version, the WEEE Directive embraced the principle of eco-design expressed by a producer's approach to using materials with less environmental impact over a product's life cycle. Accordingly, manufacturers shall give special consideration to the EEE's design for its environmental impact during its lifecycle: from the initial design project, through use, until its treatment of the waste from the product (recovery or disposal). This concept was explicitly introduced into European legislation only with the introduction of the Directive 2005/32/EC. The Directive, later revised in 2009 (Directive 2009/125/EC), established a framework for setting eco-design requirements for energy-related products. By acknowledging the importance of the eco-design principle, the WEEE Directive indirectly imposes a duty of prevention on manufacturers, importers, and distributors of EEE. This prerequisite is needed to extend the lifecycle of the EEE and reduce the e-waste and diminish the dangerousness of the substances and components used to produce EEE. Furthermore, Article 4 of the WEEE Directive urges Member States to support collaboration between manufacturers and recyclers and measures to incentivize the EEE's design and production to facilitate the reuse, dismantling, and recovery of the e-waste and its components and materials. However, as mentioned earlier, the WEEE Directive focuses mainly on the EoL stages of EEE.

A primary policy approach adopted by the EU Commission is that of Extended Producer Responsibility (EPR), which is mandatory within the context of the WEEE Directive (Pouikli, 2020). EPR requires producers, which for the aim of this policy are an enlarged category and can be manufacturers, distributors, or importers to take operational and or financial responsibility for the EoL phase of their products.

Article 12 section 1 states that "Member States shall ensure that producers provide at least for the financing of the collection, treatment, recovery and environmentally sound disposal of WEEE from private households that have been deposited at collection facilities", and section 2 continues encouraging producers "to finance also the cost occurring for collection of WEEE from private households to collection facilities. The EPR system shifts responsibility for collecting and managing e-waste up the value chain towards the producer and away from the municipalities. Depending on the case, the responsibility can be complete or partial, financial or physical. The policy is in line with the "polluter pays" principle. Those responsible for producing waste and thus damages on the environment should bear the costs of such adverse consequences (i.e. Article 10 of the Landfill Directive 99/31/EC). In this logic, the product's price must include the costs of disposal and treatment to reflect the product's environmental effects (OECD, 2016). When placing a product on the market, producers must mark their products and provide a guarantee showing that the management of all e-waste will be financed. Concerning e-waste-generated by-products placed on the market before 13 August 2005 (historical waste), the costs shall be borne by one or more management systems to which all producers that exist on the market when the respective costs occur contribute proportionately (Article 12 s 4). Each producer shall be responsible for financing the operations for all e-waste from products placed on the market after that date (Article 12 s 3). Slightly different requirements apply for e-waste from users other than private households (Article 13).

Article 14 section 2 of the WEEE Directive recites that "Member States shall ensure that users of EEE in private households are given the necessary information about (a) "the requirements not to dispose of e-waste as unsorted municipal waste and to collect e-waste separately, (b) the return and collection systems available to them [. . .], (c) (d) (e) *omissis*" (Directive 2018/849/EC). The Development of Guidance on Extended Producer Responsibility document provides a good overview of how producers can carry out their tasks individually or collectively by assigning a third-party organization called the Producer Responsibility Organization (PRO). PRO facilitates producers to collectively manage the take-back and, most often, arrange for the treatment of products (OECD, 2016; OECD, 2005). While most EU countries allow collective and individual systems to coexist, others such as Spain, France, Austria, Cyprus, Holland, and Estonia are organized into collective systems to organize the individual system. At the two opposite ends, there are Sweden and Germany. The first supports collective systems, while the second supports only individual. E-waste must be separately collected from municipal household waste as much as possible. The WEEE Directive imposes only minimum and general requirements. It is up to the states to take adequate and organizational measures for selective collection. The purpose of the collection is to separate the WEEE according to their categories and particular treatment. For this purpose, the WEEE Directive classifies EEE broadly into ten categories (Annex I), reduced to six in 2018 (Annex III). A non-exhaustive list of EEE falling within the categories listed in Annex III is contained in Annex IV.

The WEEE collected is sent to treatment companies unless the WEEE turns out to be, in the end, reusable because it is still in working order. Only recovery or elimination companies authorized by the public authority can carry out these treatment operations, which must be carried out according to the best recovery and recycling and treatment techniques available as per Article 8. The recovery of e-waste regarding Annex II of FDW consists of all operations involving the reuse of the entire equipment or its components, the recycling, and energy transformation, such as incineration with heat recovery. The same article defines *disposal* as "any operation which is not recovery even when said operation has as a secondary consequence the recovery of substances from energy". Disposal is, therefore, a residual means. This Directive mentions some examples of disposal methods in Annex I, such as landfill, incineration at sea, burial inland, or dumping at sea waste.

The WEEE Directive adopts three weight-based indicators to supervise the WEEE system's efficiency. These are the collection rate, the recycling and preparation for reuse rate, and the recovery rate. The WEEE Directive recast has undergone various changes in the collection rate target and its calculation. Initially, the target was set at 4 kg per capita of e-waste from households. From 2016 to 2018, the minimum collection rate became 45%, calculated based on the total weight of e-waste collected and the average weight of EEE placed on the market in the 3 preceding years. From 2019, it was either 65% of the EEE placed on the market and calculated as before or 85% of the total e-waste generated on the territory of that Member State. Annex V of the WEEE Directive reports all the minimum recovery targets. Finally, to reduce illegal e-waste export to non-OECD countries, the WEEE Directive includes in its Annex VI a provision enumerating minimum monitoring requirements for e-waste shipments. The provision also introduces a distinction between new, used, or waste products to tackle the false labelling of e-waste as used EEE.

From the beginning of January 2021, following the Brexit transition, new rules on waste shipments from and towards the EU Community entered into force in the UK, approved by the UK Government's Department for Environment, Food and Rural Affairs (DEFRA). Accordingly, the UK has become a non-EU country, and the EU has to treat it as any other OECD country or country part of the Basel Convention that intends to export or import waste to an EU country. It will require all the authorizations from each country they transport waste through and fill a duly reasoned request (DRR) explaining why they could not process or dispose of the waste in their territory. As the UK mainly exports waste to mainland Europe, Brexit's impact on e-waste export could disrupt shipments, delays, and congestions at ports. Alternatively, if the UK Government has secured good bilateral agreements with EU individual Member States to enable waste exports, then the flow should continue as usual.

2.4 HARMONIZATION OF EPR SCHEMES AND THE E-COMMERCE CHALLENGE

EPR schemes or Product Stewardship systems, used in the EU and most countries worldwide, are an essential environmental policy instrument, especially with EEE. Most US State laws on e-waste embrace the concept of EPR, apart from the states of California and Utah (Biedenkopf, 2020; Magalini et al., 2019). These schemes have proved to be able to generate financial resources to support the recycling and collection of e-waste and have reduced the total amount of e-waste dumped in landfills. However, the recent growth of e-commerce and online sales risk seriously compromises these results by facilitating producers' non-compliance to the EPR requirements. This effect of "free-riding" exists because online sellers often have no physical or legal entity in the country of the consumers and, therefore, are not registered with national or local EPR schemes. In such a way, they circumvent the responsibilities (take-back) and expenses (EPR fees) associated with placing products on the market in those countries. Although the available information shows that this opportunistic behaviour is contained between 5% and 10% of the value of the OECD EEE market, the phenomenon of e-commerce has rapidly grown (Hilton et al., 2019).

There is a range of strategies to correct this free-riding problem in e-commerce, raising awareness among sellers about EPR compliance, strengthening enforcement, and even harmonizing the legal framework for EPR systems. A first short-term policy would be to develop awareness among online sellers. For instance, this could be done by developing e-commerce codes of practices, including showing the PRO registration details on the websites and other information such as the logo and the legal entity address and contacts as a voluntary standard. Moreover, multiseller marketplace platforms could proactively inform sellers on the sites of their obligation towards EPR and take action against those non-compliant. Alternatively, EPR registration could be linked to other regulatory measures, such as the VAT (Value Added Tax). For example, the EPR registration could be a condition for the seller who wants to obtain a VAT number. The new technology and innovation, such as smart contracts and blockchain technologies based on general and public distributed ledgers, could facilitate EPR compliance management. Other measures that could be applied are those targeted to strengthen enforcement. If awareness-raising can help those sellers who are simply unaware of their EPR obligations, it cannot do anything for those companies which did not pay the fees associated with the EPR registration in all the jurisdictions where they operate on purpose. To tackle this problem and improve enforcement, the EU Commission has proposed to create on the competent authority's website a unique electronic register of producers for each jurisdiction along with a form to report the website trading name and/ or company's name of the unregistered producers. Harmonizing the EPR systems should support this action and coordinated enforcement across jurisdictions, for instance, in the European Union (Draft Regulation on the registration and reporting of producers of WEEE and the frequency of reporting to the register). The existence of many diverging national categories of EEE adds unnecessary layers of complexity and administrative burden. Besides, to facilitate enforcement, it would be helpful to build into the e-waste regulation to prosecute a company in another country or economic region. In the EU, it is already so, although it lacks a proper tool for prosecution since all existing forms are too slow and costly. The legislator should add enforcing powers to enable private actions to prevent illegal online selling (European Commission, 2016).

After consultations with various stakeholders via meetings, interviews and two workshops, the OECD examined the same matter and produced several proposals. One is to consider the parcel delivery companies, such as Carriers and postal services, as "producers" for the EPR laws holding a delivery contract with an online seller in the jurisdiction where the seller is not registered. Alternatively, to circumvent the difficulty of detecting and engaging online sellers, Individual Producer Responsibility systems should be established as a means to charge EPR fees back to Original Equipment Manufacturers (Hilton et al., 2019).

2.5 CONCLUSIVE CONSIDERATIONS: THE FUTURE
OF EU E-WASTE POLICY AND REGULATION

The European Union presents a robust legislative framework on e-waste on the basis of the analysis of the regulation on waste shipment (WSR) and the two directives on e-waste and hazardous substances (WEEE and RoHS). Together with the Basel Convention, these legislative tools have enormously improved e-waste management, helping develop a legal and technical framework in several countries. However, the path towards environmentally sound e-waste management is still an ongoing process. The international community must move from a linear economy characterized by a consumer society towards a circular economy, which also takes environment, resources, and people's health into account. In March 2020, the European Commission presented a new circular economy action plan to reduce e-waste (EU 'Circular Economy Action Plan', 2020). Explicit urgent objectives of the proposal are creating the right to repair, the improvement of reusability in general, introducing of a standard charger for all mobile devices, and establishing a rewards system to encourage EEE recycling. In addition, the proposal aims to incorporate circular flows in all stages of a value chain—from design to production until the consumer—to achieve sustainability (European Commission, 2019). The EU Commission has launched a public consultation to revise the ecodesign directive and increase the legislative measures to make products' placed on the EU market more sustainable (Public Consultation, 2021).

Similarly, the EU Commission has proposed and voted a revision of the WSR to explore ways to reduce the illegal shipments of e-waste by strengthening the inspection system (Waste Shipment revision of EU rules). The Revision awaits the EU Council's (and the Commission's) agreement before being adopted in a plenary session by the EU Parliament. While revising the legal framework on e-waste to keep it up-to-date with innovation is undoubtedly essential, it is also essential to develop cooperation at the supra-national level and harmonize the law and procedures to improve enforcement. In this sense, the EPR legislation could be a joint base to start with for a harmonization of the law concerning e-waste. However, regulation is only one part of the policy approach. Public awareness is also essential since all stakeholders can play a role in sustainability, starting from the public authorities. Public authorities have significant purchasing power, which can be used to give a good example. The EU Commission and some European countries have developed guidelines published in this area as national Green public procurement (GPP) criteria or green purchasing criteria. GPP is a voluntary instrument through which the public authorities can contribute to sustainable consumption and production by selecting environmentally friendly goods, services and work compliant with those criteria (GPP webpage, 2021; EU Communication, 2008).

REFERENCES

Ajibo, KI 2016 Transboundary Hazardous Wastes and Environmental Justice. *Environmental Law Review* 18(4): 267–283.

Albers, J 2015 *Responsibility and Liability in the Context of Transboundary Movements of Hazardous Wastes by Sea.* Berlin, Heidelberg: Springer, 295–304, 280ff.

Assemboni-Ogunjimi, A 2015 La problématique des déchets dangereux en Afrique: le cas du Togo. In M Faure, A Lawogni, and M Dehoumon eds. *Les mouvements transfrontières de déchets dangereux.* Bruxelles: Bruylant, 94.

Bamako Convention 1991 Bamako Convention on the Ban on the Import into Africa and the Control of Transboundary Movement and Management of Hazardous Wastes within Africa. *International Legal Materials* 30: 775.

Basel Action Network. Project of Earth Economics 2005 *The Digital Dump: Exporting Re-Use and Abuse to Africa.* Seattle. https://www.cs.swarthmore.edu/~turnbull/cs91/f09/paper/DigitalDump.pdf

Biedenkopf, K 2020 E-Waste Policies in the United States: Minimalistic Federal Action and Fragmented Subnational Activities. In MNV Prasad, M Vithanage, and A Borthakur eds. *Handbook of Electronic Waste Management International Best Practices and Case Studies.* KU Leuven, Leuven, Belgium: Butterworth-Heinemann, 577–588.

Brandt, J, Holberg, JS 2015 *Legal Possibilities to Restrict Exports of Used Electric and Electronic Equipment,* Copenhagen: Nordic Council of Ministers.

Choski, S 2001 The Basel Convention on the Control of Transboundary Movements of Hazardous Wastes and Their Disposal: 1999 Protocol on Liability and Compensation. *Ecology Law Quarterly* 28(2): 509, 518, 524–526.

Commission Decision 2005/618/EC (0.1% by Weight in Homogeneous Materials for Lead, Mercury, Hexavalent Chromium, Polybrominated Biphenyls and Polybrominated Diphenyl Ethers and of 0.01% for Cadmium).

Commission Decision 2005/717/EC for the Purposes of Adapting to Technical Progress the Annex to Directive 2002/95/EC.

Convention on the Control of Transboundary Movement of Hazardous Wastes and Their Disposal (Basel Convention) (Basel, 22 March 1989) 1673 UNTS 57.

Convention on Persistent Organic Pollutants (Stockholm Convention) (Stockholm, 22 May 2001) (2006) 40214 UNTS 119.

Council Directive 75/442/EEC of 15 July 1975 on Waste [1975] OJ L 194/39.

Council Directive 84/631/EEC on the Supervision and Control within the European Community of the Transfrontier Shipment of Hazardous Waste [1984] OJ L 326/31.

Council Directive 91/689/ EEC of 12 December 1991 on Hazardous Waste [1991] OJ L377/ 20.

Council Directive 2006/117/Euratom on the Supervision and Control of Shipments of Radioactive Waste and Spent Fuel [2006] OJ L 337/ 21.

Cox, G 2010 The Trafigura Case and the System of Prior Informed Consent Under the Basel Convention—A Broken System? *Law, Environment and Development Journal* 6(3): 273. www.lead-journal.org/content/10263.pdf (accessed 6 February 2023).

Decision COP of 6 June 26, 2013 — Institutional Arrangements for the Implementation of Convention: Establishment of a Secretariat (C1 DEC. Conference of the States Parties) https://wedocs.unep.org/handle/20.500.11822/22471

de La Fayette, L 1995 Legal and Practical Implications of the Ban Amendment to the Basel Convention, *Yearbook of International Environmental Law,* Vol. 6, 703–717.

Directive 78/319/EEC on Toxic and Dangerous Waste [1978] OJ L 84, pp. 43–48.

Directive 91/156/EEC of 18 March 1991 Amending Directive 75/442/EEC on Waste, [1975] OJ L 194/39 and [1991] OJ L 78/32.

Directive 2002/95/EC of the European Parliament and of the Council of 27 January 2003 on the Restriction of the Use of Certain Hazardous Substances in Electrical and Electronic Equipment [2003] OJ L37/19.

Directive 2005/32/EC of the European Parliament and of the Council of 6 July 2005 Establishing a Framework for the Setting of Ecodesign Requirements for Energy-Using Products and Amending Council Directive 92/ 42/ EEC and Directives 96/ 57/ EC and 2000/ 55/ EC of the European Parliament and of the Council [2005] OJ L191/29.

Directive 2006/12/EC of the European Parliament and of the Council of 5 April 2006 on Waste [2006] OJ L114/9.

Directive 2008/98/ EC of the European Parliament and of the Council of 19 November 2008 on Waste and Repealing Certain Directives [2008] OJ L312/3.

Directive 2009/125/EC of the European Parliament and of the Council of 21 October 2009 Establishing a Framework for the Setting of Eco-Design Requirements for Energy-Related Products [2009] OJ L 285.

Directive 2011/65/ EU of the European Parliament and of the Council of 8 June 2011 on the Restriction of the Use of Certain Hazardous Substances in Electrical and Electronic Equipment [2011] OJ L174/ 88.

Directive 2012/19/EU of the European Parliament and of the Council on Waste Electrical and Electronic Equipment (WEEE) [2012] OJ L 197/38.

Directive (EU) 2018/849 of the European Parliament and of the Council of 30 May 2018 Amending Directives 2000/53/EC on End-of-Life Vehicles, 2006/66/EC on Batteries and Accumulators and Waste Batteries and Accumulators, and 2012/19/EU on Waste Electrical and Electronic Equipment OJ L 150.

Directive (EU) 2018/851 of the European Parliament and of the Council of 30 May 2018 Amending Directive 2008/98/EC on Waste [2018] OJ L 150 p. 109.

Draft Proposal Regulation at https://www.europarl.europa.eu/thinktank/en/document/EPRS_BRI(2022)729330 (accessed 6 February 2023) was announced in 2020 with the European Green Deal and voted in plenary session at the EU Parliament in January 2023. It awaits the EU Council's agreement to be definitely approved in plenary session.

Draft Regulation on the Registration and Reporting of Producers of WEEE and the Frequency of Reporting to the Register at https://ec.europa.eu/info/law/better-regulation/have-your-say/initiatives/1579-Registration-and-reporting-of-producers-of-WEEE-and-the-frequency-of-reporting-to-the-register_en (accessed 6 February 2023).

EU 'Circular Economy Action Plan' 2020. Bruxelles. https://ec.europa.eu/environment/strategy/circular-economy-action-plan_en (accessed 6 February 2023).

EU Communication 2008 (COM (2008) 400) Public Procurement for a Better Environment. https://eur-lex.europa.eu/legal-content/EN/TXT/?uri=CELEX:52008DC0400 (accessed 6 February 2023).

European Commission 2016 The 'Blue Guide' on the Implementation of EU Product Rules 2016. https://eur-lex.europa.eu/legal-content/EN/TXT/?uri=CELEX%3A52016XC0726%2802%29 (accessed 6 February 2023).

European Commission 2019 New Rules Make Household Appliances More Sustainable. Brussels: Press Release. https://ec.europa.eu/commission/presscorner/detail/en/IP_19_5895 (accessed 6 February 2023).

European Parliament on 23.12.2020 E-Waste in the EU: Facts and Figures. https://www.europarl.europa.eu/news/en/headlines/society/20201208STO93325/e-waste-in-the-eu-facts-and-figures-infographic (accessed 6 February 2023). Washing Machines and Electric Stoves Count for 52.7% of the E-Waste Collected, then IT Equipment, Laptop and Printers (14.1%), Photovoltaic Panels, Video Cameras and Fluorescent Lamps (14.6%) and to Conclude Small Appliances Like Vacuum Cleaners and Toasters (10.1%).

Eze, CN 2007 The Bamako Convention on the Ban on the Import into Africa and the Control of the Transboundary Movement and Management of Hazardous Wastes Within Africa: A Milestone in Environmental Protection? African Journal of International and Comparative Law 15: 208, 219.

Forti, V, Baldé, CP, Kuehr, R, and Bel, G 2020 The Global E-Waste Monitor 2020: Quantities, Flows and the Circular Economy Potential. United Nations University/United Nations Institute for Training and Research (UNITAR), International Telecommunication Union, and International Solid Waste Association. Tokyo, Japan, 2020.

Greenhouse, S 1988 Toxic Waste Boomerang: Ciao Italy! *New York Times*, 3 September.

Grosz, M 2011 *Sustainable Waste Trade Under WTO Law: Chances and Risks of the Legal Frameworks' Regulation of Transboundary Movements of Wastes,* Leiden – Boston Martinus Nijhoff Publishers. 1–525.

Hilton, M, Sherrington, C, McCarthy, A, and Börkey, P 2019 *Extended Producer Responsibility (EPR) and the Impact of Online Sales.* OECD Environment Working Papers n. 142. Paris: OECD Publishing, 24–31, 48–51.

Ilankoon, IMSK, Ghorbani, Y, Chong, MN, Herath, G, Moyo, T, and Petersen, J 2018 E-Waste in the International Context—A Review of Trade Flows, Regulations, Hazards, Waste Management Strategies and Technologies for Value Recovery. *Waste Management* 82: 258–275.

Kummer, K 1998 The Basel Convention: Ten Years On. *Review of European Community and International Environmental Law* 7: 227–229, 232.

Kummer, K 2017 Transboundary Movement of Hazardous Waste and Chemicals from International Environmental Law. In A Nollkaemper, and I Plakokefalos eds. *The Practice of Shared Responsibility in International Law*. Cambridge: Cambridge University Press, 936–961.

Langlet, D 2009 *Prior Informed Consent and Hazardous Trade: Regulating Trade in Hazardous Goods at the Intersection of Sovereignty, Free Trade and Environmental Protection.* Alphen Aan Den Rijn [Netherlands], Austin [Texas], 2nd ed. Kluwer Law International, Chapter 5.

Langlet, D, and Mahmoudi, S 2016 *EU Environmental Law and Policy.* Oxford: Oxford University Press, 285, 301–302.

Louka, E 1993 Bringing Polluters Before Transnational Courts: Why Industry Should Demand Strict and Unlimited Liability for the Transnational Movements of Hazardous and Radioactive Wastes. *Denv. J. Int'l L. & Pol'y* 22: 63, 98–102.

Magalini, F, Sinha Khetriwal, D, and Kyriakopoulou, A July 2019 *E-Waste Policy Africa Handbook*. African Clean Energy Technical Assistance Facility Department for International Development (DFID) Africa Clean Energy Technical Assistance Facility and Coffey International Development Ltd, Nairobi – Kenya.

Maxianova, K 2008 Shipments of Electronic Waste: Providing the Right Incentives through Regulation and Enforcement, *Review of European, Comparative & International Environmental Law (RECIEL)* 17:3, 268–276.

Model National Legislation on the Management of Hazardous Wastes and Other Wastes as well as on the Control of Transboundary Movements of Hazardous Wastes and Other Wastes and their Disposal. http://www.basel.int/Portals/4/Basel%20Convention/docs/pub/modlegis.pdf (accessed 6 February 2023).

Nemeth, P 2015 The Basel Convention and the Transboundary Movements of Hazardous Waste to the Developing World: A Study of Regulatory Violations and the Problem of Legal Compliance (DPhil thesis, The State University of New Jersey), 77.

OECD 2005 For an Analysis of the Cost and Benefits of the EPR see OECD 2005. *Analytical Framework for Evaluating the Costs and Benefits of Extended Producer Responsibility Programmes*. www.oecd.org/greengrowth/tools-evaluation/34529905.pdf (accessed 6 February 2023).

OECD 2016 *Extended Producer Responsibility Updated Guidance for Efficient Waste Management*. https://www.oecd.org/environment/waste/extended-producer-responsibility-9789264256385-en.htm (accessed 6 February 2023).

Olowu, D 2006 The United Nations Special Rapporteur on the Adverse Effects of the Illicit Movement and Dumping of Toxic and Dangerous Wastes on the Enjoyment of Human Rights: A Critical Evaluation of the First Ten Years. *Environmental Law Review* 8: 199, 203–207.

On 'Green Public Procurement' see https://ec.europa.eu/environment/gpp/index_en.htm (accessed 6 February 2023).

Pasquali, L 2005 *Le régime juridique des mouvements transfrontières de déchets en droit communautaire et en droit international.* Aix-Marseille: Presses Universitaires d'Aix-Marseille.

Pouikli, K 2020 Concretising the role of extended producer responsibility in European Union waste law and policy through the lens of the circular economy. *ERA Forum* 20, 491–508.

Proposal for a Regulation of the European Parliament and of the Council on the Voluntary Participation by Organisations in a Community Eco-Management and Audit Scheme (EMAS), COM (2008) 402 final, 2 and 5.

Public Consultation Planned for the Fourth Quarter of 2021 See the Feedback Received at https://ec.europa.eu/info/law/better-regulation/have-your-say/initiatives/12567-Sustainable-products-initiative (accessed 6 February 2023).

Regulation (EC) No 850/2004 of the European Parliament and of the Council of 29 April 2004 on Persistent Organic Pollutants and Amending Directive 79/117/EEC [2004] OJ L 158/7.

Regulation (EC) No 1013/ 2006 of the European Parliament and of the Council of 14 June 2006 on Shipments of Waste [2006] OJ L190/ 1.

Regulation (EU) No 1021/2019 of the European Parliament and of the Council of 20 June 2019 on Persistent Organic Pollutants OJ L 169, 25.6.2019.

Sands, P, and Peel, J 2018 *Principles of International Environmental Law.* Cambridge: 4th ed. Cambridge University Press, 620–623.

Scotford, E 2009 The New Waste Directive—Trying to Do It All . . . An Early Assessment. *Environmental Law Review* 11(2): 75–96.

Sundram, MS 1997 Basel Convention on Transboundary Movement of Hazardous Wastes: Total Ban Amendment. *Pace International Law Review* 9(1): 13–14.

The 1972 Declaration of the United Nations Conference on the Human Environment (Stockholm Declaration) (Stockholm, 16 June 1972) (1972) 11 ILM 1416.

United Nations Environment Programme 2018a *Bamako Convention: Preventing Africa from Becoming a Dumping Ground for Toxic Wastes.* https://www.unep.org/news-and-stories/press-release/bamako-convention-preventing-africa-becoming-dumping-ground-toxic (accessed 6 February 2023).

United Nations Environment Programme 2018b *Twenty Years of the Bamako Convention: A Time for More Effective Implementation.* (UNEP/BC/COP.2/). Abidjan. https://wed-ocs.unep.org/handle/20.500.11822/22556

UN Treaty Collection, Chapter XXVII—Environment, 17. Minamata Convention on Mercury.

Vu, HQ 1994 The Law of Treaties and the Export of Hazardous Waste. *UCLA Journal of Environmental Law and Policy* 12(2): 417.

Widawsky, L 2008 In My Backyard: How Enabling Hazardous Waste Trade to Developing Nations Can Improve the Basel Convention's Ability to Achieve Environmental Justice. *Environmental Law* 38: 577.

Widmer, R, Oswald-Krapf, H, Sinha-Khetriwal, D, Schnellmann, M, and Böni, H 2005 Global Perspectives on E-Waste. *Environmental Impact Assessment Review* 25: 436–458.

3 Comparison Analysis of E-waste Management in India and Japan
Challenges and Solutions with a Focus on the COVID-19 Pandemic

Satoka Shimizu

CONTENTS

DOI: 10.1201/9781003301899-4

3.1   INTRODUCTION

The World Health Organization (WHO) has announced COVID-19 as a pandemic. The provision of evidence-based safe Waste Management (WM) and hygienic conditions is essential to protect human health during COVID-19 (WHO 2020). Large-framed management guidelines of infectious waste have already been promulgated. However, the state of WM differs among countries. Setting individual guidelines is required on the part of respective governments and municipalities (UNEP et al. 2020).

There are concerns on WM under COVID-19, e.g. suspension of recycling activities, business opportunity loss, and negative impacts on vulnerabilities in collection services and OSH (UNEP et al. 2020). According to UNEP, gender gap is deeply embedded in WM sector (UNEP et al. 2020). Although there is insufficient data existing for evaluating COVID-19 impacts on the women in the WM sector, many factors show the possibilities of higher social, economic, and health risks for women (UNEP et al. 2020).

Due to the economic collapse caused by COVID-19 pandemic, the global GDP in 2020 is estimated to be −4.3 % (The World Bank 2021). GDP in 2020 of India is estimated to be −9.6 %, and GDP in 2020 of Japan is estimated to be −5.3 % (The World Bank 2021). Women, the informal workers, youth, the poor, and those who work in contact-intensive sectors have been suffering from acute adverse effects of this collapse (IMF 2021).

However, there are also opportunities for improving WM systems under COVID-19 (UNEP et al. 2020). "Tt is very important that everyone understands that a return to the workplace is not a return to the same workplace" (British Standards Institution 2020). This study thus derives solutions to change COVID-19 risks to opportunities for both India and Japan.

According to IMF, in addition to a powerful multilateral coordination to control the global pandemic, accelerating access to a vaccine and initializing economic activity by appropriate political support will become the driving force of recovery from this collapse (IMF 2021).

3.2   METHODOLOGY

In this study, "Issue" is defined as problems occurring or expected, and "Challenge" is defined as tasks and action items to solve "Issue". "Solution" stimulates and moves "Challenge" to be enforced as real actions. "Solution" is explained in the Discussion section. Due to the continuous pandemic situation, the effectiveness of the recommended solutions remains to be assessed.

This study follows three methods, i.e., *survey method (Section 3.2.1), comparison method (Section 3.2.2), and issue and challenge identification method (Section 3.2.3).*

3.2.1 Survey Method

To explore the current circumstances of e-waste management in India and Japan, three types of surveys, i.e., field surveys, telephone/e-mail surveys, and literature/newspaper/website surveys were conducted (Table 3.1).

Field survey: To understand the holistic on-site e-waste management circumstance in India, this study includes field surveys in formal/informal sectors in India. The field survey of the formal sector was carried out from 16 to 20 March 2019 with the support of esteemed Green Waves Environmental Solutions (GWES) at Visakhapatnam in Andhra Pradesh, India.

Telephone/email survey: To avoid the infection and spreads of COVID-19, telephone and email-based surveys were conducted. This type of survey was carried out in an informal sector in Japan in March 2019 and in formal sectors in India and Japan in October 2020.

Literature/newspaper/website survey: Literature, newspaper, and website surveys were carried out in October 2020.

3.2.2 Comparison Method

The results acquired by *Survey method* are categorized in the following three categories, i.e., "E-waste management system", "E-waste management OSH", and "E-waste management under COVID-19" (Table 3.2). To extract gaps and commonalities between India and Japan, a comparison analysis was conducted. Comparison items in each category were followed as given in Table 3.2.

3.2.3 Issue and Challenge Identification Method

From the results, i.e., gaps and commonalities, acquired by *Comparison method,* India and Japan's "Issue" and its corresponding "Challenge" are identified. Corresponding "Solution" is explained in the Discussion section.

TABLE 3.1
Types of Surveys

Survey Type	Sector	Country	Period
Field	Formal/Informal	India	March 2019
Telephone/Email	Informal	Japan	March 2019
	Formal	India/Japan	October 2020
Literature/Newspaper/Website	Formal/Informal	India/Japan	October 2020

TABLE 3.2
Category and Comparison Item

Category	Comparison Item
E-waste management system	□ E-waste management regulation
	□ Restriction of hazardous substances
	□ EPR
	□ Form/Manifest
	□ The ratio of e-waste handled by the informal sector
	□ Potentiality of harming OSH and environment in other countries
	□ E-waste collection circumstance
E-waste management OSH	□ OSH regulation
	□ Compliance status of OSH regulations
	□ OSH risk
	□ Gender gap
E-waste management under COVID-19	□ Correspondence by government
	□ National guidelines for WM under COVID-19
	□ E-waste management guidelines under COVID-19
	□ WM guideline's supportive video under COVID-19
	□ Obligation of reporting employee's COVID-19 infection as occupational accidents
	□ Change of e-waste collection target
	□ Effect on e-waste management
	□ Compliance status of national guidelines under COVID-19
	□ Amount of e-waste generated
	□ Handling speed
	□ Business
	□ OSH risk
	□ Gender gap
	□ Effect on society
	□ Change of awareness on e-waste

3.3   RESULT

In this section, first, *E-waste management system (Section 3.3.1), E-waste management OSH (Section 3.3.2), and E-waste management under COVID-19 (Section 3.3.3)* in India and Japan were surveyed and compared. Next, the results of *Issue and challenge (Section 3.3.4)* in India and Japan were identified.

3.3.1 E-WASTE MANAGEMENT SYSTEM

Table 3.3 shows the results of the comparison of e-waste management systems in India and Japan. Both India and Japan have e-waste management regulations including hazardous substances, Extended Producer Responsibility (EPR), and Form/Manifest. However, both countries have not achieved their respective e-waste recycling target. Key differences between India and Japan are the ratio of e-waste handled by the informal sector and the role in the informal sector. While 95% of e-waste

TABLE 3.3

Comparison of E-waste Management System in India and Japan

Comparison Item	India*	Japan*
☐ E-waste management regulation	✓ (Since 2011)	✓ (Since 1998)
☐ Restriction of hazardous substances	✓	✓ (Designated hazardous e-wastes)
☐ EPR	✓	✓
☐ Form/Manifest	✓ (Paper)	✓ (Paper and electronic)
☐ The ratio of e-waste handled by the informal sector (and its main role)	95% (collection, dismantling, recycling, and disposal)	30%: as of 2014 (collection and export)
☐ Harming OSH and environment in other countries	-	Potential (illegal exports)
☐ E-waste collection circumstance	Formal collectors are not able to reach enough volume of e-waste	Recovery amount target of small home appliances' recycling has not been achieved

*Note: ✓: confirmed, -: no information acquired from the survey of this study.

has been treated by the informal sector in India, 30% of e-waste is illegally exported from Japan to outside the country. Illegal export has the potentiality of harming OSH and the environment in other countries if the e-waste was treated in the informal sector in the importing countries. Enforcement of the rules to the informal sector is one of the key challenges for both countries. In India, governments and authorized e-waste handlers such as GWES are struggling to connect e-waste to formal sectors with the help of awareness activities. Ensuring the profitability of the e-waste business in the formal sector is also one of the urgent challenges in India.

3.3.1.1   E-waste Management System in India

E-waste Management Regulation: Laws to manage e-waste were placed in 2011. After e-waste (Management & Handling) Rules, 2011, e-waste (Management) Rules, 2016, have come into force (MOEFCC 2016). In March 2018, the Ministry of Environment, Forest and Climate Change (MOEFCC) amended the e-waste (Management) Rules, 2016. The informal sector handles the large majority of the waste, thus, enforcing rules remains a challenge (Forti et al. 2020). The regulations support EPR schemes and scopes including informal actors for the formalization and ban the illegal import of e-waste (Pena 2012).

Restriction of Hazardous Substances (RoHS)/Extended Producer Responsibility (EPR): While provisions of RoHS have been retained from the e-waste regulations of 2011, several provisions are added in e-waste (Management) Rules, 2016. For example, CPCB examines products from the market and verifies RoHS compliance. Non-compliance products are withdrawn or recalled from the market until their compliance is assured (Das and Chatterjee 2017). Under the e-waste (Management) Amendment Rules, 2018, the cost of sampling and testing for the RoHS test is borne by the government. If the product does not comply with RoHS provisions, the cost of the test will be borne by the producers. Under the provision of EPR, not only producers who have placed goods in the market during the financial year but also the producers who have started their sales operations recently are under the scope. Producer Responsibility Organizations (PROs) are required to apply to the Central Pollution Control board (CPCB) for registration to undertake activities prescribed in the rules (MOEFCC 2018a). In the case of transfer or sale of assets by the producers, the liability under EPR also should be transferred to the buyer. The manufacturer, producer, importer, transporter, refurbisher, dismantler, and recycler are liable to pay financial penalties under the provisions of the Environment (Protection) Act, 1986 and rules for any violation of the provisions under these rules by the State Pollution Control Boards (SPCB). By following the guidelines published by the CPCB, the provisions are priorly approved by CPCB. The collection, storage, transportation, segregation, refurbishment, dismantling, recycling, and disposal of e-waste shall be following the guidelines published by the CPCB (MOEFCC 2018b). Figure 3.1 shows the structure of e-waste management in India.

Case Study (Formal Sector Authorization): Green Waves Environmental Solutions (GWES) is the first authorized company for collection, storage, and dismantling of e-waste under the e-waste (Management) Rules, 2016 by Andhra Pradesh Pollution Control Board (APPCB) in 2016 (APPCB 2016).

Case Study (E-waste Collection Circumstance/Form): The author surveyed e-waste collection by GWES at a governmental body, Visakhapatnam Port Trust (VPT) in March 2019. During the observation, e-wastes, e.g., PC monitors of the 90s, were collected from several different buildings of VPT (Figure 3.2). Personal Protective Equipment (PPE) such as globes were properly worn by the collection

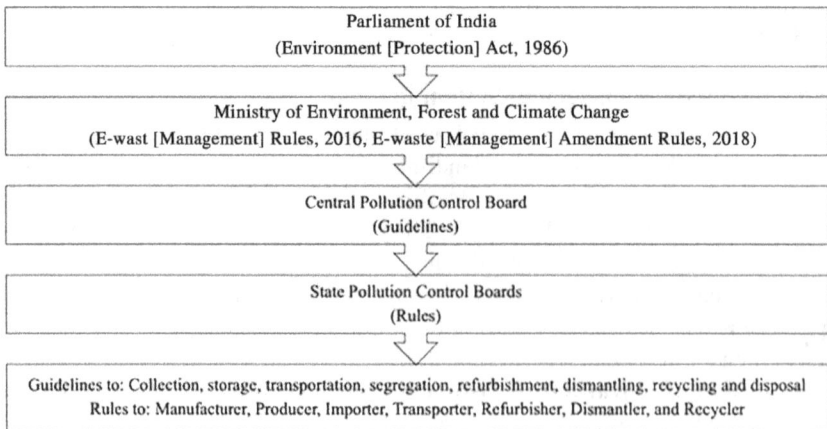

| Parliament of India |
| (Environment [Protection] Act, 1986) |

| Ministry of Environment, Forest and Climate Change |
| (E-wast [Management] Rules, 2016, E-waste [Management] Amendment Rules, 2018) |

| Central Pollution Control Board |
| (Guidelines) |

| State Pollution Control Boards |
| (Rules) |

| Guidelines to: Collection, storage, transportation, segregation, refurbishment, dismantling, recycling and disposal |
| Rules to: Manufacturer, Producer, Importer, Transporter, Refurbisher, Dismantler, and Recycler |

FIGURE 3.1   Structure of e-waste management in India.

FIGURE 3.2   Collection at VPT, 2019.

representatives. The collection process included lifting objects, going up and down the stairs, and loading objects onto the trucks. "Form", which is equivalent to "Manifest" in Japan, was properly filled and processed on paper bases.

Case Study (Formal Sector Storage): The storage process at GWES was observed by the author in March 2019 (Figure 3.3). Collected e-wastes were stored at authorized locked storage protected by the security guard. Hazardous e-waste, e.g., mercury lamps, were fully covered separately for safety and the environment.

Case Study (Formal Sector Dismantling): Dismantling process at GWES in Hyderabad was observed by the author in March 2019. After the hazardous segment checkup, the e-waste which had passed the hazardous segment test was opened and checked whether some of the parts were still working properly. If no failure was found, re-use possibilities were explored, e.g., inserting CPU or HDD to another motherboard (Figure 3.4). It takes 15 to 25 minutes to dismantle one CPU. In addition to motherboards, the author observed the dismantling process of keyboards, codes (Figure 3.5), etc. The dismantling process was mechanically conducted. Depending on the types of e-waste, the process was conducted manually, semi-manually, or automatically (Figure 3.6), and each representative was wearing proper PPEs, e.g., gloves, masks, safety specs. Dismantled e-wastes were sent to authorized recyclers.

The Ratio of the Informal Sector/E-waste Collection Circumstance: Authorized recyclers have difficulties accessing e-waste supply (Raghupathy et al. 2010). Initiatives have been taking place to integrate actors in the informal economy, who are responsible for 95% of e-waste recycling in India, into the formal economy (Chaturvedi and Bhardwak 2013). Organizing the informal

FIGURE 3.3   E-waste storage of GWES, 2019.

FIGURE 3.4   Dismantling process with PPE and opened motherboard.

economy requires knowing the actors, gaining their trust, and acquiring a good understanding of the entire recycling chain (ILO 2014). Dr. Anil Chowdary, Managing Director of GWES, an authorized company for collection, storage, and dismantling of e-waste, said in 2019, "Our key challenge is how to increase the collection amount and profitability of the business. For example, distance to carry e-waste directly affects the cost. We need to find solutions". Producers need to foster a network that can help engage the informal sector to meet their recycling targets under India's 2016 EPR-based E-Waste Rules (Radulovic 2018). "Lack of proper collection, logistics infrastructure, limited awareness of consumers on the hazards of improper disposal of e-waste, the lack of standards for the collection,

FIGURE 3.5 Code opening machine and copper removed from a cord.

FIGURE 3.6 Plastic and copper part of the electric cable was separated by cable wire crushing and segregation machine.

dismantling, treatment of e-waste, and an inefficient reporting process issues are to be tackled" (Forti et al. 2020).

Case Study (Connect E-waste to the Formal Sector): To accelerate formal collection channels in society and increase awareness of consumers, GWES supports effective pick-up service. GWES has created a smartphone app called ReByte. ReByte offers doorstep waste collection services to households, public offices, and corporates. Collection requests are easily processed via Rebyte.

Case Study (Awareness Activity): To encourage customers to continue their proper emission of e-waste through authorized dealers, GWES issues Certificate of Recycling. It certifies that "All the material has been collected and handled in an

FIGURE 3.7   ECO-BIN.

FIGURE 3.8   Educational activity in 2019.

environment-friendly manner, in accordance with the guidelines set by the CPCB".
GWES also supports St. Joseph's College for Women (SJCW) in Visakhapatnam.
GWES has installed ECO-BIN in SJCW to collect e-waste and to increase the awareness among students for the importance of formal e-waste recycling (Figure 3.7). In addition, awarding activities and educational activities (Figure 3.8) for the students in SJCW by GWES were confirmed.

3.3.1.2   E-waste Management System in Japan

E-waste Management Regulation: Japan is one of the first countries globally to implement an EPR-based system for managing e-waste. Japan relies on a strong legal framework, an advanced collection system, and a developed processing

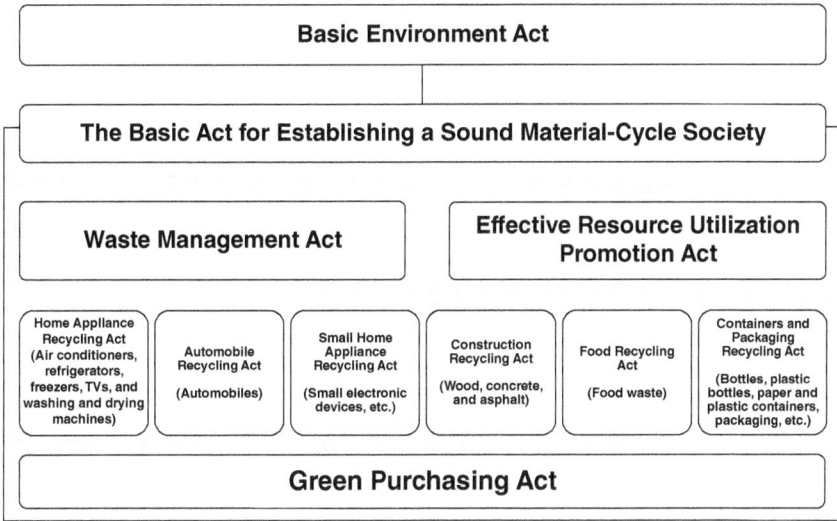

FIGURE 3.9 Legal structure to establish a "Sound Material-Cycle Society" in Japan.

infrastructure (Forti et al. 2020). Figure 3.9 shows the legal structure related to e-waste management in Japan, based on the web survey of "History and Current State of Waste Management in Japan" (MOE 2014). According to the Ministry of the Environment (MOE) in Japan, under the Basic Environment Act, the Basic Act for Establishing a Sound Material-Cycle Society was composed. Fundamental principles and guidelines for a sound material-cycle society are provided by the act. Waste Management Act protects living environments and improves public health by controlling waste generation and by treating waste appropriately. The purpose of Effective Resource Utilization Promotion Act is utilizing resources effectively, reducing waste generation, and preserving the environment. Green Purchasing Act promotes the purchase of environmentally friendly products and services. There are six types of Recycling Acts by products (MOE 2014). In Japan, Electronic and Electrical Equipment (EEE) products are collected and recycled under the Recycling Acts, such as Home Appliance Recycling Act and Small Home Appliance Recycling Act.

Effective Resource Utilization Promotion Act (Disclosure of Hazardous Substances): Under the act, business operators are required to promote the reuse of designated home appliances and personal computers, and manufacturers and importers are obligated to disclose information on designated substances contained in products (METI 2008). Other than the national regulations, as an initiative for companies in Japan, there is a growing movement to carry out so-called "green procurement" that seeks information on contained substances when procuring materials and parts (Suwa 2006).

Home Appliance Recycling Act (EPR/Manifest): Home Appliance Recycling Act was enacted for four types of waste home appliances, i.e., 1) air conditioners, 2) refrigerators and freezers, 3) TVs, and 4) washing and drying machines. The Act obligates home appliance retailers to accept them from consumers and to deliver them to manufacturers. Manufacturers are obligated to recycle them, to meet the recycling rate target, and to recover CFCs, HCFCs, and HFCs (MOE 2014). When consumers dispose of home appliances, the consumers need to cover fees of transportation, collection, and recycling (MOE 2014). A manifest system was established in which consumers can check that the waste they discharged was correctly transported and disposed (MOE 2020c).

Automobile Recycling Act (EPR): The Act defines the responsibility of automobile manufacturers to accept and recycle three items that are often illegally discarded, i.e., CFCs, shredder residues, and airbags. Automobile owners are obliged to cover disposal fees (MOE 2014).

Small Home Appliance Recycling Act (E-waste Collection Circumstance): The Act aims to make an effective use of metals and proper disposals of small home appliances since 2012 (MOE 2014). However, the recovery amount target (140,000 tons/year) set at the start of the act has not yet been achieved (MOE 2014). According to Industrial Structure Council and Central Environment Council (ISCCEC) 2020, reducing collection and transportation costs is necessary, besides, as a recent issue of Japan, the disposal costs of waste plastics increased due to the impact of China's import ban. Small home appliances are collected by consumer electronics mass retailers, by collection bases set up by certified businesses, by courier services, or by collection boxes of Municipalities. Station collection and pickup collection methods are proved highly effective to increase the collection amount by municipalities (ISCCEC 2020). The awareness of the small home appliances recycling system among consumers is about 60%, which has been flat in recent years, and it is necessary to further raise awareness (ISCCEC 2020). Dissemination and enlightenment measures linked to the "Make from Urban Mines! Everyone's Medal Project" and other dissemination and enlightenment measures by businesses, local governments, and consumers are being promoted (ISCCEC 2020).

Waste Management Act/E-manifest: The act distinguishes between industrial waste and municipal waste and also defines the responsibilities of municipalities for municipal waste management and waste-generating business operators for industrial waste management (MOE 2014). It is requested that industrial wastes are disposed of by using paper manifests or electronic ones (JW Center 2020). E-manifest reached 63% of all manifests in FY 2019 (JW Center 2020). Under the Waste Management Law Amendment 2017, someone who discharges 50 tons or more of specially controlled industrial wastes per year is obliged to use e-manifests from 1 April 2020 (JW Center 2020).

Case Study (E-manifest): The author interviewed an employee, who works for a formal industrial waste treatment company in Ibaraki-prefecture on 15 October 2020. The employee said, "E-manifesto is fully penetrated our daily work. The usage of E-manifest is very simple. By using E-manifest, we do not need to spend time inputting for writing down the information".

The Ratio of the Informal Sector/Harming OSH and Environment in Other Countries: According to Hotta et al. (2014), illegal export of e-waste was estimated at about 30% of total e-waste in Japan. There is concern that such illegally exported items will be managed improperly in the importing countries without taking into account the environmental considerations (Hotta et al. 2014). Under the Waste Management Law, when collecting, transporting, or disposing of waste are entrusted by another person, permission from the local government is required. However, there are illegal collectors in Japan, and also there are some cases where waste home appliances collected by such illegal waste collection companies are illegally dumped, scrapped, or exported (ISCCEC 2020). Illegal or unauthorized cases on the Basel Law have been monitored, and these cases have been continuing, thus, administrative guidance (strict caution) has been given (MOE 2020a). Common destinations of e-waste shipped from Japan include China, India, Pakistan, Vietnam, the Philippines, South Korea, and Thailand (Fuse et al. 2011). To prevent the illegal export of wastes, enhanced monitoring activities were undertaken by Regional Environment Offices (REO), Ministry of the Environment (MOE) in cooperation with Customs (MOE 2013). Japanese government revised the Waste Management Act and Basel Act, and both of the revised acts were enforced in 2018. Revised Waste Management Act newly covers the listed hazardous end-of-life (EoL) equipment and requests the entity who stores and/or disposes of them to notify prefectures and to follow the standards. Revised Basel Act strengthened the export control for the listed e-waste (Terazono 2019).

Case Study (Informal Sector): To understand the circumstances of e-waste export, the author held a telephone-based hearing survey with the support of a former home waste collector in Saitama prefecture on 30 March in 2019. According to the informal exporter, they said, "We used to export e-waste mainly to China through exporting companies, but not anymore". The former home waste collector said,

> E-wastes acquired from individual houses by local home waste collection companies were able to be sold to E-waste exporting companies in Japan. However, in recent years, Japanese local home waste collectors' situation has dramatically changed because the Chinese regulations have become stricter. The price of E-waste to sell to exporting companies has dropped, or these wastes acceptance have often been denied. The business is no longer viable.

The home waste collector changed the job in mid-2019.

3.3.2 E-WASTE MANAGEMENT OSH

Table 3.4 shows the comparison of e-waste management OSH in India and Japan. Although both countries have already set OSH regulations, the informal sector does exist in both countries. According to Directorate General Factory Advice Service et al., the informal sector is usually not covered by the OSH regulations. Besides, it was confirmed that the risk of the waste treatment industry is higher than the average risk of all industries in Japan. A gender gap on OSH was observed in the informal sector in India.

TABLE 3.4

Comparison of E-waste Management OSH in India and Japan

Comparison Item	India*	Japan*
☐ OSH regulation	✓	✓
☐ Compliance status of OSH regulation	Formal sector (✓) Informal sector (regulation not covered)	Formal sector (✓) Informal sector (-)
☐ OSH risk	Risk of the informal sector is higher than the risk of formal sector	Risk of waste treatment industry is higher than the average risk of all industries
☐ Gender gap	Insufficient PPE for women was observed (informal sector)	-

Note: ✓: confirmed, -: no information acquired from the survey of this study.

3.3.2.1   E-waste Management OSH in India

OSH Regulation: OHS-related statutes such as Factories Act, National OSHE Policy, Dock Workers (Safety, Health and Welfare) Regulations, Model Rules framed under Factories Act, Mines Safety Act, Indian Boilers Act, Indian Boilers Regulations, Dangerous Machines (Regulations) Act, Child Labour (Prohibition and Regulations) Act, Indian Electricity Act, Indian Electricity Rules, and Static and Mobile Pressure Vessels Rules are confirmed (Ministry of Labour, Government of India 2019).

Compliance Status of OSH Regulations: Workers in the informal economy in India are not covered under any of the OSH legislation, thus, system/mechanism supports need to be provided (Directorate General Factory Advice Service et al. 2020).

OSH Risk (General/Global): E-waste workers are exposed to safety and health hazard risks arising from their work, e.g., carrying heavy waste materials, exposure to hazardous chemicals, unsafe use of electricity and machinery, and long and irregular working hours. E-waste workers in the informal sector rarely receive adequate protection. Thus, practical support is required (IFC et al. 2019).

Gender Gap (General/Global): In the informal sector, men often have higher authority and income than women, e.g., trading wastes. On the other hand, women's roles are often limited, e.g., wastes picking and separation (UNEP et al. 2020).

Gender Gap (Informal Sector): According to the result of a field survey at an informal sector in India in 2019, Under one male director, two women without PPEs were working with bare hands under the sun (Figure 3.10). Women's insufficient occupational OSH was observed in the informal sector.

3.3.2.2   E-waste Management OSH in Japan

OSH Regulation: Industrial Safety and Health Act in Japan aims to secure, in conjunction with the Labor Standards Act, the safety and health of workers in workplaces, as well as to facilitate the establishment of a comfortable working environment by promoting comprehensive and systematic countermeasures concerning

FIGURE 3.10   Two women without PPEs at the informal sector.

the prevention of industrial accidents (Ministry of Justice 2020). In 1993, the Ministry of Labor established the "Safety and Health Management Guidelines for Cleaning Businesses" to further strengthen occupational safety and health measures in waste treatment businesses. The guideline includes safety and health measures for garbage collection work and implementation of safety and health education (MOE 1993).

Compliance Status of OSH Regulation: The survey result of *E-waste management system in Japan* confirmed the existence of the informal sector in Japan. The informal sector may potentially not follow the OSH regulations.

OSH Risk: Based on the occupational accident occurrence situation report from the Ministry of Health, the National Federation of Industrial Waste Management (NFIWM) summarized the occurrence of occupational accidents in the waste treatment industry. According to the occurrence report, the frequency rate and the intensity number of the industrial waste treatment industry and municipal waste treatment industry are higher than the average of all industries (Table 3.5) (NFIWM 2020).

Gender Gap: OSH-related gender gap in e-waste management was not confirmed from the survey. However, Japan's gender gap is by far the largest among all advanced economies and has widened over the past year. Japan has a noticeable economic and status gender gap (World Economic Forum 2020).

3.3.3 E-waste Management Under COVID-19

Table 3.6 shows the results of the comparison of e-waste management under COVID-19 in India and Japan. National guidelines of WM under COVID-19 have been issued by both countries, and each guideline is followed by the formal sector in each country. Japan created a supportive video for the guidelines. Both countries have not yet

TABLE 3.5
Occurrence of Occupational Accidents in Japan, 2019

Type of Industry (Average)	Frequency Rate	Strength Rate
All industry	1.8	0.09
Waste treatment industry	6.99	0.17
(industrial waste and municipal waste)		

TABLE 3.6
Comparison of E-waste Management under COVID-19 in India and Japan

Comparison Item	India*	Japan*
Correspondence by the government:		
☐ National guidelines for WM under COVID-19	✓	✓
☐ E-waste management guideline under COVID-19	-	-
☐ WM guideline's supportive video under COVID-19	-	✓
☐ The obligation of reporting employee's COVID-19 infection as occupational accidents	Depends on the states	✓
☐ Change of e-waste collection target	✓	-
Effect on e-waste management:		
☐ Compliance status of national guidelines under COVID-19	Compliant (formal sector)	Compliant (formal sector)
☐ Amount of e-waste generated	-	Increased
☐ Handling speed	Delayed case was reported (collection, dismantling)	-
☐ Business	Negatively affected	-
☐ OSH risk	OSH risk under COVID-19 is higher than normal time, and especially the informal sector's risk under COVID-19 is higher than the formal sector	Higher than normal time
☐ Gender gap	Suggested that the COVID-19 pandemic could put women in WM at higher social, economic, and health risk	
Effect on society:		
☐ Change of awareness on e-waste	-	Increased urban mining necessity due to the resource security reason

Note: ✓: confirmed, -: no information acquired from the survey of this study.

issued the guidelines for the e-waste management context under COVID-19. The Indian government has decided to lower the E-waste collection target of 2020–2021. While the Japanese government has an obligation to report employees' COVID-19 infection as an occupational accident, currently it depends on the states in India. The generation of the amount of e-waste in Japan was confirmed to have been increasing under COVID-19. The delay of handling speed on collection and dismantling was reported in India. There is a confirmed case in the formal sector in India that the e-waste business was negatively affected. For both countries, OSH risk under COVID-19 is higher than normal time, and especially the informal sector's risk under COVID-19 is higher than the formal sector. In general, the gender gap on WM OSH risk under COVID-19 is suggested to be increased. In Japan, the necessity of urban mining resources is increasing due to the resource security ensuring reason under COVID-19.

3.3.3.1   E-waste Management under COVID-19 in India

Correspondence by the Government: MOEFCC received requests from the electrical and electronic industry associations to reconsider the collection target for the financial year 2020–-2021 under the EPR provision of the e-waste (Management) Rules, 2016. In October 2020, MOEFCC decided to change the target from 40% to 30% (MoEFCC 2020a). CPCB issued "Guidelines for Handling, Treatment and Disposal of Waste Generated during Treatment/Diagnosis/Quarantine of COVID-19 Patients" on 25 March 2020 (MoEFCC 2020b). On occupational accident reporting rules, although there is no general obligation placed on employers to report their employees suspected to have COVID-19 to the government, certain large cities oblige employers to notify local municipal authorities (Wijekoon et al. 2020).

Effect on E-waste Management (Formal Sector): On 15 October 2020, Dr. P. Anil Chowdary of GWES, an authorized company for collection, storage, and dismantling, said,

> Under COVID-19, it has been challenging and struggling to provide service. WM requires a mass or group of people to work together. We have reduced the manpower of each shift to half to avoid the transmission of COVID-19. We have set a rotation schedule for picking up, kept thermal check for workers, made additional PPE and medical insurance available for workers. We first disinfect E-waste with 1% sodium hypochlorite before we send it to be dismantled. Speed of the recycling process has declined due to limitations in taking work schedules, and it has been negatively affecting the business.

3.3.3.2   E-waste Management under COVID-19 in Japan

Correspondence by the Government: Guidelines for waste collection and transport under COVID-19 and for household garbage disposal under COVID-19 have been published (MOE 2020d). On 20 October, the second edition of the waste treatment guidelines for the waste treatment industry was issued (JW center and Japan Environmental Sanitation Center 2020). It explains measures and caution points for waste treatment work. It contains detailed precautions, i.e., how to use toilets and rest areas, and it also encourages telework. The guidelines also recommend installing

COCOA (COVID-19 Contact-Confirming Application) app released by the Ministry of Health, Labor and Welfare (MHLW). COVID-19 infection cases at the workplace are mandatory to be reported as occupational accident occurrence situations to the MHLW (MHLW 2020). Japanese government posted a large-scale advertisement of the e-waste web portal on the Nikkei newspaper to encourage consumers to choose the formal sector (The Nikkei 2020). By using the web portal, consumers can easily search the authorized e-waste handler nearby.

Effect on E-waste Management (Formal Sector): The author interviewed an employee, who works for a formal industrial waste treatment company in Ibaraki-prefecture on 15 October 2020. The employee said, "PPE to avoid the infection of COVID-19 has been provided from the company and we keep wearing them".

Effect on Society: The temporary shutdown of refinery and mining activities as a mitigation measure due to the pandemic has impacted the supply of metals (Rajput et al. 2020). Most of the metal mines in Japan are already closed and rely on imports. Japan is required to have an idea on how to secure metal resources, which are feared to rise in price. Precious metals and rare metals contained in e-waste are called urban mines, and the business of recycling metal resources is expanding (The Nikkei 2020). Under the pandemic, the amount of Japanese e-waste has increased. According to The Nikkei, due to the influence of COVID-19, many people sorted out the disued items in their houses, and the collection of used PCs increased in 2020. "According to the PC 3R Promotion Association, 89,208 PCs were collected from homes from April to June in 2020 when the state of emergency was announced, i.e., 33.3% increase from the same period of the previous year" (The Nikkei 2020).

3.3.4   ISSUE AND CHALLENGE

In this section, "Issue" and its corresponding "Challenge" of India and Japan were identified from the results of gaps and commonalities based on *issue and challenge identification method*. As one of the key common challenges for the new normal in India and Japan, it is highlighted that there was a necessity of creating a mechanism to collect e-waste, sustainably and safely, under COVID-19.

3.3.4.1   Issue and Challenge in India

Issues and challenges in India were identified following the three categories, i.e., "E-waste management system", "E-waste management OSH", and "E-waste management under COVID-19".

E-waste Management System: Issues are inefficient reporting processes, e.g., using a paper form. In addition, the high informal sector ratio causes insufficient compliance status of e-waste management regulations and environmental pollution risk in the country. Regarding e-waste collection circumstances, formal collectors are not able to reach enough volume of e-waste. Corresponding challenges are to "Establish/implement electronic form system", "Transform informal sector to formal sector", "Enforce regulations to the informal sector", and "Connect E-waste to formal collectors".

E-waste Management OSH: Issues are that the informal sector may potentially not be compliant with the OSH regulations, the OSH risk of the informal sector is higher than the risk of the formal sector, and insufficient PPE for women were observed in the informal sector. Corresponding challenges are "Transform informal sector to the formal sector", "Reduce OSH risk of the informal sector", and "Eliminate gender gap in E-waste management".

E-waste Management Under COVID-19: Issues are the lack of e-waste management guidelines under COVID-19, lack of WM guidelines' supportive video under COVID-19, the obligation of reporting employee's COVID-19 infection as the occupational accident depends on the states. E-waste recycling target was changed due to the decrease of the e-waste recycling ratio. Handling speed of e-waste collection and dismantling was reduced. Thus, the e-waste business was negatively affected by COVID-19. Furthermore, the potential increase of the gender gap and also OSH risk especially in the informal sector are worth considering. Corresponding challenges are "Create E-waste management guidelines under COVID-19", "Create WM guidelines' supportive video under COVID-19", "Obligate reporting employee's COVID-19 infection as occupational accidents", "Avoid decrease of E-waste recycling ratio under COVID-19", "Avoid the delay of E-waste collection and reduction of dismantling speed under COVID-19", "Avoid negative effects of E-waste business under COVID-19", "Reduce OSH risk of the formal and the informal sector under COVID-19", and "Avoid gender gap increase under COVID-19".

3.3.4.2 Issue and Challenge in Japan

Issues and challenges in Japan were identified following the three categories, i.e., "E-waste management system", "E-waste management OSH", and "E-waste management under COVID-19".

E-waste Management System: Issues are that still paper manifest which requires the face to face contact is used occasionally, there is collection and export by the informal sector, the recovery amount target of small home appliances' recycling has not been achieved, and the potentiality of harming OSH and environment in other countries through the illegal export of informal sector. Corresponding challenges are "Increase electronic manifest ratio", "Transform informal sector to formal sector", "Stop illegal export", and "Increase recovery amount of small home appliances".

E-waste Management OSH: Issues are that OSH regulations will potentially not be followed by the informal sector, and the OSH risk of the waste treatment industry is higher than the average risk of all industries. Corresponding challenges are "Transform the informal sector to the formal sector" and "Reduce OSH risk".

E-waste Management Under COVID-19: Issues are the lack of e-waste management guidelines under COVID-19, OSH risk higher than normal time, and the potential gender gap increase. Corresponding challenges are "Create E-waste management guideline under COVID-19", "Reduce OSH risk under COVID-19", and "Avoid the potential gender gap increase under COVID-19"

3.4 DISCUSSION

To enforce identified challenges to result into real actions, solutions for India and Japan were argued and pointed out in this section. Solutions were categorized into three categories, i.e., "E-waste management system", "E-waste management OSH", and "E-waste management under COVID-19". Rationales are stated based on the survey results stated before.

3.4.1   SOLUTION FOR INDIA

3.4.1.1   E-waste Management System

Establish/Implement Electronic Form System: It was confirmed that E-manifest (electronic form) in Japan enhances the efficiency of the WM reporting process. Further implementation of the electronic form has the potential to improve the WM reporting efficiency in India.

Transform the Informal Sector to the Formal Sector: It was confirmed that collection boxes and pick up services are effective collection methods in Japan. MOE in Japan established a web portal of small home appliances collection to connect consumers to the formal sector. Thus, to increase the collection amount of the formal sector in India, it is recommended to establish an easy-to-access web portal for consumers and integrate it with the official collection app. Producers need to network informal sectors to meet the collection target under EPR (Radulovic 2018), thus, giving the support to guide on proper dismantling methods and designing EEE which is easy to disassemble and repair.

Enforce Regulations to the Informal Sector: It was confirmed that strengthening the punishment helped to enforce regulations on the informal sector in Japan, thus, it is recommended to tighten the monitoring/punishment for the violations of e-waste and OSH regulations in India.

Connect E-waste to Formal Collectors: To increase the e-waste supply for formal collectors, it is essential to transform the informal sector into the formal sector. As a result, the informal sector can retain its income, and the formal sector can reach more mass of e-waste (Raghupathy et al. 2010). This mutual gain will become the motivation for both sectors to be connected.

3.4.1.2   E-waste Management OSH

Transform the Informal Sector into the Formal Sector and Reduce OSH Risk of the Informal Sector: The informal sector can reach a higher OSH level by transforming into the formal sector (Raghupathy et al. 2010). It is recognized as a merit for the informal sector to transform into the formal sector. Thus, it is recommended to promote further awareness activities for the informal sector by emphasizing the merit.

Eliminate Gender Gap in E-waste Management: There was an observed circumstance of insufficient PPE of women in the informal sector. In addition to transforming the informal sector to the formal sector, women should participate in the decision-making positions in WM (UNEP et al. 2020). It gives women opportunities to access proper OHS services.

3.4.1.3   E-waste Management under COVID-19

Create E-waste Management Guidelines Under COVID-19: India has already issued the waste management guidelines under COVID-19; however, e-waste-management-specific guidelines have not yet been released. It was confirmed that e-waste management has a long and complex process. Thus, to give e-waste workers and dischargers clear directions for e-waste handling, proper guidelines to follow in the context of e-waste management under COVID-19 are required.

Create Waste Management Guidelines' Supportive Video Under COVID-19: By making municipal waste collection workers understand the caution point visually, OSH risks of e-waste handlers under COVID-19 will be reduced. Thus, creating e-waste management guidelines' supportive videos is recommended.

Obligate Reporting Employee's COVID-19 Infection as Occupational Accidents: It is essential to monitor and understand the circumstances of employee's COVID-19 infection to improve the OSH under COVID-19. Thus, it is recommended to expand the coverage of the reporting to the national level obligation in India.

Avoid the Decrease of E-waste Recycling Ratio Under COVID-19: To keep or increase collection amount under COVID-19, it is recommended to integrate collection apps into electronic form and e-waste web portal, besides adding non-face-to-face awareness activities for the informal sector. The reasons are as follows. The example of the Rebyte app created by GWES proved that the collection app supports engaging e-waste with the formal sector. Under COVID-19, it is essential to consider the collection method which minimizes face-to-face contact. Thus, it is recommended that the integrated collection system should be further integrated with the door-step collection system. These developments will create business opportunities to break the cost and technical barriers and create collaboration between industries in society. Besides, producers and consumers will also become available for tracking their e-waste efficiently without any face-to-face contact, and e-waste workers will be able to improve the work efficiency by systemization. The visualization of the e-waste management system will contribute to enhancing the responsibility awareness among stakeholders. To increase the formal recycling amount, the informal sector needs to be transformed into the formal sector. Newspaper and webinar methods will only reach people who have the access to the specific newspaper or the Internet. Thus, utilizing radio broadcasting, city speaker broadcasting, and bulletin boards methods in the local languages is recommended to reach more people in the informal sector.

Avoid the Delay of E-waste Collection and the Reduction in Dismantling Speed Under COVID-19 and Avoid Negative Affection of E-waste Business Under COVID-19: It was confirmed that one e-waste dismantling company has been disinfecting the e-waste before dismantling the e-waste to avoid the infection of COVID-19 to spread, and it has been causing the delay of the process due to the additional disinfecting work. Thus, there is a need to develop alternative methods to keep/increase the e-waste dismantling speed under COVID-19. The viruses that cause COVID-19 break down over time on the surface of an object, and the infection remains for a maximum of 72 hours according to the surface characteristics (MOE 2020e). For this reason, most of the municipalities in Japan are requesting consumers to discharge PET bottles after waiting for more than two days after they are used. Thus, it is recommended to apply this PET bottle discharging method in Japan to the e-waste management system in India. It will make it possible to dismantle e-waste without applying the disinfecting method and to keep the speed of dismantling steady. The alternative method recommended is to include the e-waste management guidelines under COVID-19. To avoid the delay of the e-waste collection and avoid the reduction of dismantling speed under COVID-19, it is recommended to make the e-waste management process more automated. The reason being not only that the automation will improve the collection and dismantling speed by continuous machine process, but also that the automation will decrease the COVID-19 infection risk of the e-waste handlers for touching the infected e-wastes and decrease the risk of infection of face-to-face contact.

Reduce OSH Risk of Formal and Informal Sectors Under COVID-19: The informal sector is not covered under the existing regulations, thus, there is a need for the informal sector to be transformed into the formal sector. All stakeholders are recommended to motivate and transform the informal sector while emphasizing clear benefits, as follows. The informal sector will reach a higher OSH level and a more efficient and sustainable e-waste management system. Consumers will have the satisfaction of fulfilling their responsibility toward ESG. Producers will achieve the national collection target and will receive more investment from investors and will acquire further sales from responsible consumers. It is confirmed that ESG-related sustainable bond issuance will increase markedly under COVID-19 (Oxford Analytica 2020), and capital flowing will affect the ESG portfolio under COVID-19 (Amanjot 2020).

Avoid Gender Gap Increase Under COVID-19: Women are potentially excluded from the process of decision-making, which could make women inaccessible to required information and services (UNEP et al. 2020). To ensure gender OSH equality in e-waste management, it is recommended that women should participate in decision-making to introduce OSH measures. Furthermore, COVID-19 infection cases at work are strongly recommended to be monitored as the aspect of occupational accidents by gender-disaggregated data.

3.4.2 Solution for Japan

3.4.2.1 E-waste Management System

Increase Electronic Manifest Ratio: It is recommended that the Japanese government integrate the web portal site, the E-manifest, and e-waste collection apps. The reasons are as follows. The current obligation of the scope of E-manifest in Japan is limited to industrial waste. By the integration of the system, e-waste from non-business consumers will be under the scope of E-manifest. The current web portal site created by the government supports consumers to search for formal collectors' contact information or the formal collection boxes' locations. By the integration of the system, consumers will be able to eliminate the process of searching and contacting the collectors and trace the proper e-waste management system. Authorized collectors will be able to eliminate the process of responding to individual inquiries by phone. MOE and JW Center will be able to increase the E-manifest scope and amount of recycling and achieve the targets.

Transform Informal Sector into Formal Sector: In addition to promoting the current e-waste web portal site, it is recommended to create the official e-waste collection app in Japan. It was confirmed that the collection app will connect e-waste to the formal sector from the informal sector. The reduction of the informal collection amount will give an opportunity to the informal sector to consider the transformation.

Stop Illegal Export: The existence of this illegal sector in Japan was confirmed. Thus, it is recommended to increase awareness among consumers (dischargers) to avoid handing its e-waste to the informal sector. To further improve the current awareness activities, it is recommended to add social networking service (SNS) and TV advertisement methods. In addition, it is recommended to flashily disclose the detection of illegal exports in a big way.

Increase Recovery Amount Target of Small Home Appliances: In addition to the continuation of the awareness activities for the consumers to connect e-waste to the formal sector, it is recommended to establish the easy-to-use-system by integrating E-manifest, the web portal, and the collection app. It will help the consumers to choose the formal sector easily.

Stop Illegal Export (How to Avoid OSH and Environmental Negative Effects on Other Countries?): There is the possibility that Japanese informal e-waste sector workers do not have an idea or knowledge of how their actions have badly affected the other countries' OSH. Thus, informal sector workers in Japan should be educated through awareness activities about the circumstances of OSH and environmental pollution in the importing countries for illegally exporting e-waste.

3.4.2.2   E-waste Management OSH

Transform Informal Sector to Formal Sector/Reduce OSH Risk: The informal sector potentially does not follow the OSH regulations. Collection and transportation processes in the informal sector potentially contain OSH risks such as lifting heavy objects and long-hour driving. According to the results of the survey of the informal sector in Japan, it was confirmed that strengthening regulations forces the informal sector workers to stop their illegal work. Thus, in addition to the increase of OSH awareness in society, it is recommended for the Japanese government to strengthen the regulations and punishments to transform the informal sector.

3.4.2.3   E-waste Management Under COVID-19

Create E-waste Management Guidelines Under COVID-19 /Reduce OSH Risk Under COVID-19: Creating e-waste management guidelines under COVID-19 and the supportive video will aid the e-waste handlers to understand the caution points and contribute to reducing the risks of the overall e-waste management system. It is recommended that the guidelines include the recommendation of the integration of the collection system and the door-step collection robots. The guidelines, videos, and integrated system will help us to reduce infection risks. Besides, the integrated efficient system will strengthen the capacity to respond to the increasing generation of e-waste under COVID-19 in Japan.

Avoid Gender Gap Increase Under COVID-19: According to the Global Gender Gap Report 2020, Japan ranks 144 out of 152 countries for the political empowerment of women in the world (World Economic Forum 2020). The increase of women's representation and positions in decision-making processes will provide more opportunities to shape important decisions for WM (UNEP et al. 2020). Thus, women in the e-waste management industry in Japan are recommended to actively participate in the decision-making processes of determining the content of e-waste management guidelines under COVID-19. It will allow us to improve both women's empowerment and sustainable e-waste management in Japan, as the "new normal".

3.5 CONCLUSION

In this study, to derive new solutions to transform COVID-19 risks to the opportunities of e-waste management in India and Japan, *survey method, comparison method,* and *issue and challenge identification method* were conducted based on the latest and practical information of e-waste management in India and Japan. It was confirmed that there are issues, challenges, and solutions, as well, to e-waste management in India and Japan. Solutions to sustainable e-waste management under COVID-19 were derived and recommended to move forward to the "new normal" e-waste management.

Key issues in India are the high ratio of the informal sector and negative affection to the e-waste collection rate and e-waste business under COVID-19. In India, higher OSH risk in the informal sector than in the formal sector was observed.

Key issues in Japan are unachieved small home appliances' recovery targets and the existence of the informal sector that exports e-wastes. In Japan, higher OSH risk in the waste treatment industry than the average industry was confirmed.

Since informal sectors are not covered by e-waste management and OSH regulations, transforming informal sectors into formal sectors is one of the key challenges to increase the formal recycling amount and reduce OSH risks in both India and Japan.

It has been confirmed that the global collapse by COVID-19 has economically adverse effects on social weak people such as informal sector workers and women and also has negative implications for contact-intensive businesses. The current forms of WM sectors in India and Japan, which contain face-to-face contact aspects in the supply chains, can be regarded as contact-intensive sectors.

In addition to the recommendations of additional awareness activities, the integration of the collection apps, the web portals, E-manifesto (electronic forms), and the door-step collection robots was suggested as one of the key solutions. The integrated collection system will contribute to transforming the informal sectors to the formal sectors in India and Japan while aiding to reduce the face-to-face contact risks under COVID-19 and providing a higher volume of e-waste and profit to the formal sectors. Besides, with the participation of women, creating e-waste management guidelines under COVID-19 and the supportive videos were recommended. It will give opportunities not only to improve OSH levels and recycling rate under COVID-19 but also to improve women's empowerment in India and Japan.

ACKNOWLEDGMENT

This research is practically and financially supported by Green Waves Environmental Solutions in India. The author gratefully acknowledges the special support by Dr. P. Anil Chowdary, Managing Director of Green Waves Environmental Solutions;

Dr. P.V. Unnikrishnan, Goa University; Andhra Pradesh Pollution Control Board; Visakhapatnam Port Trust; St Joseph's College for Women; Mr. Irie Makoto; Mr. Yuki Okano; Mrs. Mami Shimizu; Mr. Takayoshi Shimizu; and Dr. Osamu Sugiyama, Kyoto University.

REFERENCES

Andhra Pradesh Pollution Control Board. 2016. *Authorization for Collection, Storage and Dismantling of E-Waste (Under Rule 13(3) (VI) of E-Waste (Management) Rules, 2016.* https://pcb.ap.gov.in/Attachments%20070319/ewasterul/Ms.%20Green%20Waves%20 Environmental%20Solutions%20-%20Authorization.pdf.

British Standards Institution (BSI). 2020. *Special Report: Working in the 'New Normal'.* Martin Cottam, 37–38. www.bsigroup.com/en-GB/topics/novel-coronavirus-covid-19/ the-new-normal/.

Chaturvedi, B., and Bhardwak, S. 2013. *Learning to Re-E-Cycle: What Working with E-Waste Has Taught Us.* New Delhi: Chintan Environmental Research and Action Group, 6. www.chintan-india.org/sites/default/files/2019-07/chintan-study-learning-to-re-e-cycle.pdf.

Das, S. D., and Chatterjee, S. 2017. RoHS Provisions in E-Waste Rules: 2011 & 2016. *IOSR Journals* 42: Table II. doi:10.9790/1676-1201043945. www.iosrjournals.org.

Directorate General Factory Advice Service and Labour Institutes in collaboration with International Labour Organization (ILO). 2020. *National Occupational Safety and Health (OSH) Profile (Draft).* 46. https://dgfasli.gov.in/sites/default/files/service_file/ Nat-OSH-India-Draft%281%29.pdf.

Forti, V., Baldé, C. P., Kuehr, R., and Bel, G. 2020. *The Global E-Waste Monitor 2020.* 74–75, 109. www.itu.int/en/ITU-D/Environment/Documents/Toolbox/GEM_2020_def.pdf.

Fuse, M., Yamasue, E., Reck, B. K., and Graedel T. E. 2011. Regional Development or Resource Preservation? A Perspective from Japanese Appliance Exports. *Ecological Economics* 70: 788–797. ISSN 0921-8009. www.sciencedirect.com/science/article/pii/ S0921800910004702.

Hotta, Y., Santo, A., and Tasaki, T. 2014. *EPR-based Electronic Home Appliance Recycling System Under Home Appliance Recycling Act of Japan.* IGES, 21–25. www.oecd.org/ environment/waste/EPR_Japan_HomeAppliance.pdf.

IFC (International Finance Corporation), ILO (International Labour Organization) and Karo Sambhav. 2019. *Work Improvement for Safe Home—Action Manual for Improving the Safety and Health of E-Waste Workers.* International Labour Office, Decent Work Team for South Asia, 3–4. ISBN 978-92-2-133188-9 (web pdf).

ILO. 2014. *The Informal Economy of E-Waste: The Potential of Cooperative Enterprises in the Management of E-Waste.* International Labour Office, Sectoral Activities Department (SECTOR), Cooperatives Unit (COOP), 32–33. ISBN 978-92-2-129100-8 (print) ISBN 978-92-2-129101-5 (web pdf).

IMF. 2021. *World Economic Outlook Update, January 2021.* International Monetary Fund, 1. www.imf.org/-/media/Files/Publications/WEO/2021/Update/January/English/text.ashx.

Industrial Structure Council and Central Environment Council (ISCCEC). 2020. *Report on Evaluation and Examination of the Enforcement Status of the Small Home Appliances Recycling System.* 2. www.meti.go.jp/shingikai/sankoshin/sangyo_gijutsu/haikibutsu_ recycle/kogata_wg/pdf/20200807_01.pdf (accessed October 31, 2020).

Japan Economics Newspaper (The Nikkei). 2020. Urban Mine. *Invent/Planning Unit of the Nikkei.* (accessed September 23, 2020).

Japan Industrial Waste Information Center (JW Center). 2020. *About JW Center.* www.jwnet. or.jp/en/index.html (accessed October 31, 2020).

Japan Industrial Waste Information Center (JW Center) and Japan Environmental Sanitation
    Center. 2020. *New Coronavirus Guidelines for the Waste Treatment Industry* (2nd
    Edition). www.jwnet.or.jp/uploads/media/2020/06/R2coronaguideline_chousa.pdf
    (accessed October 31, 2020).
Ministry of Economy, Trade and Industry, Japan. (METI). 2008. *3R Policies: Measures to
    Provide Information on Substances Contained in Products*. www.meti.go.jp/policy/
    recycle/main/3r_policy/policy/j-moss.html.
Ministry of Environment (MOE). 1993. *Strengthening Occupational Health and Safety
    Measures in the Waste Treatment Business*. www.env.go.jp/hourei/11/000357.html.
Ministry of Environment (MOE). 2013. *Enhanced Monitoring Activities for the Prevention of
    Illegal Export and Import of Wastes*. www.env.go.jp/en/headline/2044.html.
Ministry of Environment (MOE). 2014. *History and Current State of Waste Management in
    Japan*. 2.5-7.17.23.27. www.env.go.jp/recycle/circul/venous_industry/en/history.pdf.
Ministry of Environment (MOE). 2020a. *Enforcement Status of the Law Concerning the
    Regulation of Import and Export of Specified Hazardous Waste (The First Year of Reiwa)*.
    www.env.go.jp/press/108297.html.
Ministry of Environment (MOE). 2020b. *Illegal/Unauthorized Cases*. www.env.go.jp/recycle/
    yugai/anken.html.
Ministry of Environment (MOE). 2020c. *Law for the Recycling of Specified Kinds of Home
    Appliances (Outline)*. 1. www.env.go.jp/en/laws/recycle/08.pdf.
Ministry of Environment (MOE). 2020d. *Measures for COVID-19*. www.env.go.jp/en/recycle/
    index.html.
Ministry of Environment (MOE). 2020e. *Q & A to Implement Countermeasures to
    the Novel Coronavirus Infectious Disease (COVID-19) in Waste Treatment and
    Management (29 June 2020)*. www.env.go.jp/saigai/novel_coronavirus_2020/
    COVID-19_0727QA_en.pdf.
Ministry of Environment (MOE). 2020f. *Small Home Appliance Recycling Collection Portal
    Site*. http://kogatakaden.env.go.jp.
Ministry of Environment, Forest and Climate Change (MOEFCC). 2016. G.S.R. 338 (E)
    [23-03-2016] :e-waste (Management) Rules, 2016. 1. https://moef.gov.in/en/g-s-r-338-e-
    23-03-2016-e-waste-management-rules-2016/ (accessed February 8, 2022).
Ministry of Environment, Forest and Climate Change (MOEFCC). 2018a. *E-Waste Management
    Rules Amended for Effective Management of E-Waste in the Country: Union Environment
    Minister*. 1. https://pib.gov.in/PressReleasePage.aspx?PRID=1526177.
Ministry of Environment, Forest and Climate Change (MOEFCC). 2018b. "NOTIFICATION"
    New Delhi, the 22 March 2018. http://greene.gov.in/wp-content/uploads/2019/09/
    2019091847.pdf.
Ministry of Environment, Forest and Climate Change (MoEFCC). 2020a. *E-Waste Collection
    Targets for the Financial Year 2020-2021 Under the EPR Provision of the E-Waste
    (Management) Rules, 2016 in Wake of COVID-19 Pandemic Reg*. http://moef.gov.in/wp-
    content/uploads/2017/08/e-waste-04062014215956.pdf.
Ministry of Environment, Forest and Climate Change (MoEFCC). 2020b. *Revision
    1: Guidelines for Handling, Treatment and Disposal of Waste Generated dur-
    ing Treatment/Diagnosis/Quarantine of COVID-19 Patients*. www.mohfw.gov.in/
    pdf/6394860950158855689.
Ministry of Health, Labour and Welfare (MHLW). 2020. *Occupational Accidents Caused by
    the New Coronavirus Infection must also be Submitted as a Worker Death/Injury Report*.
    1. www.mhlw.go.jp/content/10900000/000631412.pdf.
Ministry of Justice, Japan. 2020. Japanese Law Translation, 2006: Industrial Safety and
    Health Act (Act No. 57 of June 8, 1972). www.japaneselawtranslation.go.jp/law/
    detail/?id=1926&vm&re.
Ministry of Labour, Government of India. 2019. *OHS Statues*. https://dgfasli.gov.in/other-acts#.

National Federation of Industrial Waste Management (NFIWM). 2020. *Occurrence of Occupational Accidents in the Industrial Waste Treatment Industry*. 8. www.zensan pairen.or.jp/wp/wp-content/themes/sanpai/assets/pdf/disposal/safety_saigaihasei.pdf (accessed October 31, 2020).

Oxford Analytica. 2020. *Sustainable Bond Issuance Will Increase Markedly, Emerald Expert Briefings*. 1–3. ISSN:2633-304X. https://doi.org/10.1108/OXAN-DB254845.

Pena, R. 2012. *Wasting no Opportunity: The Case for Managing Brazil's Electronic Waste*. Project Report. The World Bank infoDev, 61. www.infodev.org/infodev-files/resource/ InfodevDocuments_1169.pdf (accessed March 15, 2021).

Radulovic, V. 2018. *GAP Analysis on Responsible E-Waste Management Efforts in India: Institutional, Economic, and Technological Barriers and the Potential Role of a Sustainability Standard to Build Capacity and Help Foster Solutions*. 3. https://greenel ectronicscouncil.org/wp-content/uploads/2018/11/GEC-CRB-Gap-Analysis-Report-FINAL-Oct-2018.pdf (accessed October 31, 2020).

Raghupathy, L., Krüger, C., Chaturvedi, A., Arora, R., and Henzler, M. P. 2010. *E-Waste Recycling in India—Bridging the Gap Between the Informal and Formal Sector*. 2.7–8. www.iswa.org/uploads/tx_iswaknowledgebase/Krueger.pdf (accessed October 31, 2020).

Rajput, H., Changotra, R., Rajput, P. et al. 2020. A Shock Like No Other: Coronavirus Rattles Commodity Markets. *Environ Dev Sustain* 5. doi:10.1007/s10668-020-00934-4.

Singh, A. 2020. COVID-19 and Safer Investment Bets. ELSEVIER. *Finance Research Letters* 101729: 3.5. doi:10.1016/j.frl.2020.101729. https://europepmc.org/backend/ptpmcr ender.fcgi?accid=PMC7431328&blobtype=pdf.

Suwa, K. 2006. Management of Specific Chemical Substances in Electrical and Electronic Equipment: Western and Public-Private Efforts [Japanese]. *IEEJ Journal* 126, No. 3: 144. www.jstage.jst.go.jp/article/ieejjournal/126/3/126_3_142/_pdf.

Terazono, A. 2019. *Current Issues and Challenges of E-Waste Management in Japan-After Revised Domestic Waste Regulations and China's Waste Import Ban*. Korea Waste Resource Circulation Society, 98–105. www.papersearch.net/thesis/article.asp?key= 3688090 (accessed October 31, 2020).

United Nations Environment Programme (UNEP), International Environmental Technology Centre (IETC), IGES Center Collaborating with UNEP on Environmental Technologies (CCET). 2020. *Waste Management during the COVID-19 Pandemic: From Response to Recovery*. 6-7.11.15.27.29. Table 1. www.iges.or.jp/en/pub/waste-management-during-covid-19-pandemic-response-recovery/en (accessed October 31, 2020).

WHO. 2020. *Water, Sanitation, Hygiene, and Waste Management for SARS-CoV-2, the Virus that Causes COVID-19*. Interim Guidance. www.who.int/publications/i/item/water-sanitation-hygiene-and-waste-management-for-covid-19 (accessed October 31, 2020).

Wijekoon, L., Shroff, V., and Mohapatra, A. 2020. India: COVID-19 (Coronavirus)—Employer FAQs. *JD Supra*. www.jdsupra.com/legalnews/india-covid-19-coronavirus-employer-faqs-15413/ (accessed October 31, 2020).

The World Bank. 2021. *January 2021 Global Economic Prospects*. 201. Table 1-1. ISBN (paper): 978-1-4648-1612-3 ISBN (electronic): 978-1-4648-1613-0 doi:10.1596/978-1-4648-1612-3.

World Economic Forum. 2020. *Global Gender Gap Report 2020: 13*. ISBN-13: 978-2-940631-03-2. http://www3.weforum.org/docs/WEF_GGGR_2020.pdf.

Part II

Technological Advancement towards Sustainable E-waste Management

Features Progression in the Technological Aspect of E-waste Management towards Sustainability

4 Fungal Biotechnology for the Recovery of Critical Metals from E-waste
Current Research Trends and Prospects

*Amber Trivedi, Kavita Kanaujia,
Kumar Upvan and Subrata Hait*

CONTENTS

4.1   INTRODUCTION

Owing to the rapid developments in the electrical and electronic engineering industries (EEI), continual consumer demand and accelerated obsolescence rate of electrical and electronic equipment (EEE) have led to an increase in waste electrical and electronic equipment (WEEE) or electronic waste (e-waste) generation, globally. Generation of e-waste is projected to increase to 74.7 Mt in 2030 from 53.6 Mt in 2019 at an alarming growth rate of about 40% (Forti et al. 2020). Simultaneously, the demand for critical metals for use in EEE is proliferating and thereby leading to the significant depletion of the world's primary metal reserve (Kaya 2016). However, the supply of such metals is monopolistically controlled by few countries owing to the uneven geographical distribution of metal reserves (Hou et al. 2018). The presence of valuable metals including precious metals makes printed circuit board (PCB), the essential constituent of EEE, a secondary metal reservoir. The resource recovery paradigm has focused attention on the efficient extraction of critical metals from a secondary resource like e-waste as a part of the 'circular economy' concept to ensure efficient resource circulation to maintain the worldwide supply (Iacovidou et al. 2017; Ardente et al. 2019). Recycling of e-waste is an essential aspect of waste treatment supporting environmental management and helps in the recovery of critical metals for economic development (Priya and Hait 2017a). A high content of base and heavy metals, besides the considerable presence of precious metals, makes the PCB a valuable metal reservoir (Bandyopadhyay 2008; Priya and Hait 2018). Hence, metal recycling from waste PCBs is the primary obligation for economic development as well as environmental protection (Deng et al. 2007). However, informal recycling and unscientific disposal of PCBs in landfills create environmental problems due to the presence of toxic metals (Li et al. 2007; Lim and Schoenung 2010; Noon et al. 2011). Metallurgical processes like pyro- and hydrometallurgy to recycle metals are expensive, energy-intensive, and generate secondary pollution (Ilyas et al. 2010; Priya and Hait 2017b). In the pyrometallurgy process, thermal treatment process like e-waste burning under controlled conditions emits the toxic gases, including dioxin and furans, thereby causing environmental toxicity. In hydrometallurgy, chemicals like inorganic acids and ligands are employed for metal extraction from PCBs. However, the hazardousness of these chemicals is a major concern for their applicability (Cui and Zhang 2008; Priya and Hait 2017a). The effluent generated in the hydrometallurgical processes further requires treatment before disposing into the environment (Owens et al. 2007; Verma and Hait 2019).

Nowadays, researchers are focusing on the eco-friendly and efficient metallurgical process for metal recovery to overcome the aforementioned drawbacks. Biohydrometallurgy or bioleaching technique with the integration of the hydrometallurgical process and the biotechnology employing microorganisms, mostly autotrophic bacteria, has emerged as an eco-friendly process to recycle metals from PCBs (Liang et al. 2010; Li et al. 2015). Fungal biotechnology with the application of heterotrophic fungi like *Aspergillus* spp. and *Penicillium* spp. has the potential for efficient metal dissolution from the obsolete PCBs (Srichandan et al. 2019; Arshadi et al. 2020). However, the fungal biotechnology for the extraction of critical metals from e-waste has been applied on a limited scale (Islam et al. 2020). In this context, an overview of current research trends and future prospects of fungal biotechnology for extractive metallurgy from PCBs has been provided. Various fungal bioleaching routes and schemes with the effects of process parameters for metal recovery from PCBs have been briefly discussed. Further, the underlying mechanisms and kinetics of fungi-assisted metal extraction from PCBs have been explored and summarized. Moreover, future research directions and SWOT analysis have been presented to assess the prospect of fungal biotechnology for critical metal recovery from PCBs.

4.2 METHODOLOGY

To prepare the basic structure of the chapter, a literature survey was done by exploring various search engines like Google Scholar, Google search, and Scopus Database with various relevant keywords like 'e-waste generation', 'elemental composition of e-waste', 'metal extraction from PCB', 'fungal biotechnology for metal recycling', 'kinetic study of fungal bioleaching', and 'mechanism involved in fungal bioleaching'. Data were extracted from the aforementioned sources and databases which include journal articles and national and international reports on e-waste management. About 180 articles and 10 technical reports were retrieved from the survey. To continue with the relevant dataset for the chapter, articles and reports which were unrelated to the current topic were screened manually. Finally, a total of 55 relevant documents which include 49 journal articles, 5 book chapter, and 1 report were scrutinized, consulted, and referred for the preparation of the chapter.

4.3 OVERVIEW OF FUNGAL BIOTECHNOLOGY FOR METAL RECOVERY FROM E-WASTE

Biotechnology employing fungal species has been largely limited to the production of enzymes, antibiotics, and organic acids (Rehm 1979). However, numerous fungi have been employed to leach metals from solid waste and material matrices in fungal biotechnology regarded as fungal bioleaching (Burgstaller and Schinner 1993). Autotropic bioleaching bacteria work under acidic conditions and waste PCBs are basic in nature, thereby increasing the pH of bioleaching medium upon metal leaching (Brandl et al. 2001; Priya and Hait 2020). On the contrary, fungi like *Aspergillus* spp. used in the fungal bioleaching experiments can grow under acidic-to-alkaline growth medium, unlike autotropic bioleaching bacteria (Xu et al. 2014). Fungal bioleaching can be employed in the metal leaching process from unique niches, where the autotrophic bacteria are not viable. Fungal bioleaching can be performed, given a suitable carbon source (Bindschedler et al. 2017). It has been also reported that fungi extract metals mostly by the secretion of metabolites like protons, amino acids, and organic acids (Burgstaller and Schinner 1993). The metabolic transformation of glucose to citric acid by using *A. niger* is well established. The citric acid formation pathway is completed in five steps. The pathway begins with the uptake of sugar substrate, followed by the formation of two moles of pyruvate through glycolytic catabolism of substrate in the second step and subsequently transformed to oxaloacetate and acetyl-CoA. Lastly, these two intermediate compounds are condensed and excreted as citric acid (Kubicek et al. 2011). Recent studies also reported that the fungi utilize organic carbon (OC) available in the medium and cheap organic wastes like molasses for their growth and to produce metabolites and enzymes (Arshadi et al. 2019).

Efficient strategies for e-waste management comprises collection, dismantling, recycling, and material recovery (Soo and Doolan 2014; Singh et al. 2018). The extraction of critical metals from the electronic scraps is crucial toward achieving a circular economy. Nowadays, fungal bioleaching has gained attention to recover metals from waste PCBs in the e-waste stream. A typical fungal bioleaching scheme consisting of the collection, size reduction, and separation followed by metal recovery employing fungal spp. has been shown in Figure 4.1. Fungal bioleaching of metals from a solid matrix involves primary mechanisms viz., acidolysis, complexolysis, and redoxolysis (Rasoulnia et al. 2016; Bindschedler

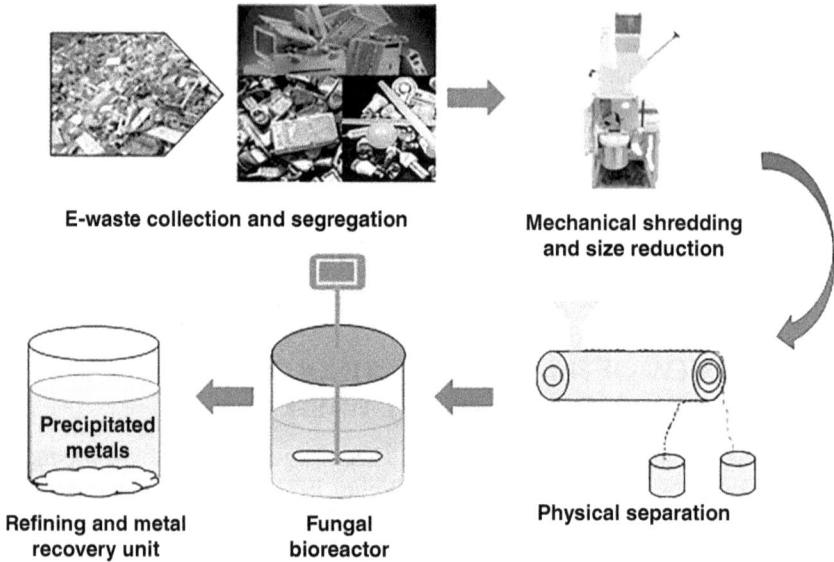

FIGURE 4.1 A typical fungal biotechnology scheme for biometallurgical metal recovery from e-waste.

et al. 2017). However, a minor role has been reported to be played by the redox processes (Burgstaller and Schinner 1993). Such leaching processes are known as direct leaching, whereas indirect leaching processes proceed by the interaction of fungal mycelia and e-waste particle, which may also increase the leaching efficiency (Burgstaller and Schinner 1993).

Different fungi-assisted biometallurgical processes, viz. one-step, two-step, and spent-media have been employed for metal extraction from PCBs and schematically presented in Figure 4.2a, b, and c, respectively. In the one-step approach, the pulverized e-waste particles are added prior to the sterilization of the bioleaching medium and subsequently added to the fungal spores followed by incubation (Horeh et al. 2016; Faraji et al. 2018). However, the metal recovery efficacy by the fungal bioleaching is affected due to toxicity induced by the metals present in the e-waste and metals adsorbed by the fungal mycelia in the one-step bioleaching process (Horeh et al. 2016; Faraji et al. 2018). To overcome the aforementioned problem, researchers have experimented with different processes like two-step and spent-media approaches. In the two-step process, sterilized PCB particles are introduced upon a substantial drop in pH of the bioleaching media during the 4th to 7th days of bioleaching. Metal bioleaching begins in a fungal-free acidic solution after 14th day of inoculation in the case of the spent-media process (Horeh et al. 2016; Faraji et al. 2018). Metal extraction efficiency from e-waste employing the fungal bioleaching approach may vary on the basis of the selection of the aforementioned processes. However, more advanced studies are required to develop a better understanding to establish and improvise the biotechnological process employing fungal bioleaching considering its limited application.

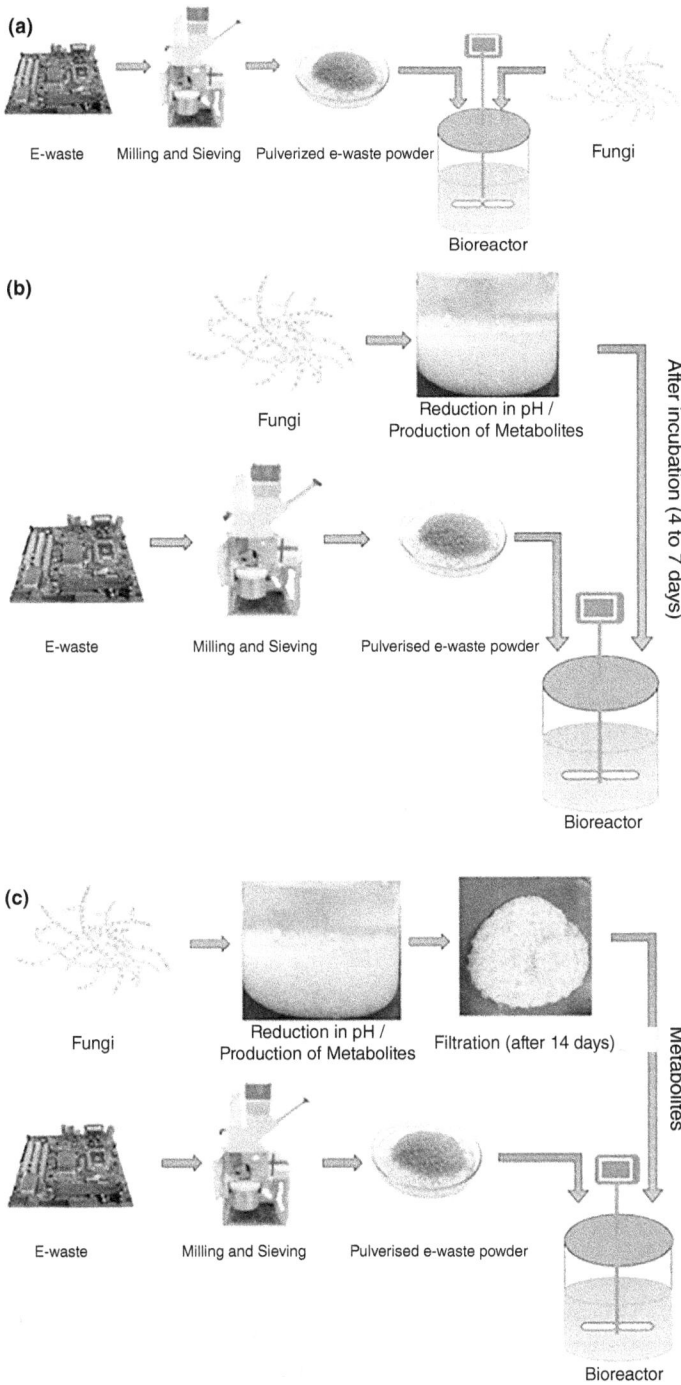

FIGURE 4.2 (a) One-step, (b) two-step, and (c) spent-media fungal bioleaching processes for extractive metallurgy from e-waste.

4.4 APPLICABILITY OF FUNGAL BIOTECHNOLOGY FOR EXTRACTIVE METALLURGY FROM E-WASTE

4.4.1 Impacts of Process Parameters on Fungal Bioleaching

E-waste comprises heterogeneous material compositions, which include metals like base, precious, and rare earth metals that are crucial to recycle for economic development as well as for environmental protection (Deng et al. 2007). It has been reported that fungi can mobilize metals from various waste streams including e-waste (Burgstaller and Schinner 1993; Brandl et al. 2001; Xu et al. 2014; Faraji et al. 2018; Srivastava et al. 2020). However, few studies have been reported on the fungal biotechnology for extractive metallurgy from electronic scraps. Fungal bioleaching is needed to be developed for a beneficial metal recycling from the metal-rich waste streams like e-waste. Based on the available scientific literature, it has been reported that the fungal bioleaching of metals from e-waste is time-consuming, which may be because of various unknown factors. The fungal bioleaching efficiency is affected by both biotic factors such as type of fungal spp. and their inoculum size as well as abiotic factors like pH, temperature, the carbon source, nitrogen-to-phosphate ratio, e-waste pulp density (PD), and metal ions content (Burgstaller and Schinner 1993; Liu et al. 2003; Ilyas et al. 2007; Santhiya and Ting 2006; Xiang et al. 2010; Xu et al. 2014). In metal bioleaching from e-waste, it has been well documented that e-waste concentration beyond a certain threshold might cause toxicity to microbial culture (Brandl et al. 2001). Further, the e-waste consists of metallic and non-metallic contents, which also add toxicity to the microbes (Brandl et al. 2001). Zhou et al. (2009) have reported that e-waste induces adverse effects on microbial growth due to its heterogeneous composition, including organic and inorganic elements. Choi et al. (2004) observed that high e-waste PD reduces the metal bioleaching rate, which might be due to the metal content beyond the threshold limit. Further, it has been also reported the inadequate aeration inhibits microbial growth and thereby decreases the bioleaching of metal. According to Faraji et al. (2018), the optimum PD of waste PCB is found to be 4.19 g/L for both one-step and two-step bioleaching for maximum metal bioleaching. Few studies have reported the optimum PCB particle size range to be between 40 and 200 μm for maximum metal recovery using bioleaching technique (Ilyas et al. 2007; Ivanus 2010).

The applicability of fungal biotechnology with process-related factors for extractive metallurgy from PCBs is summarized in Table 4.1. It is apparent that fungal bioleaching works at a broad pH range of 2 to 10, e-waste PD of 1 to 10 g/L, and temperature of 27°C to 30°C. Fungal species like *A. niger* and *P. simplicissimum* have been employed, and bioleach around 95% of Al, Ni, Pb, and Zn and 65% Cu and Sn from e-waste in 21 days duration have been used (Brandl et al. 2001). Similarly, metal extraction efficiencies of about 96% Al, 48% Cu, 25% Fe, 73% Ni, and 98% Zn from electronic scrap have been demonstrated employing *P. chrysogenum* (Ilyas et al. 2013). A study on fungal bioleaching, conducted by Kolencik et al. (2013), using *A. niger* has reported to achieve about 68.3% Cu and 27.9% Pb in 42 days from e-waste scraps at an optimized process conditions comprising pH of 5.6, temperature of 30°C, e-waste PD of 10 g/L, and shaking speed of 120 rpm. Madrigal-Arias et al.

TABLE 4.1

Summary of literature on the applicability of fungal biotechnology for metal extraction from electronic waste

E-waste	Species	Medium	Pulp Density (g/L)	Other Process Parameters	Duration (day)	Maximum Metal Extraction Efficiency (%)	References
E-waste scrap	A. niger and P. simplicissimum	Sucrose	1-10	T: 30°C; MS: 150	21	Al, Ni, Pb, and Zn: 95%; Cu and Sn: 65%	Brandl et al. 2001
Catalysts	A. niger	-	-	T: 30°C; MS: 150	70	Ni: 65% and Al: 55%	Santhiya and Ting 2005
Hydrocracking catalyst	P. simplicissimum	-	30	T: 30°C; MS: 120	14	Fe: 100%; Mo: 93%; Ni: 66% and Al: 25%	Amiri et al. 2012
Fluid catalytic cracking catalyst	A. niger	-	10	T: 30°C; MS: 120	-	Sb: 64%; V: 36%; Al: 30%; Fe: 23%; and Ni: 9%	Aung and Ting 2005
Electronic scrap	A. niger	-	10	pH: 5.6; T: 30°C; MS: 120	42	Cu: 68.3%; and Pb: 27.9%	Kolencik et al. 2013
Electronic scrap	P. chrysogenum	-	-	-	-	Zn: 98%; Al: 96%; Ni: 73%; Cu: 48%; and Fe 25%	Ilyas et al. 2013
Cellular PCBs	Mixed culture	Potato dextrose agar	13	T: 28°C, MS: 280	14	Au: 87%	Madrigal-Arias et al. 2015
Lithium-ion batteries	A. niger	Sucrose	10	pH: 6; MS: 130	30	Cu: 100%; Li: 95%; Mn: 70%; Al: 65%; Co: 45%; and Ni: 38%	Bahaloo-Horeh et al. 2016
Spent batteries	A. fumigatus, A. flavipes, A. japonicus, A. tubingensis, A. versicolor, and A. niger	Malt extract and sucrose	1	T: 27°C, MS: 150	22	Ni, Cd, and Zn: 90%	Kim et al. 2016
Mobile phone battery	A. niger	Sucrose	1	pH: 6; T: 30°C, MS: 130	30	Li: 100%; Cu: 94%; Mn: 72%, Al: 62%; and Ni: 45%	Bahaloo-Horeh et al. 2018

(Continued)

TABLE 4.1 (Continued)

Summary of literature on the applicability of fungal biotechnology for metal extraction from electronic waste

E-waste	Species	Medium	Pulp Density (g/L)	Other Process Parameters	Duration (day)	Maximum Metal Extraction Efficiency (%)	References
PCB	Mixed culture	Sucrose	1-8	pH: 5; T: 30°C, MS: 180	27	Zn: 49%; Al: 15%; Pb: 20% and Sn: 8%	Xia et al. 2018
PCB	A. niger	Sucrose	1-10	pH: 5.5; T: 28°C, MS: 120	21	Zn: 100%; Cu: 85% and Ni: 80%;	Faraji et al. 2018
Cell phone PCB	A. niger	Glucose (20 g/L)	11	pH: 4.4; T: 19-23°C (without agitation)	-	Au: 56%	Argumedo-Delira et al. 2019
Mobile phone PCB	P. simplicissimum	Sucrose, cheese whey, sugar, and sugar cane molasses	10	pH: 2-10; T: 30°C; MS: 130	-	Cu: 90%; Ni: 89%	Arshadi et al. 2019
Desktop PCB	A. niger A. tubingensis A. niger and A. tubingensis	Potato dextrose broth	1	pH: 5; T: 30°C; MS: 170	30	Zn: 79%; Cu: 71%; and Ni:32% Cu: 54%; Zn: 41%; and Ni 14% Cu: 76%; Zn: 63% and Ni: 36%	Trivedi and Hait 2019
Desktop PCB	A. tubingensis	Malt extract	2.5-10	pH: 5; T: 30°C; MS: 170	30	Zn: 54%; and Cu: 34%	Trivedi and Hait 2020

Note: T: temperature; MS: mixing speed

(2015) have reported about 87% Au extraction from mobile phone PCBs in 14 days of bioleaching, employing a mixed culture of fungal spp. using potato dextrose agar (PDA) media at 28°C of temperature, 13 g/L of e-waste PD, and 280 rpm of mixing speed. Xia et al. (2018) have reported to attain metal extraction efficiency of around 15% Al, 20% Pb, 8% Sn, and 49% Zn from desktop PCBs employing mixed fungal culture by using sucrose medium in 27 days of experimental time at 1 to 8 g/L of e-waste PD range, a pH value of 5, 30°C of temperature, and 180 rpm of mixing speed. In a study conducted by Trivedi and Hait (2019), fungal metal bioleaching of about 71% and 54% Cu, 32% and 14% Ni, and 79% and 41% Zn from desktop PCBs employing the pure culture of *A. niger* and *A. tubingensis*, respectively, has been reported. However, a slight improvement in the recovery for Cu and Ni has been reported employing the mixed culture of *Aspergillus* spp.

4.4.2 Kinetics of Fungal Bioleaching

Kinetics development is a critical aspect of metal dissolution from electronic scraps from the process control perspectives. It has been reported that kinetics through fungi-mediated metal dissolution from electronic scraps generally follow the shrinking core model (Rasoulnia and Mousavi 2016). Considering the PCB particles as spherical and heterogeneous in nature in the solid-liquid extraction system, this model assumes that the major fraction of the metals (solute) is present inside at the core of the particle rather than the particle surface, and the core shrinks over the time as the metals (solute) are extracted (Goto et al. 1996). The three main sequential steps occur in fungal bioleaching: (1) diffusion of reactant through the bulk bioleaching media, (2) diffusion of reactant through the solid particles, and (3) chemical reaction (Rasoulnia and Mousavi 2016). Since vigorous shaking is applied during bioleaching, diffusion of reactant through the bulk bioleaching media layer is usually not regarded to be rate-limiting. Faraji et al. (2018) have reported that fungal bioleaching as assisted by organic ligands is controlled by the diffusion at the product layer for Zn in one-step bioleaching, while the mixed control mechanisms work for two-step and spent-media processes. In one-step and two-step fungal bioleaching processes, it has been reported that Cu extraction is controlled by the diffusion, whereas it is controlled by the chemical reaction in spent-media processes. Moreover, the chemical reaction has been reported to control the Ni extraction in all three approaches of fungal bioleaching. From the available literature, it can be inferred that the chemical reaction controls the Cu, Ni, and Zn extraction in the spent-medium fungal bioleaching approach. In contrast, the diffusion through the bulk media controls the Cu and Zn extraction in one-step fungal bioleaching. However, the mixed control mechanisms govern in the case of two-step fungal bioleaching of metals from e-waste.

4.4.3 Mechanisms of Fungal Bioleaching

Aspergillus and *Penicillium* spp. are the most common fungal species employed to solubilize metals from electronic scraps. During fungal bioleaching, the

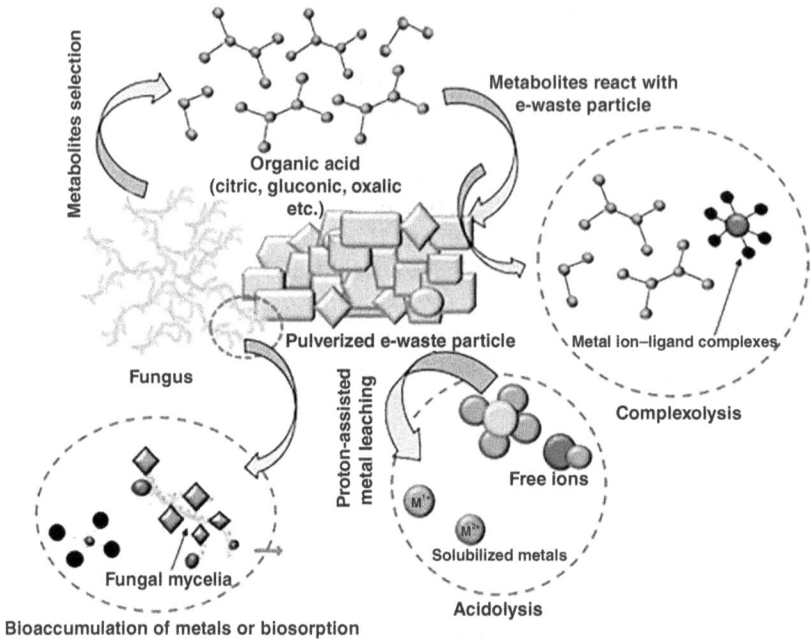

FIGURE 4.3 Overall mechanism of fungal biotechnology for metal extraction from electronic scraps

interaction of metals with fungi alters their physicochemical characteristics such as their mobility in the solution from the solid particles (Gadd 2007, 2010). The recovery of critical metals from e-waste is mainly associated with the secretion of fungal metabolites including organic acids. The prevalent mechanisms behind the fungal biotechnology for metal extraction from electronic scraps involve acidolysis, complexolysis, and redoxolysis (Burgstaller and Schinner 1993; Rasoulnia and Mousavi 2016; Bindschedler et al. 2017). However, it has been reported that the redox processes play a minor role (Burgstaller and Schinner 1993). The overall mechanism of fungal biotechnology for metal extraction from electronic scraps is presented in Figure 4.3. Fungi secrete metabolites like protons, amino acids, and organic acids to mobilize metals in the solution. As depicted in Figure 4.3, protons excreted by microbes in acidolysis convert the metal dissolution through the deterioration of critical bonds, which causes mobilization of metals from the solid particle matrix. The organic ligands, like citric acid, tartaric acid, and oxalic acid produced by the fungal species during their growth, involve in the complexolysis. Any substance with the capability of complexation and chelation can potentially facilitate for metal dissolution (Bindschedler et al. 2017). Besides, Burgstaller and Schinner (1993) have reported that the fungi mycelia facilitate the biosorption and adsorption of metals from the solid matrices including PCB particles.

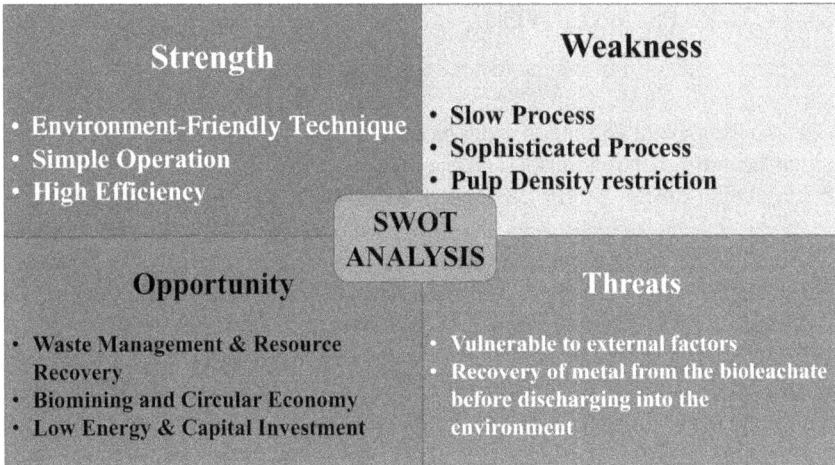

Strength

- Environment-Friendly Technique
- Simple Operation
- High Efficiency

Weakness

- Slow Process
- Sophisticated Process
- Pulp Density restriction

SWOT ANALYSIS

Opportunity

- Waste Management & Resource Recovery
- Biomining and Circular Economy
- Low Energy & Capital Investment

Threats

- Vulnerable to external factors
- Recovery of metal from the bioleachate before discharging into the environment

FIGURE 4.4 SWOT analysis of fungal biotechnology for metal recovery from e-waste

4.5 SWOT ANALYSIS

The strengths, weaknesses, opportunities, and threats (SWOT) analysis of the fungal biotechnology, based on the available scientific literature, to assess the viability of the industry-level applications for potential cost-effective benefiBciation from e-waste is presented in Figure 4.4. The major strengths of fungal bioleaching technique include an environmentally friendly approach with miniBmum or no secondary pollution, limited energy requirement, ease of operation, and usage of waste containing organic carbon (OC) like whey, molasses as the growth media instead of commercial media, and high metal extraction efficacy. The weaknesses of the fungal biotechnology are extensive time requirement and e-waste PD restriction, limiting the metal extraction efficiency. However, recent scientific studies have shown a considerable reduction in the time requirement employing two-step and spent-media fungal biotechnology (Faraji et al. 2018). The increase in pulp density inhibits microbial growth and reduces metal bioBleaching efficiency. Further, the usage of OC in the heterotrophic fungal bioleBaching process increases the chance of biological contamination and thereby is likely to hinder metal dissolution from electronic scraps. Fungal biotechnology has the potential to extract valuable metals from electronic scraps, leading to resource recovery and helping to address the growing problem of e-waste. This approach aligns with the goal of transitioning to a circular economy.. Like any other biotechnological process, the fungal bioleaching technique is also vulneraBble to the external factors in hampering the overall process. Furthermore, it must be ensured that all metals are recovered from the bioleachate in a downstream operation before discharging into the environment and assessing the biodegradBability to avoid any environmental pollution.

4.6   FUTURE PERSPECTIVES

The current scenario of fungal biotechnology for the extraction of critical metals from electronic scraps is promising. However, there are several limitations to using fungal bioleaching for metal extraction from electronic waste. Firstly, the research in this area has been limited and mostly confined to laboratory-scale studies. Additionally, different fungal species require different process conditions to extract the same metal from e-waste, which complicates the optimization of the process. There are also contradictions in the literature regarding the effect of various process parameters on metal mobilization during fungal bioleaching. These issues highlight the need for more research to improve the process and make it more practical at an industrial scale. Currently, fungal bioleaching is not a cost-effective method for extracting metals from electronic waste on a large scale. Hence, to improve the metal recovery process through fungal bioleaching to an industrial level, a joint effort is needed to study and optimize the process parameters through experiments and interventions. Fungal bioleaching with a suitable alternative medium like dairy and agro-industrial waste which contains OC instead of commercial media like sucrose, is one of such innovative interventions for metal recovery from e-waste including PCBs. Fungal bioleaching technique has demonstrated an enhanced metal recovery from waste PCBs albeit it being a slow process. It is a prerequisite to develop a suitable fungal bioleaching approach to overcome the weakness of long leaching time. Moreover, the metal bioleaching efficiency is also inhibited as the e-waste PD increases. Research is warranted to overcome the PD restriction during the extraction of metals from e-waste using fungal bioleaching. To recover the precious and rare earth metals from e-waste, fungal biotechnology can be modified suitably. Furthermore, it can also be employed to extract critical metals from other secondary reserves comprising metal-rich waste and industrial sludge. It is necessary to develop the fungal bioleaching process in scale-up reactors in semi-continuous and continuous modes for metal recovery from e-waste to move towards the industrial-scale applications.

4.7   CONCLUSIONS

Managing of metals as a resource in the developing world is crucial for the advancement of emerging technologies. Recovery of critical metals while managing e-waste for environmental protection is an essential aspect of the circular economy for the sustainable material supply. Bioleaching employing heterotopic microorganisms like fungi for metal dissolution from electronic scraps is time-consuming despite being efficient. Fungal bioleaching employing fungi using commercial media or alternative substrates like wastes containing OC has shown as a promising green and efficient approach for metal recovery from electronic scraps. The SWOT analysis indicates that the fungal biotechnique is efficient and eco-friendly with minimum or no secondary pollution. However, the impediments like higher time requirement and e-waste PD restriction during fungal bioleaching need to be addressed through further research. Further, the challenge lies in advancing the fungal biotechnology to scale-up from the lab-scale to the industry level in a semi-continuous or continuous mode of operation for commercialization for achieving a circular economy.

REFERENCES

Amiri, Farnaz, Seyyed M. Mousavi, Soheila Yaghmaei, and Mansoor Barati. 2012. "Bioleaching Kinetics of a Spent Refinery Catalyst Using *Aspergillus Niger* at Optimal Conditions." *Biochem. Eng.* 67: 208–217. doi:10.1016/j.bej.2012.06.011.

Ardente, Fulvio, Cynthia E. L. Latunussa, and Gian A. Blengini. 2019. "Resource Efficient Recovery of Critical and Precious Metals from Waste Silicon PV Panel Recycling." *Waste Manag.* 91: 156–167. doi:10.1016/j.wasman.2019.04.059.

Argumedo-Delira, Rosalba, Mario J. Gómez-Martínez, and Brenda J. Soto. 2019. "Gold Bioleaching from Printed Circuit Boards of Mobile Phones by *Aspergillus Niger* in a Culture Without Agitation and with Glucose as a Carbon Source." *Metals* 9 (5). doi:10.3390/met9050521.

Arshadi, Mahdokht, Alireza Esmaeili, and Soheila Yaghmaei. 2020. "Investigating Critical Parameters for Bioremoval of Heavy Metals from Computer Printed Circuit Boards using the Fungus *Aspergillus Niger*." *Hydrometallurgy* 197: 105464.

Arshadi, Mahdokht, Sheida Nili, and Soheila Yaghmaei. 2019. "Ni and Cu Recovery by Bioleaching from the Printed Circuit Boards of Mobile Phones in Non-Conventional Medium." *J. Environ. Manage.* doi:10.1016/j.jenvman.2019.109502.

Aung, Khin M. M., and Yen P. Ting. 2005. "Bioleaching of Spent Fluid Catalytic Cracking Catalyst Using *Aspergillus Niger*." *J. Biotechnol.* 116 (2): 159–170. doi:10.1016/j.jbiotec.2004.10.008.

Bahaloo-Horeh, Nazanin, Seyyed M. Mousavi, and Mahsa Baniasadi. 2018. "Use of Adapted Metal Tolerant *Aspergillus Niger* to Enhance Bioleaching Efficiency of Valuable Metals from Spent Lithium-Ion Mobile Phone Batteries." *J. Clean. Prod.* 197: 1546–1557.

Bahaloo-Horeh, Nazanin, Seyyed M. Mousavi, and Seyed A. Shojaosadati. 2016. "Bioleaching of Valuable Metals from Spent Lithium-Ion Mobile Phone Batteries Using *Aspergillus Niger*." *J. Power Sources* 320: 257–266. Elsevier B.V. doi:10.1016/j.jpowsour.2016.04.104.

Bandyopadhyay, Amitava. 2008. "A Regulatory Approach for E-Waste Management: A Cross-National Review of Current Practice and Policy with an Assessment and Policy Recommendation for the Indian Perspective." *Int. J. Environ. Waste Manag.* 2 (1–2): 139–186. doi:10.1504/IJEWM.2008.016998.

Bindschedler, Saskia, T. Q. T. V. Bouquet, Daniel Job, Edith Joseph, and Pilar Junier. 2017. "Fungal Biorecovery of Gold from E-Waste." *Advances in Applied Microbiology* 99: 53–81. Elsevier.

Brandl, Helmut, R. Bosshard, and M. Wegmann. 2001. "Computer-Munching Microbes: Metal Leaching from Electronic Scrap by Bacteria and Fungi." *Hydrometallurgy* 59 (2–3): 319–326. Elsevier.

Burgstaller, Wolfgang, and Franz Schinner. 1993. "Leaching of Metals with Fungi." *J. Biotechnol.* 27 (2): 91–116. doi:10.1016/0168-1656(93)90101-R.

Choi, Moon-Sung, Kyung-Suk Cho, Dong-Su Kim, and Dong-Jin Kim. 2004. "Microbial Recovery of Copper from Printed Circuit Boards of Waste Computer by *Acidithiobacillus Ferrooxidans*." *J. Environ. Sci. Heal. Part A* 39 (11–12): 2973–2982. Taylor & Francis.

Cui, Jirang, and Lifeng Zhang. 2008. "Metallurgical Recovery of Metals from Electronic Waste: A Review." *J. Hazard. Mater.* 158 (2–3): 228–256. doi:10.1016/j.jhazmat.2008.02.001.

Deng, Wen-Jing, Jinshu Zheng, Xinhui Bi, Jiamo M. Fu, and Ming H. Wong. 2007. "Distribution of PBDEs in Air Particles from an Electronic Waste Recycling Site Compared with Guangzhou and Hong Kong, South China." *Environ. Int.* 33 (8): 1063–1069. doi:10.1016/j.envint.2007.06.007.

Faraji, Fariborz, Rabeeh Golmohammadzadeh, Fereshteh Rashchi, and Navid Alimardani. 2018. "Fungal Bioleaching of WPCBs Using *Aspergillus Niger*: Observation, Optimization and Kinetics." *J. Environ. Manage.* 217: 775–787. Elsevier Ltd. doi:10.1016/j.jenvman.2018.04.043.

Forti, Vanessa, Cornelis P. Baldé, Ruediger Kuehr, and Garam Bel. 2020. *The Global E-Waste Monitor 2020: Quantities, Flows, and the Circular Economy Potential*. United Nations University (UNU)/United Nations Institute for Training and Research (UNITAR)— Co-Hosted SCYCLE Programme, International Telecommunication Union (ITU) & International Solid Waste Association (ISWA).

Gadd, Geoffrey M. 2007. "Geomycology: Biogeochemical Transformations of Rocks, Minerals, Metals and Radionuclides by Fungi, Bioweathering and Bioremediation." *Mycol. Res.* 111 (1): 3–49. Elsevier.

Gadd, Geoffrey M. 2010. "Metals, Minerals and Microbes: Geomicrobiology and Bioremediation." *Microbiology* 156 (3): 609–643. Microbiology Society.

Goto, Motonobu, Bhupesh C. Roy, and Tsutomu Hirose. 1996. "Shrinking-Core Leaching Model for Supercritical Fluid Extraction." *J. Supercrit. Fluids* 9: 128–133.

Horeh, N. B., S. M. Mousavi, and S. A. Shojaosadati. 2016. "Bioleaching of Valuable Metals from Spent Lithium-Ion Mobile Phone Batteries Using *Aspergillus Niger*." *J. Power Sources* 320: 257–266. doi:10.1016/j.jpowsour.2016.04.104

Hou, Wenyu, Huifang Liu, Hui Wang, and Fengyang Wu. 2018. "Structure and Patterns of the International Rare Earths Trade: A Complex Network Analysis." *Resour. Policy.* 55 (November 2017): 133–142. Elsevier Ltd. doi:10.1016/j.resourpol.2017.11.008.

Iacovidou, Eleni, Costas A. Velis, Phil Purnell, Oliver Zwirner, Andrew Brown, John Hahladakis, Joel Millward-Hopkins, and Paul T. Williams. 2017. "Metrics for Optimising the Multi-Dimensional Value of Resources Recovered from Waste in a Circular Economy: A Critical Review." *J. Clean. Prod.* 166: 910–938. Elsevier Ltd. doi:10.1016/j.jclepro.2017.07.100.

Ilyas, Sadia, Munir A. Anwar, Shahida B. Niazi, and Muhammad A. Ghauri. 2007. "Bioleaching of Metals from Electronic Scrap by Moderately Thermophilic Acidophilic Bacteria." *Hydrometallurgy* 88 (1–4): 180–188. doi:10.1016/j.hydromet.2007.04.007.

Ilyas, Sadia, Jae-Chun Lee, and Ru-An Chi. 2013. "Fungal Leaching of Metals from Electronic Scrap." *Mining, Metallurgy & Exploration* 30: 151–156.

Ilyas, Sadia, Chi Ruan, Haq N. Bhatti, Muhammad A. Ghauri, and Munir A. Anwar. 2010. "Column Bioleaching of Metals from Electronic Scrap." *Hydrometallurgy* 101 (3–4): 135–140. Elsevier B.V. doi:10.1016/j.hydromet.2009.12.007.

Islam, Aminul, Tofayal Ahmed, Md Rabiul Awual, Aminur Rahman, Monira Sultana, Azrina Abd Aziz, Minhaj Uddin Monir, Siow Hwa Teo, and Mehedi Hasan. 2020. "Advances in Sustainable Approaches to Recover Metals from E-Waste-A Review." *J. Clean. Prod.* 244: 118815.

Ivanus, Radu C. 2010. "Bioleaching of Metals from Electronic Scrap by Pure and Mixed Culture of *Acidithiobacillus Ferrooxidans* and *Acidithiobacillus Thiooxidans*." *Metal. Int.* 15 (4). EDITURA STIINTIFICA FMR CALEA GRIVITEI, NR 83, SECTOR 1, OP 12, BUCHAREST: 62–70.

Kaya, Muammer. 2016. "Recovery of Metals and Nonmetals from Electronic Waste by Physical and Chemical Recycling Processes." *Waste Manag.* 57: 64–90. Elsevier Ltd. doi:10.1016/j.wasman.2016.08.004.

Kim, Min-Ji, Ja-Yeon Seo, Yong-Seok Choi, and Gyu-Hyeok Kim. 2016. "Bioleaching of Spent Zn-Mn or Ni-Cd Batteries by *Aspergillus* Species." *Waste Manag.* 51: 168–173. Elsevier Ltd. doi:10.1016/j.wasman.2015.11.001.

Kolencik, Marek, Martin Urik, Slavomir Cernansky, Marianna Molnarova, and Peter Matus. 2013. "Leaching of Zinc, Cadmium, Lead and Copper From Electronic Scrap Using Organic Acids and the *Aspergillus Niger* Strain." *Fresenius Environ. Bull.* 22 (12): 3673–3679.

Kubicek, Christian P., Peter Punt, and Jaap Visser. 2011. "Production of Organic Acids by Filamentous Fungi." In *Industrial Applications*, 215–234. Springer.

Li, Huiru, Liping Yu, Guoying Sheng, Jiamo Fu, and Ping'an Peng. 2007. "Severe PCDD/F and PBDD/F Pollution in Air around an Electronic Waste Dismantling Area in China." *Environ. Sci. Technol.* 41 (16): 5641–5646. doi:10.1021/es0702925.

Li, Jingying, Changjin Liang, and Chuanjing Ma. 2015. "Bioleaching of Gold from Waste Printed Circuit Boards by *Chromobacterium Violaceum*." *J. Mater. Cycles Waste Manag.* 17 (3): 529–539. Springer. doi:10.1007/s10163-014-0276-4.

Liang, Guobin, Yiwei Mo, and Quanfa Zhou. 2010. "Novel Strategies of Bioleaching Metals from Printed Circuit Boards (PCBs) in Mixed Cultivation of Two Acidophiles." *Enzyme Microb. Technol.* 47 (7): 322–326. Elsevier Inc. doi:10.1016/j.enzmictec.2010.08.002.

Lim, Seong-Rin, and Julie M. Schoenung. 2010. "Toxicity Potentials from Waste Cellular Phones, and a Waste Management Policy Integrating Consumer, Corporate, and Government Responsibilities." *Waste Manag.* 30 (8–9): 1653–1660. Elsevier Ltd. doi:10.1016/j.wasman.2010.04.005.

Liu, Jian-Zhong, Li-Ping Weng, Qian-Ling Zhang, Hong Xu, and Liang-Nian Ji. 2003. "Optimization of Glucose Oxidase Production by *Aspergillus Niger* in a Benchtop Bioreactor Using Response Surface Methodology." *World J. Microbiol. Biotechnol.* 19 (3): 317–323. doi:10.1023/A:1023622925933.

Madrigal-Arias, Jorge E., Rosalba Argumedo-Delira, Alejandro Alarcón, Ma R. Mendoza-López, Oscar García-Barradas, Jesús S. Cruz-Sánchez, Ronald Ferrera-Cerrato, and Maribel Jiménez-Fernández. 2015. "Bioleaching of Gold, Copper and Nickel from Waste Cellular Phone PCBs and Computer Goldfinger Motherboards by Two *Aspergillus Niger* Strains." *Brazilian J. Microbiol.* 46 (3): 707–713. doi:10.1590/S1517-838246320140256.

Noon, Michael S., Seung-Jin Lee, and Joyce S. Cooper. 2011. "A Life Cycle Assessment of End-of-Life Computer Monitor Management in the Seattle Metropolitan Region." *Resour. Conserv. Recycl.* 57: 22–29. Elsevier B.V. doi:10.1016/j.resconrec.2011.09.017.

Owens, Clyde V., Christy Lambright, Kathy Bobseine, Bryce Ryan, L. Earl Gray, Brian K. Gullett, and Vickie S. Wilson. 2007. "Identification of Estrogenic Compounds Emitted from the Combustion of Computer Printed Circuit Boards in Electronic Waste." *Environ. Sci. Technol.* 41 (24): 8506–8511. doi:10.1021/es071425p.

Priya, Anshu, and Subrata Hait. 2017a. "Comparative Assessment of Metallurgical Recovery of Metals from Electronic Waste with Special Emphasis on Bioleaching." *Environ. Sci. Pollut. Res.* 24 (8): 6989–7008. doi:10.1007/s11356-016-8313-6.

Priya, Anshu, and Subrata Hait. 2017b. "Qualitative and Quantitative Metals Liberation Assessment for Characterization of Various Waste Printed Circuit Boards for Recycling." *Environ. Sci. Pollut. Res.* 24 (35): 27445–27456. doi:10.1007/s11356-017-0351-1.

Priya, Anshu, and Subrata Hait. 2018. "Extraction of Metals from High Grade Waste Printed Circuit Board by Conventional and Hybrid Bioleaching Using *Acidithiobacillus Ferrooxidans*." *Hydrometallurgy* 177 (November 2017): 132–139. Elsevier. doi:10.1016/j.hydromet.2018.03.005.

Priya, Anshu, and Subrata Hait. 2020. "Biometallurgical Recovery of Metals from Waste Printed Circuit Boards Using Pure and Mixed Strains of *Acidithiobacillus Ferrooxidans* and *Acidiphilium Acidophilum*." *Process Saf. Environ. Prot.* 143: 262–272. Institution of Chemical Engineers. doi:10.1016/j.psep.2020.06.042.

Rasoulnia, Payam, and Seyyed M. Mousavi. 2016. "V and Ni Recovery from a Vanadium-Rich Power Plant Residual Ash Using Acid Producing Fungi: *Aspergillus Niger* and *Penicillium Simplicissimum*." *RSC Adv.* 6 (11): 9139–9151. Royal Society of Chemistry. doi:10.1039/c5ra24870a.

Rasoulnia, Payam, Seyyed M. Mousavi, Seyed O. Rastegar, and H. Azargoshasb. 2016. "Fungal Leaching of Valuable Metals from a Power Plant Residual Ash Using *Penicillium Simplicissimum*: Evaluation of Thermal Pretreatment and Different Bioleaching Methods." *Waste Manag.* 52: 309–317. Elsevier Ltd. doi:10.1016/j.wasman.2016.04.004.

Rehm, Hans-Jürgen. 1979. Industrielle Mikrobiologie. *Carbohydr. Res.* 95. doi:10.1016/s0008-6215(00)85590-1.

Santhiya, Deenan, and Yen-Peng Ting. 2005. "Bioleaching of Spent Refinery Processing Catalyst Using *Aspergillus Niger* with High-Yield Oxalic Acid." *J. Biotechnol.* 116 (2): 171–184. doi:10.1016/j.jbiotec.2004.10.011.

Santhiya, Deenan, and Yen-Peng Ting. 2006. "Use of Adapted *Aspergillus Niger* in the Bioleaching of Spent Refinery Processing Catalyst." *J. Biotechnol.* 121 (1): 62–74. Elsevier.

Singh, Narendra, Huabo Duan, Fengfu Yin, Qingbin Song, and Jinhui Li. 2018. "Characterizing the Materials Composition and Recovery Potential from Waste Mobile Phones: A Comparative Evaluation of Cellular and Smart Phones." *ACS Sustain. Chem. Eng.* 6 (10): 13016–13024. ACS Publications.

Soo, Vi K., and Matthew Doolan. 2014. "Recycling Mobile Phone Impact on Life Cycle Assessment." *Procedia CIRP.* 15: 263–271. Elsevier B.V. doi:10.1016/j.procir.2014.06.005.

Srichandan, Haragobinda, Ranjan Kumar Mohapatra, Pankaj Kumar Parhi, and Snehasish Mishra. 2019. "Bioleaching Approach for Extraction of Metal Values from Secondary Solid Wastes: A Critical Review." *Hydrometallurgy.* 189: 105–122.

Srivastava, Rajiv R., Sadia Ilyas, Hyunjung Kim, Sowon Choi, Ha B. Trinh, Muhammad A. Ghauri, and Nimra Ilyas. 2020. "Biotechnological Recycling of Critical Metals from Waste Printed Circuit Boards." *J. Chem. Technol. Biotechnol.* 95 (11): 2796–2810.

Trivedi, Amber, and Subrata Hait. 2019. "Bioleaching of Selected Metals from E-Waste Using Pure and Mixed Cultures of *Aspergillus* Species." *Measurement, Analysis and Remediation of Environmental Pollutants.* 271.

Trivedi, Amber, and Subrata Hait. 2020. "Efficacy of Metal Extraction from Discarded Printed Circuit Board Using *Aspergillus Tubingensis*." In *Bioresource Utilization and Bioprocess*, edited by Agarwal, A. Gupta, T., Singh, S., and Rajput, P., 167–175. Springer. doi:10.1007/978-981-15-0540-9_13.

Verma, Auchitya, and Subrata Hait. 2019. Chelating Extraction of Metals from e-waste Using Diethylene Triamine Pentaacetic Acid. *Process Saf. Environ. Prot.* 121: 1–11. Elsevier B.V. https://doi.org/10.1016/j.psep.2018.10.005

Xia, Mingchen, Peng Bao, Ajuan Liu, Mingwei Wang, Li Shen, Runlan Yu, Yuandong Liu, Miao Chen, Jiaokun Li, Xueling Wu, Guanzhou Qiu, Weimin Zeng. 2018. "Bioleaching of Low-Grade Waste Printed Circuit Boards by Mixed Fungal Culture and Its Community Structure Analysis." *Resour. Conserv. Recycl.* 136 (May): 267–275. Elsevier. doi:10.1016/j.resconrec.2018.05.001.

Xiang, Yun, Pingxiao Wu, Nengwu Zhu, Ting Zhang, Wen Liu, Jinhua Wu, and Ping Li. 2010. "Bioleaching of Copper from Waste Printed Circuit Boards by Bacterial Consortium Enriched from Acid Mine Drainage." *J. Hazard. Mater.* 184 (1–3): 812–818. Elsevier B.V. doi:10.1016/j.jhazmat.2010.08.113.

Xu, Tong-Jiang, Thulasya Ramanathan, and Yen-Peng Ting. 2014. "Bioleaching of Incineration Fly Ash by *Aspergillus Niger*—Precipitation of Metallic Salt Crystals and Morphological Alteration of the Fungus." *Biotechnol. Reports* 3: 8–14. Elsevier B.V. doi:10.1016/j.btre.2014.05.009.

Zhou, Hong-Bo, Wei-Min Zeng, Zhi-Feng Yang, Ying-Jian Xie, and Guan-Zhou Qiu. 2009. "Bioleaching of Chalcopyrite Concentrate by a Moderately Thermophilic Culture in a Stirred Tank Reactor." *Bioresour. Technol.* 100 (2): 515–520. Elsevier. https://doi.org/10.1016/j.biortech.2008.06.033

5 Nanotechnology as a Cleaner and Greener Approach toward E-waste Recycling and Valorization

*Tirthankar Mukherjee, Shweta Mitra,
Sunil Herat and Prasad Kaparaju*

CONTENTS

5.1 INTRODUCTION

The growing demand for electronic and electrical appliances in our day-to-day life and their end-of-life (EoL) waste have created an additional environmental and health challenge (Robinson and Brett 2009). However, the appropriate reuse of this waste

DOI: 10.1201/9781003301899-7

can preserve natural resources and also prevent air and water pollution. Electronic waste, which is typically known as e-waste, comprises waste generated from electrical/ electronic industry or household, which have been discarded due to functional or quality defects while in production or use (Chen et al. 2010). These consist of parts of electrical or electronic hardware like mobile phones, videotape recorders, electronic scanners, fax machines, laser printers, electronic tablets, DVDs, microwave oven, x-ray machinery, and some laboratory or scientific equipment (Zoeteman et al. 2010). Huge amounts of e-waste are produced due to unceasing consumption and/or constant technological advancement and replacement, especially in computer and mobile phones' industry (Ha et al. 2009). In 2016, 44.7 million tons of e-waste was produced and is anticipated to touch 57.4 million tons by the year 2021 at a yearly growth rate of 3–4% (Balde et al. 2015). Further, e-waste generation from obsolete computers in both South Africa and China is projected to peak by 200% to 400% (Baldé et al. 2017).

Globally, the cumulative e-waste generation has progressively increased and also threatens human health and environment due to its hazardous and highly toxic constituents. Based on the number of mobile phones and/or personal computer (PC) manufactured, it is projected that 100 million cell phones and also 17 million PCs are to be disposed yearly (Breivik et al. 2014). Table 5.1 demonstrates the major electrical and electronic appliances used in various sectors. Developing a profitable and eco-friendly recycling method involves categorizing and quantifying the valuable resources and hazardous materials. This will enable us to comprehend the physical features of e-waste and increase the metal salvage opportunity and thereby preserve the natural resources and deliver an ecological solution of e-waste management (Barbara and Graedel, 2012). Conceptualizing the use of the EoL by-products as precursor raw materials for manufacturing innovative high-end and merchantable goods could stimulate recycling and generate economic revenues (Khin et al. 2012).

Nanoparticles (NP) are extremely small particles that can range between 1 and 100 nanometers (nm) in size. These fragments are undetectable by the human eye and can exhibit significantly different physical and also chemical characteristics due to their vastly large surface-area-to-volume ratio and reactivity. NPs are small enough to confine their electrons and produce quantum effects. There is an increasing focus on the function of metal-based NPs in manufacturing synthetic compounds and environmental remediation such as wastewater treatment, air pollution (SO_x, CO, NO_x, etc.), and disinfection of wastewater and air containing fungi, bacteria, and viruses (Stark et al. 2015). Due to their unique surface chemistry, NPs offer the potential to be altered or grafted with functional groups that can target specific pollutant of interest for efficient remediation. Further, changes in physical characteristics of NPs such as size, morphology, and porosity can offer additional characteristics that can improve the remediation properties.

Silver (Ag), iron (Fe), gold (Au), titanium oxide (TiO_2), iron oxide (FeO), and zinc oxide (ZnO) are a few of the consistent nanoscale metallic elements and metal oxides that are commonly used in environmental remediation (Shapira et al. 2015). NPs can be manufactured in several structures such as cylinders, wires, filaments, and

TABLE 5.1

Major Source of Electrical and Electronic Waste Generated in the World

Number	Category	Chief Electrical and Electronic Equipment
1	Bulky household equipment	Refrigerator, washing machines, cooking devices, power-driven fans, and air conditioner (AC).
2	Small household equipment	Vacuum cleaners, iron boxes, toasters, electrical choppers, coffee apparatuses, electrical knife, hair clippers, electrical toothbrushes, wall clocks and wrist watches, weighing scales, etc.
3	Information technology and telecommunications appliances	Centralized servers, PCs, workstations, notebooks, adding machines, printers, replicating gear, mobile phones, data cables and modems.
4	Consumer products	Television (TV), radio, camcorders, recorders, headphones, earphones, digital box, Digital Versatile Disc (DVD) player, compact disc player, digital camera, and digital media players.
5	Light appliances	Luminaires for fluorescent lamps, straight or conservative fluorescent lamps, focused energy release lamps, sodium lamps, and Light Emitting Diode (LED) lamps.
6	Electrical tools	Drilling devices, electrical saws, and sewing gears
7	Toys, recreational, and also sports appliances	Electric trains or vehicle dashing sets, computer games or consoles, PCs for bike, plunging, running, paddling, and so on, coin gaming machines
8	Medical tools	Radiotherapy gear, cardiology and dialysis tools, pneumonic ventilators
9	Observing and control apparatuses	Smoke alarms, warming controllers, indoor regulators
10	Programmed dispensers	Programmed dispensers for both hot and cold drinks, solid products, etc.

sheets. Silver NPs have shown to be effective antimicrobial agents in the treatment of wastewater containing bacteria, viruses, and fungi. Similarly, TiO_2 was shown to kill bacteria and disinfect water. Gold NPs have been used in treating water containing chlorinated organic compounds, pesticides, and inorganic mercury (Pati et al. 2016). Finally, the combination of TiO_2 with gold NPs has shown to remediate toxic air pollutants such as sulfur dioxide to sulfur.

Carbon nanotubes (CNT) are employed to remediate a broad range of pollutants by absorbing bacteria, heavy metals like Cd, and biological contaminants like benzene and also 1,2-dichlorobenzene (Uyguner-Demirel et al. 2017). There is limited or no literature that summarizes the scope and production of different NPs. Thus, a detailed review on different clean and green production methods of NPs from e-waste is presented in this chapter. Further, a comprehensive analysis on the application of e-waste NPs for environmental remediation processes is also presented.

TABLE 5.2
Major E-waste and Its Composition

E-Wastes and its Constituents	Explanation
Batteries	Heavy metals like lead (Pb), mercury (Hg), and also Cd are present in the batteries.
Cathode ray tubes (CRTs)	Pb is present inside the cone glass and fluorescent covering over within the panel glass.
Components comprising mercury, (switches)	Hg is primarily used in indoor regulators, sensors, transfers, and switches. Furthermore, it is also used in clinical gear, information transmission, telecom, and mobile phones.
Toner cartridges, liquid and color toner	Toner and toner cartridges consist of synthetic compounds that are hazardous to human health.
Printed Circuit Boards (PCB)	In PCBs, Pb and Cd are present; Cd is present in certain parts, e.g., SMD chip-resistors, infrared locators, and semiconductors.
Capacitors containing Polychlorinated Biphenyl (PCB)	Capacitors that contain PCB must be removed before demolition.
Liquid Crystal Displays (LCDs)	Pb and Hg are present in LCDs, especially in old models. LCDs with surface area > 100 cm^2 have to be excluded from e-waste.
Plastics comprising halogenated fire retardants	During the combustion of plastics, halogenated fire retardants can yield harmful contaminants.
Refrigeration equipment with Chloroflourocarbons (CFCs), Hydrochloroflourocarbons (HCFCs), or Hydroflourocarbon (HFC)	CFCs, HCFCs, or HFCs are cooling agents in refrigeration and foam. These must be removed and destroyed.
Gas emission lamps	Hg must be removed.

5.2   CONSTITUENTS OF E-WASTE

Table 5.2 illustrates the major components present in e-waste. It is generally made up of precious metals and dormant toxic constituents. The configuration of the e-waste alters on the category of electrical or electronic device/model, its producer, and manufacturing date including the time of scrap (Cui et al. 2011).

Scrap materials from media transmission frameworks and IT contain large volumes of important metals than scrap from the assembly parts of an electronic or electrical equipment. For example, a cell phone is built with more than 40 distinct modules and contains common metals (copper and tin); extraordinary metals (lithium, cobalt, indium, and antimony); and significant metals (silver, gold, and palladium) (Zhang et al. 2012). A separate treatment technology for each e-waste must be developed and used in order to prevent wasting of precious resources and rare elements. More efficient extractions of materials like palladium (Pd) and gold present in e-waste can be employed in contrast to the extraction from their respective ores (Hsu et al. 2019). Paradoxically, e-waste includes a combination of fire retardants, plastics, and different segments (PBDEs). Circuit boards in all electronic equipment contain arsenic (As), cadmium (Cd), chromium (Cr), lead (Pb), mercury (Hg), and other

TABLE 5.3
Mass Constituents of Metals for Diverse E-waste Samples

Electronic Waste	Mass (%)					Mass (mg/kg)		
	Iron (Fe)	Copper (Cu)	Aluminum (Al)	Lead (Pb)	Nickel (Ni)	Silver (Ag)	Gold (Au)	Palladium (Pd)
Television board	28	10	10	1.0	0.3	280	20	10
Computer board	7	20	5	1.5	1	1000	250	110
Cell phone	5	13	1	0.3	0.1	1380	350	210
Audio player	23	21	1	0.14	0.03	150	10	4
DVD player	62	5	2	0.3	0.05	115	15	4
Calculator	4	3	5	0.1	0.5	260	50	5
PC mainboard	4.5	14.3	2.8	2.2	1.1	639	566	124
PCBs	12	10	7	1.2	0.85	280	110	–
TV (CRTs detached)	–	3.4	1.2	0.2	0.038	20	<10	<10
Nokia cell phones	8	19	9	0.9	1	9000	–	–

poisonous synthetic substances (Xu et al. 2016). Outdated coolers, refrigerators, and other cooling units contain CFCs (Guo et al. 2009). Major metals like barium (Ba), Cd, copper (Cu), Pd, zinc (Zn), and additionally other uncommon metals are present in EoL of CRTs present in PC screens and TVs. Table 5.3 shows the different mass composition of metals from various e-wastes. Most developed countries have prohibited the disposal of CRTs in landfills due to their toxic properties (Hageluken and Christian 2006). However, the challenge in electronic industry is to develop technologies to recover and recycle e-waste by altering its composition using the advanced technology (Babu et al. 2007).

5.3 ENVIRONMENTAL AND HEALTH IMPACT ON TRADITIONAL E-WASTE RECYCLING PROCESSES

E-waste recycling includes scrap sorting, grinding and thermal, and/or hydrometallurgical, methods and thereby emit toxic compounds (Zielonka et al. 2017). In general, polybrominated diphenyl ethers (PBDEs) and also organophosphate esters have been extensively consumed in polystyrene as fire-resistive supplements. Nevertheless, these supplements can escape to the environment during the grinding of the e-waste because these compounds are chemically not combined with the polymer network (Zada et al. 2018). Sjödin et al. (2004) reported that both brominated compounds and also phosphorus in the polymeric substances are released to the environment in the places associated with recycling (Sjödin et al. 2004). Eight PBDE compounds comprising decabromodiphenyl ether, decabromobiphenyl, 1,2-bis(2,4,6-tribromo phenoxy) ethane, tetrabromobisphenol A, and five arylated and six alkylated organophosphate esters were measured in the atmospheric samples collected from the disassembly hall and the shredder room of an electronics recycling plant in Sweden (Industry Council for Electronic Equipment Recycling 2000). Several researchers have investigated the contact of lethal materials in some unofficial recycling centers in developing countries like China, Nigeria, and India (Darby et al. 2005). Due to the physical methods engaged in the resource recovery methods, the concentration of toxins like acids and dioxins discharged to the environment is significant in some unofficial recycling centers. For instance, e-waste recycling in Guiyu, China, has led to lead build up and pollution of water, air, and soil, thereby affected both biota of that district (Cooper and Tim 2000).

5.4 EXTRACTION OF NANOPARTICLES FROM E-WASTE

Nanomaterials are materials ranging from 1 to 100 nm, which have invited major attention in the current time because of their chemical properties and beneficial functions. Both the bottom-up and top-down approach are employed to produce NPs. In the bottom-up approach, the nanostructure is developed through self-organization, with each atom, and with each molecule, while during the top-down approach, the raw materials are utilized in the form of bulk and then reduced to nano-size (Dutta et al. 2018; Rienzie et al. 2019). The manufacture of NPs by green or biological methods follow the bottom-up strategy. The most striking opportunities for e-waste valorization were noticed during the retrieval of precious metals. Interestingly, electronic wastes can also be a reliable source of silicon and carbon for the manufacturing of nanostructured materials (Ahmed et al. 2018).

5.4.1 Copper Nanoparticles from E-waste

Copper (Cu) nanoparticles are potentially used as disinfectants and are considered as the most important application. They find application in medicine and food industry due to their antimicrobial properties. Cu is generally found in significant quantity in the PCBs. Sobiya et al. (2017) synthesized Cu NPs in an eco-friendly method by employing *Cassia auriculate*. The leaf extract of *Cassia auriculate* has reducing properties and can perform as an improving agent when the surface-area-to-volume ratio is increased (Sobiya et al. 2017). The manufactured Cu NPs were shown to salvage Cu from PCBs and produce Cu NPs (Sobiya et al. 2017). Majumder (2012) demonstrated the bioremediation of e-waste for extracting valuable metals. Microorganisms such as *Pseudomonas* spp. and *Fusarium oxysporum* were capable of extracting Cu (84–130 nm) from integrated circuits (ICs) on electronic boards under ambient environment (Majumder 2012). Majumder (2012) also reported that *Lantana camara*, a common weed in India, can leach Cu. Cu NPs were found to be effective against hospital strain *Escherichia coli* 2065 (Majumder 2012). Martins et al. (2019) focused their work on the extraction of Cu metal from PCBs through the physical route of NP production. Results showed that the PCBs had 43% match with the metallic segment of which 21.5% was Cu (Martins et al. 2019).

5.4.2 Carbon-based Nanoparticles' Synthesis from E-waste

E-wastes such as PCBs, waste toner powder, and printer ink contain numerous carbon-based compounds. Several studies have been reported of synthesizing carbon-based NPs from e-waste. In a simple and sustainable technology, Saini et al. (2019) carbonized the waste black toner ink at around 600°C in a muffle furnace and with supplemental oxidation process transformed it into a functional iron-oxide nanocarbons (f-FeO-NC). The produced f-FeO-NC was then employed for the degradation of Congo Red toxic azo dye via photocatalytic process. FeO-doped NC improved the photocatalytic action and was effective in the disposal of the waste black toner ink (Saini et al. 2019). Similarly, Bhongade et al. (2019) synthesized carbon nanotubes (CNTs) from waste toner powder by thermal breakdown of waste toner into hydrocarbons, which were used as a foundation for manufacturing 40–50 nm multi-walled carbon-nanotubes (MWCNTs) via chemical vapor deposition (CVD) technique. Xu et al. (2019) prepared graphene oxide quantum dots (GOQDs) from waste toner. In the aforementioned study, one-pot hydrothermal process was engaged for the fabrication of GOQDs (Xu et al. 2019). The aforementioned researchers used waste toner as the precursor and hydrogen peroxide as the oxidant for the production of GOQDs. The process did not involve any complex purification steps or strong acids and did not generate toxic metal ions. The GOQDs were employed to compute targeted DNA sequences extracted from genetically modified plant tissues (Xu et al. 2019).

5.4.3 Gold Nanoparticles from E-waste

One of the efficient methods to resolve the predicament concerning resource deficiency along with increasing demands is to foster a cost-efficient method to reuse and recover the valuable metals, particularly Au and Ag originating from the e-wastes. Tian et al. (2020) demonstrated a simple and innovative method to retrieve Au from e-wastes by exploiting hollow polyaniline nanospheres (P(Van-g-PANI)). In the aforementioned process, PANI and its byproducts can reduce Au^{3+} to Au^0 from the metal salts and stabilize the obtained polymer nanospheres-supported Au NPs (AuNP(VAn-g-PANI)) (Tian et al. 2020). In the aforementioned study, *in-situ* recycle and use of Au for the construction of electronic devices by using polymer nanospheres were demonstrated (Tian et al. 2020). Panda et al. (2020) described an innovative hybrid method for the complete retrieval of Au from small abandoned components of e-waste. In the aforementioned study, ICs and connectors from PCBs were crushed and treated for leaching Au by using 10 g/L sodium cyanide at 40°C. More than 95% Au was recovered in a single stage. From the leachate, metallic Au was retrieved by adsorption using charcoal, followed by a heat-treatment process. The leftover raffinate was discovered to comprise around 10 mg Au/L, which was further refined by employing 25 mL/g Amberlite IRA 400Cl at pH 9.6 for 30 minutes in aqueous/resin (A/R) (Panda et al. 2020). The concentration of raffinate was increased to 882.41 mg/L to recover metal/salt by cementation/evaporation processes. The obtained leachate was then treated for non-ferrous metals. The effluent left after leaching was treated in effluent treatment plant by employing standard environmental practice (Panda et al. 2020). Das and Ting (2017) investigated the leaching of ultrasound-assisted thiourea for the retrieval of Au from powered PCBs (Das and Ting 2017). Cyanidation is usually used for the extraction of metallic Au but is an extremely toxic procedure. Thiourea might be also employed as a substitute lixiviant for the extraction of metallic Au from electronic scrap materials (ESMs) due to higher leaching rates and low toxicity. In the aforementioned study, Au was recovered through in-vitro reduction of the thiourea leachate by the bacterium *Delftia acidovorans* in the form of Au NPs and is considered as a green process (Das and Ting 2017).

5.4.4 Silver Nanoparticles from E-waste

Metallic Ag is the most significant component of e-waste and has been applied in the medical sector, food, agriculture sector, textiles, plastics, and cosmetics sector and in the manufacture of catalysts. Further, Ag NPs possess antibacterial characteristics and are extensively used in industries. Although Ag in soluble form can produce harm to the human beings, its nontoxic and noncarcinogenic nature was well documented (Byeon et al. 2012; Sharma et al. 2009). Cerchier et al. (2017) synthesized Ag NPs from PCBs. In the aforementioned study, silver chloride (AgCl) was reduced to metallic Ag NPs in ammonia by utilizing glucose syrup as a reducing and a capping agent. This reaction was done by using a traditional heating, microwave irradiation, and also ultrasound treatment. The obtained Ag NPs were characterized by various analytical instruments. Although there are numerous green methods to prepare Ag NPs, the production of Ag NPs from e-waste is extremely rare (Cerchier et al. 2017).

5.4.5 Metal and Nonmetal Nanoparticles from E-waste

Nanoparticles and micro-plastics are the major contaminants in our environment. The printer toner mainly comprises micro-plastics and other nano pollutants. It contains micro polyacrylate styrene and also nano-Fe_3O_4 particles. Nanometals and polyacrylate styrene were shown to cause irreversible toxicity to all-natural cells. Hence, toners are classified as an ecological hazard and harmful to health. Ruan et al. (2018) adopted the vacuum-gasification-condensation process to recover the nano-pollutants and micro-plastics present in the toner. Nekouei et al. (2018) used solid-state mechanical alloying to straightaway alter PCBs to a standardized nano-structured alloy (Cu79-Zn13-Fe3-Sn3-Ni1). Chemical analyses confirmed that waste PCBs can directly be converted to metallic alloys devoid of employing any heat source or solutions (Nekouei et al. 2018). Babar et al. (2019) reported an eco-friendly and sustainable method to reuse the waste toner powder reinforced in the organic residues and also magnetic Fe_3O_4 (Babar et al. 2019). In the aforementioned study, waste toner powder was heated at 600°C to ensure a total breakdown of the organic remains and also the conversion of Fe_3O_4 into valuable magnetic Fe_2O_3 with no chemical usage. Babar et al. (2019) also demonstrated Fe_2O_3 as an iron precursor and also thiourea for the production of g-C_3N_4-Fe_2O_3 photocatalyst via one-step calcination method. The photocatalytic activity of the fabricated g-C_3N_4-Fe_2O_3 was assessed by the decomposition of methyl orange dye and also other textile effluents under sunlight. The resulting g-C_3N_4-Fe_2O_3 displayed improved photocatalytic activity than the normal Fe_2O_3 and g-C_3N_4. With outstanding steadiness and magnetic separation capability, g-C_3N_4-Fe_2O_3 composite exhibited splendid recyclability without the loss of substantial photocatalytic activity up to five repetitive runs, and more (Babar et al. 2019). Thus, we can conclude that this novel e-waste recovering approach may deliver innovative potentials for a low-cost and large-scale conversion process of waste toner powder into a photocatalyst which is magnetically separable for waste-water treatment.

Waste multilayer ceramic capacitors (MLCCs), comprising $BaTiO_3$, Ag, Pd, Ni, and Sn, etc., are treasured secondary resources. The current recovering methods have countless disputes when contemplating environmentally friendly and also effective separation and retrieval processes of resources (Babar et al. 2019). Niu et al. (2019) directly applied the composite constituents of waste MLCCs as Nb\\Pb co-doped and Ag–Pd–Sn–Ni-loaded $BaTiO_3$ nano-photocatalyst via a one-step ball milling process (Niu et al. 2019). The prepared photocatalyst exhibited high photocatalytic activity. The recycled material also displayed outstanding reusability and photostability. The outstanding photocatalytic activity was qualified to the Nb\\Pb co-doping and Ag–Pd–Sn–Ni loading, which boosted the visible light absorption and also the effective charge separation. On comparison with previous research, Niu et al. (2019) demonstrated a sustainable and simple method to alter e-waste into enhanced and also an efficient photocatalyst for judicious waste consumption and also environmental protection (Niu et al. 2019).

E-waste has also been a precursor for the development of nanocomposites utilized for photocatalytic degradations. They used recycled vanadium nitrate from redox battery to develop a reduced graphene oxide–vanadium pentoxide nanocomposite for oxytetracycline degradation in photocatalytic route (Mohan et al. 2019). Liquid Crystal Display (LCD) monitors has been used to develop silicon carbon nanowires with glass serving as the precursor of silica and the polymer frame as the carbon source. A combined method of pyrolysis and carbothermal reduction was employed to achieve the desired product (Assefi et al. 2019). These can be used as photocatalysts in wastewater treatment (Rodríguez-Padrón et al. 2020).

Ruan et al. (2017) presented an eco-friendly and also a valuable recovery process of waste metallic film resistors (Ruan et al. 2017). The recovery procedure comprises shearing, magnetic separation, and finally milling. The ceramic material was milled into nano-Al_2O_3. The nano-Al_2O_3 has a particle size which extended from 100 to 500 nm, and it is a chief powder-coating material. The commercial benefits of retrieving waste metallic film resistors were evaluated on the basis of energy costs, apparatus devaluation cost, employment cost, and also economic revenue (Ruan et al. 2017). Nayak et al. (2019) demonstrated the production of metal oxide nanocomposites from PCBs (Nayak et al. 2019). Metals in the electronic apparatuses of the waste memory slots were stripped out by means of nitric acid (HNO_3). The leached-out metal salt solutions were exposed to alkaline hydrothermal treatment to produce nanocomposites. Two nanoparticles were synthesized, one devoid of any stabilizing agent and another with just PVP as the stabilizing agent. The ZnO/CuO nanocomposite particles acquired via this method conducted by Nayak et al. contained fine ZnO nanostructures precipitated on CuO cores. The nanocomposites demonstrated decent visible light photo-Fenton methyl orange dye degradation by using pseudo-zero order kinetics (Nayak et al. 2019).

5.5 APPLICATION OF NANOPARTICLES OBTAINED FROM E-WASTE

Given the exceptional features of NPs, they can be utilized in various applications. A few significant applications of these are given in the next sections.

5.5.1 DRUGS AND MEDICATIONS

Iron oxide nanoparticles like magnetite (Fe_3O_4) or its oxidized state maghemite (Fe_2O_3) constitutes the frequently exploited nanomaterials for biomedical purposes. Choosing the correct NPs for attaining an effective variation in cell and biological imaging use and for the photothermal treatments has been found concerning the photosensitive characteristics of NPs. Most of the semiconductors and metallic NPs have the enormous ability for cancer identification and treatment because of their surface plasmon resonance (SPR) heightened light scattering and also absorption. Gold NPs effectively change the intense absorbed light into confined heat which is used for the selective laser photothermal (SLPT) therapy of cancer. Alongside these benefits, the antineoplastic effect of NPs is successfully utilized to prevent the growth of tumors (Loureiro et al. 2016). The multi-hydroxylated NPs display antineoplastic function with decent efficacy and also lower harmfulness. Silver NPs have been used progressively in wound coverings, catheters, and also numerous household products due to their antimicrobial property. In industries like the textile, medicine, water purification, and food packaging, antimicrobial agents are extremely crucial. Thus, the anti-infective features of inorganic NPs enhance effectiveness to show the aforementioned significant characteristics, in contrast to the organic compounds, which are lethal as compared to biotic systems. Such NPs are functionalized with some groups to overpower the selectively of microbial species. TiO_2, ZnO, $BiVO_4$, Ni, and Cu-centered NPs have been employed for this intention because of their appropriate antimicrobial abilities (Zhang et al. 2013).

5.5.2 MANUFACTURING AND MATERIALS

NPs emphasize the property, engineering, and designing of biological and non-biological assemblies that are less than 100 nm, which also exhibit exclusive and innovative properties. The latent advantages of nanomaterials can be recognized by numerous producers at both low and high stages, and also profitable harvests have now been produced massively in the forms of microelectronics, pharmaceutical, and aerospace manufacturing units (Todescato et al. 2016). Among the nanotechnology purchaser goods till date, the fitness commodity forms the greatest class, then comes the computer and electronic sector including garden and home department. Nanotechnology has been accepted as a new development in several manufacturing sectors consisting of food processing and packing. Metal NPs like the noble metals, comprising Au and Ag, have several colors in the visible section centered on plasmon resonance because of the combined oscillations of the electrons present on the surface of the NPs. The strength of the wavelength during resonance is governed by the size and also the shape of the NPs, dielectric characteristic, and the interparticle distance of the neighboring medium. The exclusive plasmon absorbance properties of these noble metals NPs have been developed for an extensive variability of implementation together with sensors (Unser et al. 2015).

5.5.3 ENVIRONMENTAL

The presence of elevated surface-area-to-mass ratio helps the NPs to be a vital part of the solid contaminant absorption from waste streams. The interface of the pollutants with NPs is reliant on the NP properties like the dimension, chemical constitution, surface morphology, porosity, accumulation, and the aggregation. NPs embedded in activated carbon and polymers enhance the adsorption capacity of pollutants from waste streams, both liquid and gaseous (Santra et al. 2001; Mitra et al. 2021; Mukherjee et al. 2019; Bose et al. 2021). Maximum sustainable functions of NPs fall into the following three classes.

1. Ecologically benevolent sustainable products (e.g., pollution prevention or green chemical approach)
2. Recovering constituents polluted with hazardous materials
3. Biosensors for different sustainable operations

5.5.4 ENERGY HARVESTING

Keeping in mind the nonrenewable nature of conventional energy resources, research has warned about the probable scarcity to be experienced globally very soon. Hence, multiple researches have been going on for cost-effective manufacturing of non-conventional energies with readily available substances. Following these researches, large specific surface area (SSA), optical nature, and also catalytic behavior make NPs the best choice for the aforementioned purpose (Shaalan et al. 2016). Particularly in photocatalytic purposes, NPs are extensively used to produce energy from photo-electrochemical (PEC) and also electrochemical water splitting. Alongside water splitting, electrochemical CO_2 reduction to fuels predecessors, solar cells, and also piezoelectric generators, NPs can be fabricated to provide endless possibilities to produce energy (Holzinger et al. 2014). Nanoscale-level storage energy into diverse configurations is also offered by NPs in energy packing sectors. Lately, nanogenerators are used to translate mechanical energy into electrical energy utilizing piezoelectricity, which stands as an alternative strategy in power generation (Li et al. 2016).

5.5.5 MECHANICAL INDUSTRIES

As discovered from the mechanical characteristics of NPs, it can be concluded that they possess outstanding young modulus, strain, and also stress properties, and, thus, they are used for various purposes in mechanical productions particularly in coating, adhesive, and also lubricant applications. Besides, this characteristic of NPs is supposed to be beneficial in attaining high mechanical strength of nanodevices for varied systems (Kot et al. 2016). The regulation of tribological characteristics at the nanoscale level can be achieved by implanting NPs in a metal or polymer medium to enhance their machine-driven properties. This is due to the rolling property of the nanoparticles in the area of lubricated contact, which provides minimum friction, wear, and also tears. Moreover, NPs provide decent delamination and sliding properties, which cause little friction and henceforth elevate the effect of lubrication. On coating, many mechanically durable characteristics are steered, because it upgrades the strength and also corrosion resistance. Carbon-based, titania, and also alumina NPs are effectively proved to instill the appropriate mechanical characteristics in coatings (Guo et al. 2013).

5.6 FUTURE IMPLICATIONS AND SCOPE OF NANOPARTICLE DERIVED FROM E-WASTE

The economic viability of e-waste NP production and application compared with conventional chemicals is crucial to scale up this technology. E-waste generation will increase in the future and thereby become a major resource for the production of valuable NPs. However, the impact of additional metals and the need to increase the purity of the produced NPs without mercury and lead, which are the constituents of e-waste, play a crucial role in the economics of NP production (Debnath et al. 2020). Currently, studies on microbe-mediated production of NPs from e-waste are very limited. Thus, more research is required to examine the potential use of e-waste for the production of precious metals (Debnath 2020; Debnath et al. 2019). Regardless of the methods for the production in general, NPs have numerous properties because of the diversity of capping chemicals. The production of NPs from e-waste is relatively a new research area started in the last few years. Table 5.4 gives a detailed character analysis of different NPs produced from various e-wastes. New technologies are yet to be discovered which can be scaled up for NP production from e-waste on a large scale.

5.7 CONCLUSION

Nanoparticles exhibit exclusive properties like high surface area and also reactivity, which allow NPs to detect and remove miscellaneous targets, together with hazardous chemicals; toxic gases like SO_x, CO, and NO_x; and also biological matter from diverse media like natural waters, air, and wastewaters. Synthesis of NPs from e-waste is a cleaner and greener approach toward e-waste recycling and valorization. The production of nanoparticles from e-waste is relatively a new research area. New technologies are yet to be discovered, which can be scaled up for NP production from e-waste on a large scale. The production of gold, silver, carbon, copper, metal, and non-metal nanoparticles from e-waste has been addressed in this chapter. However, there are numerous green methods for the preparation of Ag nanoparticles, but the production of Ag NPs from e-waste is extremely rare. Thus, we can conclude that this sector needs further investigation in the near future.

TABLE 5.4
Characterization of Nanoparticles Obtained from E-waste

E-Waste Source	Nanoparticle Type	Characterization	References
PCB	Copper	UV–Visible spectroscopy Fourier Transform Infrared (FTIR) X-Ray Diffraction Scanning Electron Microscopy (SEM) EDAX	Sobiya et al. (2017)
Electronic boards	Copper	X-ray diffraction energy-dispersive spectroscopy Scanning electron microscopy Fourier Transform Infrared Transmission electron microscopy Thermogravimetric analysis Cyclic voltammetry	Majumder (2012)
Waste toner powder	CNT	XRD Raman spectroscopy FESEM HRTEM	Bhongade et al. (2019)
PCB	Silver	Scanning electron microscope (SEM) XRD Infrared (IR)-spectroscopy Transmission electron microscope (TEM) Ultraviolet (UV)-spectroscopy Laser diffraction particle size analyzer	Cerchier et al. (2017)
PCB	Nanostructured alloy	Electron microscopy analysis XRD analysis XPS UV-Visible analysis pH and conductivity analysis	Nekouei et al. (2018)
Waste toner powder	$g\text{-}C_3N_4\text{-}Fe_2O_3$ photocatalyst	XRD, TGA, SEM-EDS, TEM, BET, XPS, UV–Visible, DRS, FT-IR, PL, EIS, and VSM	Babar et al. (2019)
PCB	ZnO/CuO nanocomposites	XRD, HRXRD, HR-TEM, UV-DRS, UV–Visible spectroscopy	Nayak et al. (2019)

REFERENCES

Ahmed, Elaf, Shafeer Kalathil, Le Shi, Ohoud Alharbi, and Peng Wang. 2018. Synthesis of ultra-small platinum, palladium and gold nanoparticles by Shewanella loihica PV-4 electrochemically active biofilms and their enhanced catalytic activities. *J. Saudi Chem. Soc.* 22, no. 8: 919–929.

Assefi, Mohammad, Samane Maroufi, and Veena Sahajwalla. 2019. Recycling of the scrap LCD panels by converting into the $InBO_3$ nanostructure product. *Environ. Sci. Pollut. Res.* 26, no. 36: 36287–36295.

Babar, Santosh, Nana Gavade, Harish Shinde, Anil Gore, Prasad Mahajan, Ki Hwan Lee, Vijaykumar Bhuse, and Kalyanrao Garadkar. 2019. An innovative transformation of waste toner powder into magnetic $g\text{-}C_3N_4\text{-}Fe_2O_3$ photocatalyst: sustainable e-waste management. *J. Environ. Chem. Eng.* 7, no. 2: 103041.

Babu, Balakrishnan Ramesh, Anand Kuber Parande, and Chiya Ahmed Basha. 2007. Electrical and electronic waste: a global environmental problem. *Waste Manag. Res.* 25, no. 4: 307–318.

Baldé, C. P., Vanessa Forti, Vanessa Gray, Ruediger Kuehr, and Paul Stegmann. 2017. *The Global e-waste Monitor–2017.* United Nations University (UNU), International Telecommunication Union (ITU) & International Solid Waste Association (ISWA), Bonn/Geneva/Vienna. ISBN Electronic Version, 978–992.

Balde, C. P., Ruediger Kuehr, K. Blumenthal, S. Fondeur Gill, M. Kern, P. Micheli, E. Magpantay, and Jaco Huisman. 2015. *E-waste Statistics-Guidelines on Classification, Reporting and Indicators.* United Nations University, IAS-SCYCLE, Bonn (Germany), 51.

Barbara, K.R. and Graedel, T.E., 2012. Challenges in metal recycling. *Science,* 337, no. 6095: 700.

Bhongade, Tejas, Gogaram, Deepak M. Gautam, and R. P. Vijayakumar. 2019. Synthesis of MWCNTs using waste toner powder as carbon source by chemical vapor deposition method. *Fuller. Nanotub. Car. N.* 27, no. 11: 864–872.

Bose, S., Mukherjee, T. and Rahaman, M., 2021. Simultaneous adsorption of manganese and fluoride from aqueous solution via bimetal impregnated activated carbon derived from waste tire: Response surface method modeling approach. *Environmental Progress & Sustainable Energy,* 40, no. 4: p. e13600.

Breivik, Knut, James M. Armitage, Frank Wania, and Kevin C. Jones. 2014. Tracking the global generation and exports of e-waste. Do existing estimates add up? *Environ. Sci. Technol.* 48, no. 15: 8735–8743.

Byeon, Jeong Hoon, and Young-Woo Kim. 2012. A novel polyol method to synthesize colloidal silver nanoparticles by ultrasonic irradiation. *Ultrason Sonochem* 19, no. 1: 209–215.

Cerchier, Pietrogiovanni, Manuele Dabalà, and Katya Brunelli. 2017. Synthesis of SnO$_2$ and Ag nanoparticles from electronic wastes with the assistance of ultrasound and microwaves. *JOM* 69, no. 9: 1583–1588.

Chen, Lei, Chunna Yu, Chaofeng Shen, Congkai Zhang, Lei Liu, Kaili Shen, Xianjin Tang, and Yingxu Chen. 2010. Study on adverse impact of e-waste disassembly on surface sediment in East China by chemical analysis and bioassays. *J. Soils Sediments.* 10, no. 3: 359–367.

Cooper, Tim. 2000. WEEE, WEEE, WEEE, WEEE, all the way home? An evaluation of proposed electrical and electronic waste legislation. *Eur. Environ.* 10, no. 3: 121–130.

Cui, Jirang, and Hans Jørgen Roven. 2011. Electronic waste. In *Waste,* 281–296.

Darby, Lauren, and Louise Obara. 2005. Household recycling behaviour and attitudes towards the disposal of small electrical and electronic equipment. *Resour Conserv Recycl.* 44, no. 1: 17–35.

Das, S., and Ting, Y. P. 2017. In-vitro synthesis of gold nanoparticles using thiourea leachate of electronic waste. In *Sixteenth International Waste Management and Landfill Symposium.* CISA Publisher, Italy.

Debnath, Biswajit. 2020. Towards sustainable e-waste management through industrial symbiosis: a supply chain perspective. In *Industrial Symbiosis for the Circular Economy.* Springer, Cham, 87–102.

Debnath, Biswajit, Ranjana Chowdhury, and Sadhan Kumar Ghosh. 2019. Urban mining and the metal recovery from e-waste (MREW) supply chain. In *Waste Valorisation and Recycling.* Springer, Singapore, 341–347.

Debnath, Biswajit, Ranjana Chowdhury, and Sadhan Kumar Ghosh. 2020. Recycling of polymers from WEEE: issues, challenges and opportunities. In *Urban Mining and Sustainable Waste Management.* Springer, Singapore, 69–80.

Dutta, Tanushree, Ki-Hyun Kim, Akash Deep, Jan E. Szulejko, Kowsalya Vellingiri, Sandeep Kumar, Eilhann E. Kwon, and Seong-Taek Yun. 2018. Recovery of nanomaterials from battery and electronic wastes: a new paradigm of environmental waste management. *Renew. Sust. Energ. Rev.* 82: 3694–3704.

Guo, Dan, Guoxin Xie, and Jianbin Luo. 2013. Mechanical properties of nanoparticles: basics and applications. *J. Phys. D: Appl. Phys.* 47, no. 1: 013001.

Guo, Jiuyong, Jie Guo, and Zhenming Xu. 2009. Recycling of non-metallic fractions from waste printed circuit boards: a review. *J. Hazard. Mater.* 168, no. 2–3: 567–590.

Ha, Nguyen Ngoc, Tetsuro Agusa, Karri Ramu, Nguyen Phuc Cam Tu, Satoko Murata, Keshav A. Bulbule, Peethmbaram Parthasaraty, Shin Takahashi, Annamalai Subramanian, and Shinsuke Tanabe. 2009. Contamination by trace elements at e-waste recycling sites in Bangalore, India. *Chemosphere* 76, no. 1: 9–15.

Hageluken, Christian. 2006. Improving metal returns and eco-efficiency in electronics recycling-a holistic approach for interface optimisation between pre-processing and integrated metals smelting and refining. In *IEEE International Symposium on Electronics and the Environment.* 218–223.

Holzinger, Michael, Alan Le Goff, and Serge Cosnier. 2014. Nanomaterials for biosensing applications: a review. *Front. Chem.* 2: 63.

Hsu, Emily, Katayun Barmak, Alan C. West, and Ah-Hyung A. Park. 2019. Advancements in the treatment and processing of electronic waste with sustainability: a review of metal extraction and recovery technologies. *Green Chem.* 21, no. 5: 919–936.

Industry Council for Electronic Equipment Recycling. 2000. *UK Status Report on Waste from Electrical and Electronic Equipment.*

Khin, Mya, A. Sreekumaran Nair, V. Jagadeesh Babu, Rajendiran Murugan, and Seeram Ramakrishna. 2012. A review on nanomaterials for environmental remediation. *Energy Environ. Sci.* 5, no. 8: 8075–8109.

Kot, M., Ł. Major, J. M. Lackner, K. Chronowska-Przywara, M. Janusz, and W. Rakowski. 2016. Mechanical and tribological properties of carbon-based graded coatings. *J. Nanomater.*

Li, Da, Habib Baydoun, Cláudio N. Verani, and Stephanie L. Brock. 2016. Efficient water oxidation using CoMnP nanoparticles. *J. Am. Chem. Soc.* 138, no. 12: 4006–4009.

Loureiro, Ana, Nuno G. Azoia, Andreia C. Gomes, and Artur Cavaco-Paulo. 2016. Albumin-based nanodevices as drug carriers. *Curr. Pharm. Des.* 22, no. 10: 1371–1390.

Majumder, D. R. 2012. Bioremediation: copper nanoparticles from electronic-waste. *Int. J. Eng. Sci. Technol.* 4, no. 10.

Martins, Thamiris Auxiliadora Gonçalves, Karen Espina Gomes, Carlos Gonzalo Alvarez Rosario, Denise Crocce Romano Espinosa, and Jorge Alberto Soares Tenório. 2019. Characterization of printed circuit boards of obsolete (PCBs) aimed at the production of copper nanoparticles. In *Characterization of Minerals, Metals, and Materials.* 543–551.

Mitra, S., Mukherjee, T. and Kaparaju, P., 2021. Prediction of methyl orange removal by iron decorated activated carbon using an artificial neural network. *Environmental Technology*, 42, no. 21: 3288–3303.

Mohan, Harshavardhan, Dhanakumar Selvaraj, Shanthi Kuppusamy, Janaki Venkatachalam, Yool-Jin Park, Kamala-Kannan Seralathan, and Byung-Taek Oh. 2019. E-waste based V2O5/RGO/Pt nanocomposite for photocatalytic degradation of oxytetracycline. *Environmental Progress & Sustainable Energy* 38, no. 4: 13123.

Mukherjee, T., Rahaman, M., Ghosh, A. and Bose, S., 2019. Optimization of adsorbent derived from non-biodegradable waste employing response surface methodology toward the removal of dye solutions. *International Journal of Environmental Science and Technology*, 16, 8671–8678.

Nayak, Pritish, Sunil Kumar, Indrajit Sinha, and Kamalesh Kumar Singh. 2019. ZnO/CuO nanocomposites from recycled printed circuit board: preparation and photocatalytic properties. *Environ. Sci. Pollut. Res.* 26, no. 16: 16279–16288.

Nekouei, Rasoul Khayyam, Farshid Pahlevani, Ravindra Rajarao, Rabeeh Golmohammadzadeh, and Veena Sahajwalla. 2018. Direct transformation of waste printed circuit boards to nano-structured powders through mechanical alloying. *Mater. Des.* 141: 26–36.

Niu, Bo, and Zhenming Xu. 2019. Innovating e-waste recycling: from waste multi-layer ceramic capacitors to NbPb codoped and Ag-Pd-Sn-Ni loaded $BaTiO_3$ nano-photocatalyst through one-step ball milling process. *Sustain. Mater. Techno.* 21: e00101.

Panda, Rekha, Om Shankar Dinkar, Manis Kumar Jha, and Devendra Deo Pathak. 2020. Recycling of gold from waste electronic components of devices. *Korean. J. Chem. Eng.* 37, no. 1: 111–119.

Pati, Paramjeet, Sean McGinnis, and Peter J. Vikesland. 2016. Waste not want not: life cycle implications of gold recovery and recycling from nano waste. *Environ. Sci. Nano* 3, no. 5: 1133–1143.

Rienzie, Ryan, Amarasinghage Tharindu Dasun Perera, and Nadeesh Madusanka Adassooriya. 2019. Biorecovery of precious metal nanoparticles from waste electrical and electronic equipments. In *Electronic Waste Management and Treatment Technology.* 133–152.

Robinson, Brett H. 2009. E-waste: an assessment of global production and environmental impacts. *Sci. Total Environ.* 408, no. 2: 183–191.

Rodríguez-Padrón, Daily, Zeid Alothman, Sameh M. Mahmoud, and Rafael Luque. 2020. Recycling electronic waste: prospects in green catalysts design. *Current Opinion in Green and Sustainable Chemistry* 100357.

Ruan, Jujun, Jiaxin Huang, Lipeng Dong, and Zhe Huang. 2017. Environmentally friendly technology of recovering nickel resources and producing nano-Al_2O_3 from waste metal film resistors. *ACS Sustain. Chem. Eng.* 5, no. 9: 8234–8240.

Ruan, Jujun, Baojia Qin, and Jiaxin Huang. 2018. Controlling measures of micro-plastic and nano pollutants: a short review of disposing waste toners. *Environ. Int.* 118: 92–96.

Saini, Deepika, Ruchi Aggarwal, Satyesh Raj Anand, and Sumit Kumar Sonkar. 2019. Sunlight induced photodegradation of toxic azo dye by self-doped iron oxide nano-carbon from waste printer ink. *Sol. Energy* 193: 65–73.

Santra, Swadeshmukul, Peng Zhang, Kemin Wang, Rovelyn Tapec, and Weihong Tan. 2001. Conjugation of biomolecules with luminophore-doped silica nanoparticles for photo-stable biomarkers. *Anal. Chem.* 73, no. 20: 4988–4993.

Shaalan, Mohamed, Mona Saleh, Magdy El-Mahdy, and Mansour El-Matbouli. 2016. Recent progress in applications of nanoparticles in fish medicine: a review. *Nanomedicine: NBM* 12, no. 3: 701–710.

Shapira, Philip, and Jan Youtie. 2015. The economic contributions of nanotechnology to green and sustainable growth. In *Green Processes for Nanotechnology.* 409–434.

Sharma, Virender K., Ria A. Yngard, and Yekaterina Lin. 2009. Silver nanoparticles: green synthesis and their antimicrobial activities. *Adv. Colloid Interface Sci.* 145, no. 1–2: 83–96.

Sjödin, Andreas, Richard S. Jones, Jean-François Focant, Chester Lapeza, Richard Y. Wang, Ernest E. McGahee 3rd, Yalin Zhang et al. 2004. Retrospective time-trend study of polybrominated diphenyl ether and polybrominated and polychlorinated biphenyl levels in human serum from the United States. *Environ. Health Perspect.* 112, no. 6: 654–658.

Sobiya, S., S. M. Akshayalakshmi, S. Umamaheshwari, and M. Priyadharshini. 2017. Phyto-mediated synthesis of copper nanoparticles by Cassia auriculata and its characteriza-tion with reference to e-waste management. *Int. J. Environ., Agri. Biotechnol.* 2, no. 2: 238733.

Stark, Wendelin J., Philipp R. Stoessel, Wendel Wohlleben, and A. J. C. S. R. Hafner. 2015. Industrial applications of nanoparticles. *Chem. Soc. Rev.* 44, no. 16: 5793–5805.

Tian, Xiangyu, Bin Zhang, Jie Hou, Minchao Gu, and Yu Chen. 2020. In situ preparation and unique electrical behaviors of gold@ hollow polyaniline nanospheres through recovery of gold from simulated e-waste. *Bull. Chem. Soc. Jpn.* 93, no. 3: 373–378.

Todescato, Francesco, Ilaria Fortunati, Alessandro Minotto, Raffaella Signorini, Jacek J. Jasieniak, and Renato Bozio. 2016. Engineering of semiconductor nanocrystals for light emitting applications. *Mater.* 9, no. 8: 672.

Unser, Sarah, Ian Bruzas, Jie He, and Laura Sagle. 2015. Localized surface plasmon resonance biosensing: current challenges and approaches. *Sensors* 15, no. 7: 15684–15716.

Uyguner-Demirel, Ceyda Senem, Burak Demirel, Nadim K. Copty, and Turgut T. Onay. 2017. Presence, behavior and fate of engineered nanomaterials in municipal solid waste landfills. In *Nanotechnology for Environmental Remediation*. 311–325.

Xu, Qiang, Yan Gong, Zhifeng Zhang, Yanming Miao, Dongxia Li, and Guiqin Yan. 2019. Preparation of graphene oxide quantum dots from waste toner, and their application to a fluorometric DNA hybridization assay. *Microchimica Acta* 186, no. 7: 483.

Xu, Yan, Jinhui Li, and Lili Liu. 2016. Current status and future perspective of recycling copper by hydrometallurgy from waste printed circuit boards. *Procedia Environ. Sci.* 31: 162–170.

Zada, Shah, Aftab Ahmad, Sikandar Khan, Arshad Iqbal, Shahbaz Ahmad, Hamid Ali, and Pengcheng Fu. 2018. Biofabrication of gold nanoparticles by Lyptolyngbya JSC-1 extract as super reducing and stabilizing agents: synthesis, characterization and antibacterial activity. *Microb. Pathog.* 114: 116–123.

Zhang, Junwei, and Mark Saltzman. 2013. Engineering biodegradable nanoparticles for drug and gene delivery. *Chem. Eng. Prog.* 109, no. 3: 25.

Zhang, Kai, Jerald L. Schnoor, and Eddy Y. Zeng. 2012. E-waste recycling: where does it go from here? 10861–10867.

Zielonka, Aleksandra, and Magdalena Klimek-Ochab. 2017. Fungal synthesis of size-defined nanoparticles. *Adv. Nat. Sci-Nanosci.* 8, no. 4: 043001.

Zoeteman, Bastiaan C. J., Harold R. Krikke, and Jan Venselaar. 2010. Handling WEEE waste flows: on the effectiveness of producer responsibility in a globalizing world. *Int. J. Adv. Manuf. Technol.* 47, no. 5–8: 415–436.

Part III

Sustainability Aspects of E-waste Management

Features Three Pillars of Sustainability, i.e., Environment, Economic, and Social with Respect to E-waste Management

6 Health and Environmental Effects of E-waste Recycling Processes
Issues, Challenges, and Solutions

Anisha Modak and Pousali Bhattacharjyya

CONTENTS

DOI: 10.1201/9781003301899-9

6.1 INTRODUCTION

E waste is an abbreviated term which denotes electronic waste. The electronics industry is a multi-billion dollar—newfangled—fastest growing manufacturing industry that has provided a determinant and strong influence on the socio-economic and technological demand of a developing country for the past few decades (Joseph, 2007); let's say electronic industries are one of the major tributaries in the field of developmental process. E-waste, which is known as Waste Electrical and Electronic Equipment (WEEE) in the Western world, is basically discarded non-functional appliances that run on electricity or an electric power source (Robinson, 2009). An unrevealed amount of electronics is imported and shipped to other countries. In developing nations, like Bangladesh and India, a good number of users usually choose to buy second-hand products due to economic constraints (Sinha-Khetriwal et al., 2005; Wath et al. 2011). There is a very lucrative market for second-hand electronic products in India, which is often located at the outskirts (Ongondo et al., 2011). However, the consequence of its consumerd-dependent production and growth with increasing technological advances leads to rapid product obsolescence or wastage which is a major environmental threat (Vats and Singh, 2014). As a part of WEEE after the usage, these electronics become complex waste matter which contains certain valuable materials (Ag, Au, Pb, Pt) which can be profitable if recovered and also contains many hazardous waste metals, nondegradable materials (plastic and plastic products, etc.), toxic chemicals and substances, acids, and toxic materials such as chromium, mercury, cadmium, lead, and polychlorinated biphenyl (PCBs) which are harmful to the environment unless they are properly treated and disposed (Tanskanen, 2013; Jhariya et al. 2014). Many waste e-products are either openly burnt and dumped or recycle, and some of them are refurbished as well (Sivaramanan, 2013). According to a study by Huisman et al. (2016), nearly 0.6 Mt of e-waste is thrown in the bins in the EU countries. While refurbishing any product, some parts of the waste are remanufactured or reused. While doing dissembling or dismantling, informal recyclers are also exposed to hazardous substances or chemicals via dust particles and fumes. Some of these hazardous materials are carcinogenic in nature. These toxic hazardous materials can cause inflammation and lesions to lungs and skin whereas carcinogenic substances may cause skin and lung cancer (Bridgen et al. 2008).

India is the third electronic waste generator in the world. In 2019, nearly 53.6 Mt of electronics was produced worldwide, and it is estimated that, by 2030, the figure will exceed 74 Mt, which is alarming. According to UN global e-waste monitor 2020, quantity of e-waste is growing nearly at 2 Mt per year (Forti et al., 2020). Unorganized or informal sector and scrap dealers deal with the recycling of more than 95% of e-waste generated in India (Chaturvedi et al., 2007). The typical composition of e-waste includes printed circuit board, radios, cathode ray tubes, LCD/plasma televisions, mobile phones, air conditioners, discarded computer monitors, refrigerators, headphones, motherboards, chargers, discs, microwaves, etc. According to the study on 'E-Waste Management in India', electronic items and appliances such as computer equipment account for almost 70% of e-waste material for Indians, followed by telecommunication equipment (12%), electrical equipment (8%), and medical equipment (7%). The remaining 4% are accounted for by household e-crap. Poverty is the main reason for Third- World countries to consume e-wastes from developed countries.

Research has shown that around 80% of PCs are installed in the business sector. The most ideal way to reduce the number of electronic waste products is to replace hazardous electronic gadgets by non-hazardous substances. Survey has shown that almost 1.3 million personal computers are only from domestic and business sectors. According to the CII (Confederation of Indian Industry) report, a humungous amount of e-waste is generated in India from its 1.46 million tons of electronics usage per year (Udhayakumar, 2017).

6.2   METHODOLOGY

The chapter is grounded on information accumulated from various published research papers to identify several issues and challenges focusing on pollutants from recycling e-waste, impact of heavy metals from e-waste on environment and human health, e-waste processing, possible prevention techniques, and management strategies. We have explored different scientific article databases such as ScienceDirect, Springer, Wiley, and Google Scholar employing keywords, e.g.—'e-waste', 'health effects', 'personal protection', 'dismantling factories', and 'lung manifestation'. An in-depth assessment of the articles allows acquiring knowledge of the current situation in a comprehensive manner. The references cited in each relevant literature and in this chapter are reviewed and referred properly.

6.3   DISPOSAL OF E-WASTE AND ITS HEALTH EFFECTS

E-waste is said to be the fastest growing waste stream in the world. In countries with high labor costs, recycling mobile phones, televisions, and radios has not been much profitable. Hence, less developed countries such as China, India, Ghana, and Nigeria perform the disposal, storage, or illegal import of the aforementioned wastes where they are recycled informally using low-tech methods such as open burning, manual dismantling, and acid leaching for the recovery of gold, copper, and other valuable metals (Lucier and Gareau, 2019). The Partnership on Measuring ICT for development (Baldé et al., 2015) measures the most salient characteristics of the e-waste dynamics in a steady manner. E-waste collection, sorting, transportation, and recycling are the major components of e-waste management.

6.3.1   E-WASTE DISPOSAL TECHNIQUES

E- Waste is recycled in mainly two sectors: Informal and formal sectors. The informal sector is the unregistered and unauthorized sector that collects e-waste from doorsteps and recycles them in a very simple way and ends up polluting and having an effect on the environment including flora and fauna (Debnath et al., 2019). It forms the Base of Pyramid as the livelihood of the informal sector depends on the recycling of e-waste and other scraps (Wilson et al.,2006). The formal sector is the authorized sector that is making business out of e-waste in a comparatively environmentally benign way, registered and permitted to operate by the pollution control board (Ghosh et al., 2014). India as a developing country needs simpler economical and low-cost techniques for a proper implementation of e-waste management. In industries, management of e-waste is done by waste-minimizing techniques and by sustainable product design.

6.3.1.1 Landfilling

Landfilling is the method where trenches and pits are dug on plain surface, and waste products are buried in it, covered by a thick layer of soil. It is the most common procedure in India. Landfills contribute to very complicated and time-consuming degradation process that involves additional environmental risks due to leaching of metals (Bhutta et al., 2011). In addition, when plastic components are burned, dioxins, cadmium, mercury are diffused or released via the landfill gas combustion plant. These hazardous types of materials can percolate or leach out from the landfills into groundwater causing water pollution, and these gases can cause air pollution (Vats and Singh, 2014; Ismail and Hanafiah, 2020). However, landfills contain a mixture of various waste streams so the landfill gas is a complex mixture of different gases produced by the action of microorganisms in a landfill. These microorganisms decompose organic waste. Landfill gas contains mostly methane, with the remainder being mostly carbon dioxide. CO_2 and methane are greenhouse gases. Methane is 25 times more potent than carbon dioxide at trapping heat in the atmosphere hence causing unavoidable environmental and climate changes. There can be several years of delay in emission of pollutants from landfills. For these reasons, a landfill site is very different from native soil, and it is very much difficult to quantify environmental impacts of e-waste in landfills (Nagajothi and Felix Kala, 2015). The landfills are combination of volatile and non-biodegradable gases like Cd, Hg, and CFC and persistent (PCB) substances. Hence, it is difficult to exclude long-term environmental risks (Weldeslassie et al. 2018).

6.3.1.2 Incineration

Incineration is a complete combustion process in which the waste material is burned at a very high temperature. By incineration, e-waste volume can be reduced and the combustible materials can be utilized form of energy. But incineration is associated with a major risk for producing and disseminating notorious toxic substances and contaminants including hazardous gases escaping out those combustion plants and from flue gas cleaning and combustion fly ash. Seelampur in metropolis is the largest e-waste disassembly center of India. Waste combustion plants contribute considerably to the annual emissions of cadmium and mercury (Nagajothi and Felix Kala, 2015). However, heavy metals are not emitted directly in the atmosphere, but they are transported into slag and exhaust gas residues. These pollutants can enter into the environment in a different route if not disposed of properly. Hence, it is important to reduce these heavy metals in the feedstock.

6.3.1.3 3R (Recycling, Reuse, and Recovery)

E-waste contains a lot of different materials including basic and rare earth metals, polymers, electronic components (ECs), and some hazardous elements. If these hazardous substances are not handled properly, they can create a severe impact on environment and human health (Sankhla et al., 2016). Therefore, recycling and recovery are done to cut back the concentration of those perilous chemicals of e-waste. E-waste is a potential source of secondary raw materials which can be recovered via efficient recycling technologies. The concept of 'urban mining' came into focus due to the high concentration of metals present in the e-waste, and the term has become synonymous with e-waste material recovery since then (Cossu, 2013). The concentration of metals in e-waste is much greater than that of in natural ores (Xavier et al., 2019). Due to this reason, coupled with low mining cost and availability of feedstocks, i.e., e-waste, urban mining is becoming cheaper compared to virgin mining (Zeng et al., 2018). Hence, 3R can serve as a more meaningful instrument from the perspective of circular economy.

6.3.2 Health Effects

E waste and its constituents such as heavy metals have enormous impact on the human body and health as shown in Figure 6.1. There are chances of accidents like cuts and burns while shredding, dismantling, and acid baths, and incineration process and exposure to toxic chemical substances for long time may have many serious ill effects. Some e-waste products and their health effects are shown in Table 6.1.

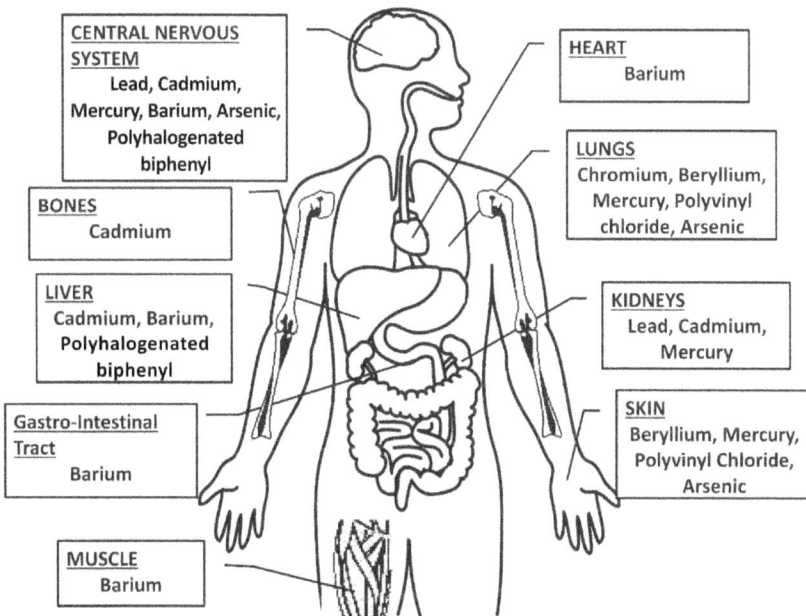

FIGURE 6.1 Heavy metal effects on human body.

TABLE 6.1
E-waste and Associated Health Effects

E-Waste Sources	Constituents	Health Effects	Reference
PCBs, panel and funnel glass, cathode ray tubes	Lead (Pb)	Damage to central and peripheral nervous systems, blood systems, renal damage, affect fetal development.	Vats and Singh, 2014
Chip resistors and semiconductors, switches, computer batteries and mobile phones, CRTs	Cadmium (Cd)	Cd is a known carcinogen. Long period of exposure causes irreversible damage to the kidney, liver, and bone density and also causes neural damage.	Abba et al., 2018
Corrosion protection of untreated and galvanized steel plates, decorator or hardener for steel housing, landfills, and incineration	Hexavalent chromium (Cr)	DNA damage, asthmatic bronchitis, or strong allergic reactions	Sankhla et al., 2016
Motherboards	Beryllium (Be)	Carcinogenic can cause lung cancer, chronic beryllium disease, or beryllicosis due to long-term inhalation of fumes and dust—symptoms of chronic beryllium disease are breathing problems, coughing, chest pain, general weakness, and skin disease (warts)	Abba et al., 2018
Switches and flat screen monitors, printed circuit boards	Mercury (Hg)	Chronic exposure through inhalation and ingestion can cause CNS damage, respiratory damage and kidney damage; respiratory and skin disease due to bio accumulation in fishes	Sankhla et al., 2016
Cable insulation or coating and computer housing	Plastic including PVC	PVC produces large amounts of hydrogen chloride gas when burned, and this gas reacts with water to produce hydrochloric acid (HCL); inhalation of HCL can cause respiratory damage. PVC also produces dioxins when burned which causes skin disease, immune system damage, interference with regulatory hormones, reproductive and developmental problems	Sankhla et al., 2016
CRT panel	Barium (Ba)	Brain swelling, damage to heart, liver, spleen, GIT, hypertension, muscle weakness	Vats and Singh, 2014
Arsenic gas used in tech manufacturing	Arsenic (As)	Lung cancer, skin disease, nerve damage. It is poisonous and has serious impacts on fetus and early childhood—can cause serious cognitive impairment and deaths in young adults	Sankhla et al., 2016
Circuit boards, plastic housing of electronic equipment. BFRs are used to make materials more flame resistant	Brominated flame retardants (BFRs)	Endocrine system function disrupter; hormonal problems cause growth and sexual developmental problems and harm the reproductive system	Abba et al., 2018
Transformers, capacitors, voltage regulators, switches, fluorescent light ballasts, cable insulation	Polychlorinated biphenyl (PCBs)	Immune suppression, liver damage, nervous damage, cancer promotion, reproductive damage, and behavioral changes	Vats and Singh, 2014; Abba et al., 2018

6.4 ENVIRONMENTAL POLLUTION DUE TO UNSUSTAINABLE E-WASTE RECYCLING

6.4.1 AIR POLLUTION

Heavy metals (Pb, Cd, Zn, Cr, Ni); brominated flame retardants (BFRs); neuro-toxins such as Pb, Hg, Cd, Cr, PBDEs, and PCBs are produced by informal recycling of e-waste which includes burning, improper dismantling, incinerating, acid bath, etc. (Grant et al., 2013). These unscientific activities pose notorious impact on human and animal health by releasing toxic pollutants in the ambient air. Heavy metals are resistant to decomposition in natural condition and become persistent pollutants in environment. Air pollution levels (PM_{10}) and heavy metal concentration were monitored and compared by several researchers in recycling sites with other unpolluted sites (Lau et al. 2014; Amoabeng et al. 2020; Gangwar et al. 2019). Air samples were also collected with the help of RDS (Respirable Dust Sampler). Lung function tests by using spirometer were assessed among workers in many epidemiological studies. Particulate matter (PM) of 2.5, 2.5–10, and 10 µm exposures affected the lung parameters such as forced vital capacity (FVC) and forced expiratory volume in one second (FEV_1); ratio between FVC and FEV_1 was measured, and less than 80% and more than 120% of predictive value is considered as detrimental. During metals' combustion, fumes can be inhaled and can cause acute lung damage and cardiovascular disease in workers. Hypoxemia and hypertension are most prevailing diseases that can affect major organs such as the heart, kidney, and brain. Low blood oxygen saturation level (SpO_2) can cause symptoms like blue discoloration of skin, shortness of breath, cough, wheezing, and choking sensation (Cao et al. 2020; Chen et al. 2011). Prolonged chronic exposure to PM can result in severe damage in lung tissue known as fibrosis, remodeling of the tissue, scarring of tissue, etc. These damages are often irreversible and can lead to chronic conditions in lungs like bronchitis, asthma, chronic obstructive pulmonary disease (COPD), and even lung cancer.

6.4.2 WATER POLLUTION

E-waste components disposed outside are drenched routinely or flooded by rainfall and run off to near local waterways; such exposures limit the diversity of aquatic environment. Heavy metals released from open burning and salvaging useful materials pollute not only water but also air and soil. When e-waste is irresponsibly dumped, it can leach into soil and water and can thereafter render water unfit for consumption. Organic pollutants (PCDD/Fs and PBDEs) were detected in the sediments in such quantities that put a heavy impact on life cycle of aquatic organisms (Zhao et al., 2010). Several researchers checked underground water quality for potential contaminates by using Atomic Absorption Spectrometry (AAS) around dumping and recycling sites. Drinking water was assessed for mean concentration of Pb, Cr, Ni, Zn, and Cd, which, at the dump site was generally lower, but well water was higher in such concentration in dry season near exposed sites (Ouabo et al. 2019). Studies show that Pb toxicity results in miscarriages and potent mortality among female animals and can cause symptoms such as abdominal discomfort, lassitude, irritation, and anemia in local people including neurological problems in fetus (Zhang et al., 2019). Ni and Zn can cause skin symptoms and asthma in 10% to 20% of population that has direct contact with it. Exposure to Cr for prolonged period causes high blood pressure, kidney disease, and DNA damage. Mn level from exposed and reference water sites respectively was 14 times and 7 times higher than ideal Mn concentration in drinking water. Mercury is only detected where open burning, acid digestion activities are performed (Zhanga et al., 2018).

6.4.3 Soil Pollution

The harmful pollutants that remain in soil may consequently cause unseen threat to the environment and human health as wealthy countries have a tendency to landfill or export their e-waste to developing countries of Asia and Africa illegally instead of recycling them properly due to the lack of facilities, high labor cost, and to recover economically valuable resources by rudimentary processing methods like manual disassembly, mechanical shredding, heavy acid digestion, and open burning leading to soil pollution at local areas (Soetrisno and Delgado-Saborit, 2020). The soil microorganisms are negatively affected by the incremental metal content which may eventually have direct harmful outcome on soil fertility (Ahmad et al., 2019; Panhwar et al., 2016). Leaching of heavy metals through the soil prevents plants from absorbing their required essential nutrients (Kladsomboon et al. 2020), such as potassium, calcium, magnesium, and nitrates. It was found that lead concentration was very high from soil samples because processing of lead-acid batteries and the assessment of mercury levels were also conducted by investigators in those areas. The level of soil contamination is estimated by pH values and Maximum Allowable Concentration (MAC) and hence the pH values of the soil were associated with the phylogenesis and bioavailability of the heavy metals and MAC in mixed soil with various values of pH. Cd, Cu, and Hg exceeded the MAC levels according to researchers (Fu et al., 2008). Local inhabitants who consume harvesting crops like rice grain, etc., are exposed to heavy metal concentration. TDI (Tolerable Daily Intake) and EDI (Estimated Daily Intake) are calculated to determine the exact value intervening the health of local people.

6.4.4 Metals in Drinking Water, Food, Blood

Negligible act toward remediation of polluted site enhances the probability of risk of contamination, and the hazard of pollution in soil is often overlooked. Due to these high levels of metals (Pb, Cd, Cu, Zn, Cr, and Ni) existing in drinking water and food, blood and risk assessments of exposure (biomonitoring and environment monitoring approaches) give a wide-ranging understanding on the body load of metals, exposure sources, and health threat caused by metals (Panhwar et al.,2016). Increasing concentration of metals in body develops severe and chronic toxicity leading to harming of kidneys and bones and enhanced risk of cancers (Kladsomboon et al., 2020). In addition, oxidative stress is the possible mechanism of metal-induced disease. Animal and in-vitro studies disclose that redox active metals such as Cr, Cu, Co, and other metals go through redox cycling reaction and generate reactive oxygen species (ROS), which encourages DNA damage by overpowering antioxidant properties (Bibi and shah, 2020). Micronutrients like Cu, Se, Zn, and vitamin (A, C and E) act as antioxidants that affect the defenses of a human body against $PM_{2.5}$ exposures (Takyi et al., 2020). Sb and Sn both can activate a variety of health crises by taking excess of them through rice grains; Sb in its organic form can be the reason of dermatitis and impaired immune system, and Sn organic compounds can cause headaches, cerebral edema, and dermal irritation (Song et al. 2015). According to WHO information, evidence of lead concentration in blood of e-waste informal workers is present and standard blood lead levels (BLL) surpassed the safe limit (BLL limit =10 µg/dL). The contamination of halogenated flame retardants (HFRs) and PBDEs (polybrominated diphenyl ethers) in poultry and sequential consumption are also confirmed (Ge et al. 2020).

6.4.5 ROUTES OF EXPOSURE AND CONTAMINATION

Uncontrolled recycling process gives rise to chronic heterogeneous exposure to various sources of e-waste, which are released in the environment in workshops and nearby neighborhoods (Soetrisno and Delgado-Saborit, 2020). Local people are barely exposed to these toxicants through contaminated air dust, soil, food, and drinking water. Expected routes are ingestion, inhalation, dermal contact and, in the case of children and infants, mostly through transplacental and lactational depending upon pollutants and their recycling process (Cai et al., 2020). Emission products and their disposal explain the attributed metal contamination with soil, food, water bodies, and so on (Thu Ngo et al., 2020). High concentration of exposure rate does not directly considerate with the effects of it within the body because it depends on certain factors such as exposure duration, synergistic effects of other chemicals present in ambience, individual susceptibility, and metabolic rate of xenobiotics (Grant et al., 2013; Sepulveda et al., 2010).

6.5 EFFECTS OF ENVIRONMENTAL DEGRADATION

6.5.1 EFFECTS ON FOOD CHAIN

Food chain is a linear network between different categories of organisms and species as shown in Figure 6.2, indicating how they relate with each other by the food they eat. It will pose a grave danger to every element of food chain if any contamination with e-waste occurs (Nkwunonwo et al., 2020). The food chain pyramid is basically an intricate relationship among primary, secondary, and tertiary consumers; so, contamination with any level can really endanger livelihood of that species and also affect the consecutive levels (Afonne and Ifediba, 2020). The equilibrium among different organisms and environment is necessary to ensure their survival, existence, and stability, and it can be disrupted due to ecological imbalance caused by e-waste contamination.

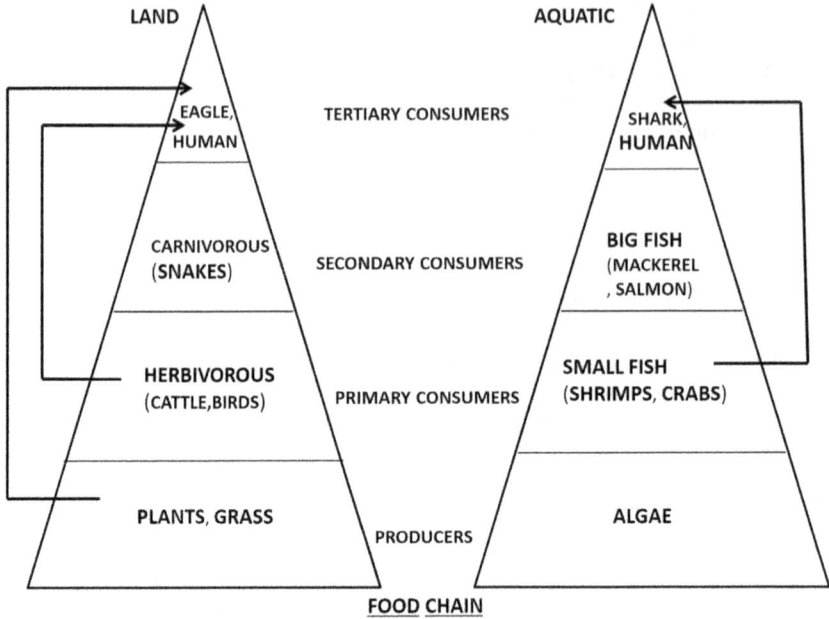

FIGURE 6.2 Effect of e-waste contamination on food chain.

6.5.2 EFFECTS ON PREGNANT WOMAN AND NEONATES

Researchers after investigating thoroughly have found out a rising concentration of heavy metals among pregnant woman by analyzing collected samples of maternal blood, cord blood, and maternal urine (N Singh et al., 2020). Certain metals like Pb, Cd, Cr, and Mn; flame retardants; PBDEs; and also some harmful by-products produced from e-waste recycling like dioxins and polycyclic aromatic hydrocarbons (PAHs) have the ability to cross placenta and have a vital task to play in order to understand the exposure mechanism in the neonates and infants (Frazzoli et al. 2010; Kim et al., 2019). Pb is a potent neurotoxicant that deposits into brain, liver, kidney, and bones; Cr and Cd are known to act as carcinogens in our body; and Cd, which can be harmful at low concentration, is related with learning disabilities and at high level it can lead to low birth length, birth weight, and decreased intelligence quotient in children (Yang et al., 2013). Also, cumulative exposure in humans can ultimately lead to renal and skeletal toxicity with Cd and have adverse effects on respiratory, hepatic, gastrointestinal, cardiovascular, and reproductive system with Cr as well. Mn interferes with neurodevelopmental outcomes in infants though it is an essential nutrient for usual growth, and also workers who met with high exposure experienced Parkinson's-like symptoms (Zhang et al. 2019; Chen et al. 2011). Human breast milk is the most significant pathway of exposure to breastfed infants. Effects on children are directly proportional to the amount of milk consumed every day. Estimated daily intake of PBDE values also is a factor that can affect the children. Body burdens of chemicals like OH-PBDEs (hydroxylated polybrominated diphenyl ethers) elicit a variety of concerns like interruption of thyroid hormones homeostasis, changed estradiol synthesis, and oxidative phosphorylation (Singh et al. 2022; Grant et al. 2013; Cao et al. 2020; Li et al., 2017).

6.5.3 Developmental Problems

Neurotoxicants work through many complicated mechanisms to develop neurological problems which may interfere with daily activity of children (Schantz et al. 2003; Chen et al., 2011). Several molecular-level mechanisms including oxidative stress, neurotransmission (glutaminergic and dopaminergic), and calcium homeostasis, neuroendocrine disruption, changes in gene function by methylation of DNA, histone modifications, microRNAs that defect posttranscriptional regulation can lead to detrimental effects (Baccarelli and Bollati, 2009). Glutamate is excitatory neurotransmitter in the brain tissues associated with learning and long-term memory. Selective binding of Pb with the NMDA (N-methyl D-aspartate) receptor initiates Ca^{2+} influx and thus interferes with synaptic function (Toscano and Guilarte, 2005; Chen et al., 2011). Current studies show that children exposed to Mn had deficit in attention and social interaction (Bauer et al., 2020). Impaired cognitive function can develop in children exposed to lead and behavioral disturbances, hyperactivity, and conduct problems in children with cadmium toxicity. PCBs (Polychlorinated biphenyls) are known to be responsible for affecting motor functions, visual spatial function, executive function (Boucher et al., 2009; Schantz et al., 2003; Chen et al., 2011). Another known carcinogen is hexavalent chromium, but as per studies it does not show much affect in children after occupational inhalation exposure (Pellerin and Booker, 2000). Hg is released in its elemental vapor form due to improper recycling of printed circuit boards, laptop monitor, and cell phones. It also affects neuropsychological functions like cognitive, language, motor function, and attention (Ramesh et al., 2007; Chen et al., 2011).

6.6 PREVENTION AND MANAGEMENT OF E-WASTE

Indiscriminate or improper disposal of e-waste in countries like India arises from the lack information how to handle those wastes, lack of acceptable facilities, weak social control, lack of regulation, and low level of awareness in the society (Tengku-Hamzah and Chibunna, 2011; Chibunna et al., 2012). In developing nations, the importation of electronic products and electronic waste from first-world countries is a major drawback as the consignment consists mostly of noncurrent, hazardous hardware (Hicks et al., 2005). This management practice impacts health and environment critically. Reuse of second-hand products, repairing broken items instead of buying new, and designing refillable or reusable products are some important aspects of waste management (Ghosh et al., 2014). A fraction of e-waste can be used as secondary raw materials for the recovery of valuables. Lead, cadmium, PCBs, and brominated flame retardants are some of the diverse toxic constituents of e-waste, which need proper handling. As a result of incompatible recycling practices in informal sectors and improper handling of these materials, the environment experiences uncontrolled releases of such toxic substances. Unit operations like segregation of ferrous metals and non-ferrous metals, dismantling, refurbishment, and reuse are essential for recycling of dangerous waste and recovery of valuable items. Other than these methods, thermal radiation is used in the case of certain e-waste samples with CFCs and underground landfills are often used for recycling or disposing of PCBs and mercury. In view of the ill-effects of hazardous wastes to both environment and health, several countries have agreed to address the problems and challenges posed by hazardous waste (Borthakur and Singh, 2012).

6.6.1 POLICY-LEVEL INITIATIVES

An immediate initiation of complete national-level policy including environmentally sound technology for the recycling of electronic waste covering all the cities and sectors which will address all issues regarding production to final disposal is a mandate. The government should also consider various other processes adapted by other countries, which can enhance the efficiency of collection and recycling of e-waste (Manish and Chakraborty, 2019). The deposition of e-waste in public landfills must be prohibited strictly. Policies must emphasize the management of restricted substances through awareness among producers and manufacturers in new product development (Nagajothi and Felixkala, 2015). The government should encourage new entrepreneurs to take technological guidance into consideration pertaining to the adverse impacts of untreated e-waste on land, water, and air (Kitila et al. 2019; Kumar et al. 2019). Start-ups associated with e-waste recycling and disposal, when granted a special allowance, help to form a well-established collection network in unorganized sector. Hence, the waste gathered by the unorganized sector may be handed over to a consistent capital-intensive organized sector to be processed in an environment-friendly way. The illegal trade of e-waste can be brought down by implementing collection schemes and management practices. A public–private participatory forum in e-waste management must be developed. According to the environmental policy of extended producer responsibility (EPR), a producer's responsibility for a product is extended to the post-consumer stage of the product's life cycle, including its final disposal (Joseph 2007). This policy has highly improved the e-waste management by promoting sustainable development in manufacturing products that are acquiescent with environmental requirements, leading to higher degree of recycling and reuse of equipment. This has reduced manufacturing and integrating environmental costs in the product price (Rudăreanu, 2013). EPR is designed to make the manufacturers internalize the external costs associated with the end-of-life disposal of their products (Sach, 2006). A portion of burden of waste management has been shifted by EPR from the local governments to the upstream producers. It is also expected to provide incentives for producers to take environmental considerations into their product design (Turga et al., 2019). India's first e-waste regulations, known as e-waste (Management and Handling) Rules, 2011, used EPR approach and suggested the producers of electronic products to set up collection centers as they have the caliber (i.e., physical responsibility) and apprise the consumers (i.e., information responsibility) on how the used electronic products can be returned to the collection centers. Indian e-waste sector has been observing several changes such as significant efforts by producers and extension of formal waste management sector pertaining to the regulations. To process and recover different components or parts of e-waste, indigenous technologies have developed. The old electronic devices should be sent for recycling or reuse, collection centers should be established to ensure proper collection and transportation directly to the recycle unit, and authorized recyclers should only be given the permission. The workers should have proper equipment and training because e-waste management not only depends on the effectiveness of the policy level initiatives but also on the attitude of buyers. Lack of awareness about the segregation of e-waste from municipal waste is increasing the immensity of e-waste problem in India. Awareness-raising programs and activities on issues related to the e-waste management, health, and safety aspects of e-wastes may be conducted in order to encourage better management practices among different target groups. Consumers need to be educated to buy only necessary products that utilize emerging technologies such as use of lead-free and halogen-free products to be identified through eco-labeling. Environmental management system (EMS) is the most significant and important driving enabler to influence all the other existing enablers (Sharma et al., 2020). EMS of e-wastes can be well organized with the help of technical guidelines which emphasize on better collection, recycling, and disposal options. However, the informal sector in India has taken under control the bulk of e-waste despite this successful development (Turaga et al., 2019).

6.6.2 COVID-19 Scenario

Given the epidemic of the COVID-19, everybody is staying at their homes and been attached to their digital devices, which proved to be providing boundless reasons for procuring more electronic products thus increasing more electronic gadget production and usage, and this will finally lead to more and more product obsolescence and will ultimately lead to increased number of e-waste products and exposure to them. The novel coronavirus (SARS-CoV2) has a huge negative impact on 219 countries and territories around the world and two international conveyances with around 58,886,225 infections and caused 1,392,165 confirmed deaths till the initial time of writing this piece (Worldmeters.info 2020). Owing to this exceptional calamity, the primary priorities and actions of government become the security of lives and livelihoods at each and every stage. In order to beat this public health crisis, improvement of medical norms, massive testing campaigns, and the development of public policies have been the approach so far (WHO, 2020a, 2020b). The epidemic has created an immense effect on the policymakers and sanitation workers, producing anguish among them (Mallapur, 2020). Since the time when rapid person-to-person transmission of coronavirus has been in the worldwide news, an abrupt and rapid demand for masks, gloves, hand sanitizers, and other important supplies has emerged. Now at time of an epidemic, it is quite reasonable that a mixture of medical and dangerous wastes along with infected masks, gloves, and additional protective equipment is being generated (BASEL 2020; UNEP, 2020). Thus, there is a huge risk of transmission of this disease to the sanitation workers as a result of inappropriate management which could subsequently direct to contamination of normal municipal solid waste with the virus-laden biomedical wastes. Hence, the essential factor that leads to a successful emergency response should be the discarding of this waste with secure and protected handling along with active civilian contribution and assistance (Sharma et al., 2020). Proper classification, compilation, partition, storage, movement, handling, and dumping of the biomedical and healthcare waste should become a vital part to make the management more efficient, and also disinfection, personnel security, and education on how to deal with the wastage should be of utmost importance (UNEP, 2020). Throughout the entire lockdown around the world, the demand for home delivery services of food and groceries has immensely enhanced, and as an outcome of that, the generation of common packaging plastic waste such as Polypropylene, Low-Density and High-Density Polyethylene, and Polystyrene has also accelerated as the recycling activities are diminished (Tenenbaum, 2020) (Kaufman and Chas an, 2020). Further to get accustomed with the elevated urge for necessary universal medical logistics, there is an anticipated surge in plastic packaging waste from medical industries (WHO, 2020c; 2020d). This mounting intricacy in the management of plastic waste has become an enormous challenge as well for the waste management industry during the Corona virus epidemic (Ferronato and Torretta, 2019).

6.7 CONCLUSION

In this chapter, we gathered information that how both an occupationally exposed population and non-occupationally exposed population around an e-waste disposal site have been affected by illegitimate methods of many informal sectors. Unsophisticated methods of e-waste processing have significantly contributed to the high levels of heavy metals in soil, water, air, and food. Chronic exposure to this contaminated environment is associated with many comorbid conditions involving predominantly cardiovascular, respiratory, and nervous systems. Health prevention strategies require preventing, especially children and pregnant women, to have an excessive exposure to toxicants. Easily traceable sources should address in a proper way for sustainable environmental remediation through campaigns, detoxification systems, and also protective dietary factors. Both developed and developing countries should take the responsibility in regulating electronic device rudimentary dismantling process and its manufacturing to avoid open burning practices in rotary incinerators. It should be made mandatory for workers to be aware of e-waste recycling activities and take training about preventive measures. Policymakers are suggested to take further interest in encouraging scientific advancements to perform recycling activities and new technology adoption by the industries.

ACKNOWLEDGMENT

The authors would like to acknowledge the Consortium of Researchers for Sustainable Development (CRSD) and Prof. Ankita Das, Visiting Researcher, Aston University, UK, for her help and guidance.

REFERENCES

Abba, Abdullahi Muhammad, Ahmed TujanniGubio, and O. S. Yekeen. "A conceptual paper on critical review of recycling of electronic waste (e-waste)." *International Journal of Research Technology in Engineering and Management* 2 (2018): 83–89.
Afonne, Onyenmechi Johnson, and Emeka Chinedu Ifediba. "Heavy metals risks in plant foods–need to step up precautionary measures." *Current Opinion in Toxicology* 22 (2020): 1–6.
Ahmed, Sirajuddin, Rashmi Makkar Panwar, and Anubhav Sharma. "Heavy metal contamination of soil and groundwater due to e-waste handling in Mandoli industrial area of Delhi,

India." In *Contamination in Soil Environment*, pp. 93–104 (2019). ISBN: 978-93-5124-950-4 (HB).

Amoabeng Nti, Afua Asabea, John Arko-Mensah, Paul K. Botwe, Duah Dwomoh, Lawrencia Kwarteng, Sylvia Akpene Takyi, Augustine Appah Acquah et al. "Effect of particulate matter exposure on respiratory health of e-waste workers at Agbogbloshie, Accra, Ghana." *International Journal of Environmental Research and Public Health* 17, no. 9 (2020): 3042.

Baccarelli A, Bollati V. 2009. Epigenetics and environmental chemicals. *Curr Opin Pediatr* 21: 243–251.

Balde, C. P., Kuehr, R., Blumenthal, K., Fondeur Gill, S., Kern, M., Micheli, P., ... & Huisman, J. (2015). E-waste statistics-Guidelines on classification, reporting and indicators", United Nations University, IAS-SCYCLE, Bonn (Germany) 2015, 51 pages (ISBN Print: 978-92-808-4553-2) (with CP Balde, K. Blumenthal, S. Fondeur Gill, M. Kern, P. Micheli, E. UNU).

BASEL. *Waste management an essential public service in the fight to beat COVID-19.* (2020). www.basel.int/Implementation/PublicAwareness/PressReleases/Wastemanagementand COVID19/tabid/8376/Default.aspx.

Bauer, Julia A., Katrina L. Devick, Jennifer F. Bobb, Brent A. Coull, David Bellinger, Chiara Benedetti, Giuseppa Cagna et al. "Associations of a metal mixture measured in multiple biomarkers with IQ: evidence from Italian adolescents living near ferroalloy industry." *Environmental Health Perspectives* 128, no. 9 (2020): 097002.

Bhutta, M. Khurrum, S., Adnan Omar, and Xiaozhe Yang. "Electronic waste: a growing concern in today's environment." *Economics Research International* 2011 (2011).

Bibi, Kalsoom, and Munir H. Shah. "Appraisal of metal imbalances in the blood of thyroid cancer patients in comparison with healthy subjects." *Biological Trace Element Research* (2020): 1–13.

Borthakur, Anwesha, and Pardeep Singh. "Electronic waste in India: problems and policies." *International Journal of Environmental Sciences* 3, no. 1 (2012): 353–362.

Boucher O, Muckle G, Bastien CH. 2009. Prenatal exposure to polychlorinated biphenyls: a neuropsychologic analysis. *Environ Health Perspect* 117: 7–16.

Brigden, K., Labunska, I., Santillo, D., & Johnston, P. (2008). Chemical contamination at e-waste recycling and disposal sites in Accra and Korforidua, Ghana, Greenpeace Toxic Tech 10, August 2008.

Cai, Kaihan, Qingbin Song, Wenyi Yuan, JujunRuan, Huabo Duan, Ying Li, and Jinhui Li. "Human exposure to PBDEs in e-waste areas: a review." *Environmental Pollution* (2020): 115634.

Cao, Peiqing, Takashi Fujimori, Albert Juhasz, Masaki Takaoka, and Kazuyuki Oshita. "Bioaccessibility and human health risk assessment of metal (loid) s in soil from an e-waste open burning site in Agbogbloshie, Accra, Ghana." *Chemosphere* 240 (2020): 124909.

Chaturvedi, Ashish, Rachna Arora, V. Khatter, and Jaspreet Kaur. "E-waste assessment in India–specific focus on Delhi." *MAIT-GTZ Study* (2007).

Chen, Aimin, Kim N. Dietrich, Xia Huo, and Shuk-mei Ho. "Developmental neurotoxicants in e-waste: an emerging health concern." *Environmental Health Perspectives* 119, no. 4 (2011): 431–438.

Chibunna, John Babington, ChamhuriSiwar, Rawshan Ara Begum, and Ahmad Fariz Mohamed. "The challenges of e-waste management among institutions: a case study of UKM." *Procedia-Social and Behavioral Sciences* 59 (2012): 644–649.

Cossu, Raffaello. "The urban mining concept." *Waste Management (New York, NY)* 33, no. 3 (2013): 497.

Debnath, Biswajit J. K. P., N. Karaikal, S. Thomas, and A. Nzhiou. "Sustainability of WEEE recycling in India." In *Reuse and Recycling of Materials: New Headways*. River Publishers, Netherlands, p. 15 (2019).

Ferronato, Navarro, and Vincenzo Torretta. "Waste mismanagement in developing countries: a review of global issues." *International Journal of Environmental Research and Public Health* 16, no. 6 (2019): 1060.

Forti, V., Balde, C. P., Kuehr, R., and Bel, G. (2020). The Global E-waste Monitor 2020: Quantities, flows and the circular economy potential.

Frazzoli, Chiara, Orish Ebere Orisakwe, Roberto Dragone, and Alberto Mantovani. "Diagnostic health risk assessment of electronic waste on the general population in developing countries' scenarios." *Environmental Impact Assessment Review* 30, no. 6 (2010): 388–399.

Fu, Jianjie, Qunfang Zhou, Jiemin Liu, Wei Liu, Thanh Wang, Qinghua Zhang, and Guibin Jiang. "High levels of heavy metals in rice (*Oryzasativa* L.) from a typical e-waste recycling area in southeast China and its potential risk to human health." *Chemosphere* 71, no. 7 (2008): 1269–1275.

Gangwar, Charu, Ranjana Choudhari, Anju Chauhan, Atul Kumar, Aprajita Singh, and Anamika Tripathi. "Assessment of air pollution caused by illegal e-waste burning to evaluate the human health risk." *Environment International* 125 (2019): 191–199.

Ge, Xiang, Shengtao Ma, Xiaolan Zhang, Yan Yang, Guiying Li, and Yingxin Yu. "Halogenated and organophosphorous flame retardants in surface soils from an e-waste dismantling park and its surrounding area: distributions, sources, and human health risks." *Environment International* 139 (2020): 105741.

Ghosh, S. K., R. Baidya, B. Debnath, N. T. Biswas, and Lokeswari M. De D. "E-waste supply chain issues and challenges in India using QFD as analytical tool." In *Proceedings of International Conference on Computing, Communication and Manufacturing, ICCCM*. Spinger, pp. 287–291 (2014).

Grant, Kristen, Fiona C. Goldizen, Peter D. Sly, Marie-Noel Brune, Maria Neira, Martin van den Berg, and Rosana E. Norman. "Health consequences of exposure to e-waste: a systematic review." *The Lancet Global Health* 1, no. 6 (2013): e350–e361.

Hicks, Charlotte, Rolf Dietmar, and Martin Eugster. "The recycling and disposal of electrical and electronic waste in China—legislative and market responses." *Environmental Impact Assessment Review* 25, no. 5 (2005): 459–471.

Huisman, J., Habib, H., Brechu, M.G., Downes, S., Herreras, L., Løvik, A.N., Wäger, P., Cassard, D., Tertre, F., Mählitz, P. and Rotter, S., 2016, September. ProSUM: Prospecting Secondary raw materials in the Urban mine and Mining wastes. In 2016 electronics goes green 2016+(EGG) (pp. 1–8). IEEE.

Ismail, H., and Hanafiah, M. M. (2020). A review of sustainable e-waste generation and management: Present and future perspectives. *Journal of Environmental Management*, 264, 110495.

Jhariya, M. K., Sahu, K. P., and Raj, A. E-waste, A New Challenge to the Environmentalists. *Nature Environment and Pollution Technology* 13, no. 2 (2014): 333.

Joseph, K. (2007). Electronic waste management in India–issues and strategies. In *Eleventh international waste management and landfill symposium*, Sardinia. S. Margherita di Pula, Cagliari, Italy; 1–5 October 2007. CISA, Environmental Sanitary Engineering Centre, Italy.

Kaufman, L., and E. Chasan. "Cities wonder whether recycling counts as essential during the virus." *Bloomberg Green* (2020).

Kim, Stephani, Xijin Xu, Yuling Zhang, Xiangbin Zheng, Rongju Liu, Kim Dietrich, Tiina Reponen et al. "Metal concentrations in pregnant women and neonates from informal electronic waste recycling." *Journal of Exposure Science & Environmental Epidemiology* 29, no. 3 (2019): 406–415.

Kitila, Abenezer Wakuma, and Solomon Mulugeta Woldemikael. "Waste electrical and electronic equipment management in the educational institutions and governmental sector offices of Addis Ababa, Ethiopia." *Waste Management* 85 (2019): 30–41.

Kladsomboon, Sumana, Chakkaphop Jaiyen, Chalisa Choprathumma, Thitaporn Tusai, and Amara Apilux. "Heavy metals contamination in soil, surface water, crops, and resident blood in Uthai District, Phra Nakhon Si Ayutthaya, Thailand." *Environmental Geochemistry and Health* 42, no. 2 (2020): 545–561.

Kumar, Sandeep, and Vinti Singh. "E-waste: generation, environmental and health impacts, recycling and status of e-waste legislation." *Journal of Emerging Technologies and Innovative Research* 6, no. 1 (2019): 592–600.

Lau, Winifred Ka Yan, Peng Liang, Yu Bon Man, Shan Chung, and Ming Hung Wong. "Human health risk assessment based on trace metals in suspended air particulates, surface dust, and floor dust from e-waste recycling workshops in Hong Kong, China." *Environmental Science and Pollution Research* 21, no. 5 (2014): 3813–3825.

Li, Xinghong, Yuan Tian, Yun Zhang, Yujie Ben, and Quanxia Lv. "Accumulation of polybrominated diphenyl ethers in breast milk of women from an e-waste recycling center in China." *Journal of Environmental Sciences* 52 (2017): 305–313.

Lucier, C. A., & Gareau, B. J. (2019). Electronic waste recycling and disposal: an overview. *Assessment and Management of Radioactive and Electronic Wastes*, 1–12.

Mallapur, C. "Sanitation workers at risk from discarded medical waste related to COVID-19." *India Spend* (2020). www.indiaspend.com/sanitation-workersat-risk-from-discarded-medical-waste-related-tocovid-19/ (accessed 4.26.20).

Manish, Akanksha, and Paromita Chakraborty. "E-waste management in India: challenges and opportunities". *Teriin.Org.* (2019). www.teriin.org/article/e-waste-management-india-challenges-and-opportunities

Nagajothi, P. Gomathi, and T. Felixkala. "Electronic waste management: a review." *International Journal of Applied Engineering Research* 10, no. 68: (2015).

Ngo, Hien Thi Thu, Pensri Watchalayann, Diep Bich Nguyen, Hai Ngoc Doan, and Li Liang. "Environmental health risk assessment of heavy metal exposure among children living in an informal e-waste processing village in Vietnam." *Science of The Total Environment* (2020): 142982.

Nkwunonwo, Ugonna C., Precious O. Odika, and Nneka I. Onyia. "A review of the health implications of heavy metals in food chain in Nigeria." *The Scientific World Journal* 2020 (2020).

Ongondo, Francis O., Ian D. Williams, and Tom J. Cherrett. "How are WEEE doing? A global review of the management of electrical and electronic wastes." *Waste Management* 31, no. 4 (2011): 714–730.

Ouabo, Romaric Emmanuel, Mary B. Ogundiran, Abimbola Y. Sangodoyin, and Babafemi A. Babalola. "Ecological risk and human health implications of heavy metals contamination of surface soil in e-waste recycling sites in Douala, Cameroun." *Journal of Health and Pollution* 9, no. 21 (2019): 190310.

Panhwar, Abdul Haleem, Tasneem Gul Kazi, Hassan Imran Afridi, Salma Aslam Arain, Mariam Shahzadi Arain, Kapil Dev Brahaman, and Sadaf Sadia Arain. "Correlation of cadmium and aluminum in blood samples of kidney disorder patients with drinking water and tobacco smoking: related health risk." *Environmental Geochemistry and Health* 38, no. 1 (2016): 265–274.

Pellerin C, Booker SM. 2000. Reflections on hexavalent chro-mium: health hazards of an industrial heavyweight. *Environ Health Perspect* 108: A402–A407.

Ramesh BB, Parande AK, Ahmed BC. 2007. Electrical and electronic waste: a global environmental problem. *Waste Manag Res* 25: 307–318.

Robinson, Brett H. "E-waste: an assessment of global production and environmental impacts." *Science of the Total Environment* 408, no. 2 (2009): 183–191.

Rudăreanu, C. "WEEE legislation and its role in Romanian WEEE management system." *Contemporary Readings in Law and Social Justice* no. 2 (2013): 889–903.

Sachs, Noah. "Planning the funeral at the birth: extended producer responsibility in the European Union and the United States." *Harv. Envtl. L. Rev.* 30 (2006): 51.

Sankhla, M. S., Kumari, M., Nandan, M., Mohril, S., Singh, G. P., Chaturvedi, B., & Kumar, R. (2016). Effect of electronic waste on environmental & human health-a review. *IOSR J. Environ. Sci. Toxicol. Food Technol*, 10(09), 98–104.

Schantz SL, Widholm JJ, Rice DC. Effects of PCB exposure on neuropsychological function in children. *Environ Health Perspect* 111 (2003): 357–576.

Sharma, Manu, Sudhanshu Joshi, and Ashwani Kumar. "Assessing enablers of e-waste management in circular economy using DEMATEL method: an Indian perspective." *Environmental Science and Pollution Research* (2020): 1–14.

Singh, Narendra, Oladele A. Ogunseitan, and Yuanyuan Tang. "Systematic review of pregnancy and neonatal health outcomes associated with exposure to e-waste disposal." *Critical Reviews in Environmental Science and Technology* (2020): 1–25.

Singh, V., Cortes-Ramirez, J., Toms, L. M., Sooriyagoda, T., & Karatela, S. (2022). Effects of Polybrominated Diphenyl Ethers on Hormonal and Reproductive Health in E-Waste-Exposed Population: A Systematic Review. *International Journal of Environmental Research and Public Health* 19, no. 13 (2022): 7820.

Sinha-Khetriwal, Deepali, Philipp Kraeuchi, and Markus Schwaninger. "A comparison of electronic waste recycling in Switzerland and in India." *Environmental Impact Assessment Review* 25, no. 5 (2005): 492–504.

Sivaramanan, Sivakumaran. "E-waste management, disposal and its impacts on the environment." *Universal Journal of Environmental Research & Technology* 3, no. 5 (2013).

Soetrisno, FitriaNurbaidah, and Juana Maria Delgado-Saborit. "Chronic exposure to heavy metals from informal e-waste recycling plants and children's attention, executive function and academic performance." *Science of The Total Environment* 717 (2020): 137099.

Song, Qingbin, and Jinhui Li. "A review on human health consequences of metals exposure to e-waste in China." *Environmental Pollution* 196 (2015): 450–461.

Takyi, Sylvia A., NiladriBasu, John Arko-Mensah, Paul Botwe, Afua Asabea Amoabeng Nti, Lawrencia Kwarteng, Augustine Acquah et al. "Micronutrient-rich dietary intake is associated with a reduction in the effects of particulate matter on blood pressure among electronic waste recyclers at Agbogbloshie, Ghana." *BMC Public Health* 20, no. 1 (2020): 1–14.

Tanskanen, Pia. "Management and recycling of electronic waste." *Acta Materialia* 61, no. 3 (2013): 1001–1011.

Tenenbaum, L. "The amount of plastic waste is surging because of the coronavirus pandemic." *Forbes* (2020). www.forbes.com/sites/lauratenenbaum/2020/04/25/plastic-waste-during-the-time-of-covid-19.

Tengku-Hamzah, Tengku, and Adura Adeline Chibunna, "Making sense of environmental governance: a study of e-waste in Malaysia." PhD diss., Durham University (2011).

Toscano CD, Guilarte TR. 2005. Lead neurotoxicity: from exposure to molecular effects. *Brain Res Brain Res Rev* 49: 529–554.

Turaga, R. M. R., Bhaskar, K., Sinha, S., Hinchliffe, D., Hemkhaus, M., Arora, R., ... & Sharma, H. (2019). E-waste management in India: issues and strategies. *Vikalpa*, 44(3), 127–162.

Udhayakumar, T. Disposal methods of e-waste in India Survey conducted in Chennai. *International Journal of Applied Environmental Sciences*, 12, no. 3 (2017), 505–512.

UNEP. (2020). Waste management an essential public service in the fight to beat COVID-19. Retrieved from https://buff.ly/39oKjdi.

Vats, M. C., and S. K. Singh. "E-waste characteristic and its disposal." *International Journal of Ecological Science and Environmental Engineering* 1, no. 2 (2014): 49–61.

Wath, Sushant B., P. S. Dutt, and Tapan Chakrabarti. "E-waste scenario in India, its management and implications." *Environmental Monitoring and Assessment* 172, no. 1–4 (2011): 249–262.

WHO. *COVID-19 strategy update*. (2020a). www.who.int/publications/m/item/covid-19-strategy-update

WHO. *Strategic preparedness and response plan*. (2020b). www.who.int/publications/i/item/strategic-preparedness-and-response-plan-for-the-new-coronavirus

WHO. "Advice for the public on COVID-19—World Health Organization." *Who. Int.* (2020c). www.who.int/emergencies/diseases/novel-coronavirus-2019/advice-for-public.

WHO. *Water, sanitation, hygiene and waste management for COVID-19: technical brief, 03 March 2020.* No. WHO/2019-NcOV/IPC_WASH/2020.1. World Health Organization (2020d).

Weldeslassie, T., Naz, H., Singh, B., & Oves, M. (2018). Chemical contaminants for soil, air and aquatic ecosystem. *Modern age environmental problems and their remediation*, 1–22.

Wilson, David C., Costas Velis, and Chris Cheeseman. "Role of informal sector recycling in waste management in developing countries." *Habitat International* 30, no. 4 (2006): 797–808.

Worldmeterd.info. "Coronavirus update (live): 58,886,225 cases and 1,392,165 deaths from COVID-19 virus pandemic—worldometer". *Worldometers.Info.* (2020). www.world ometers.info/coronavirus/?utm_campaign=homeAdvegas1?.

Xavier, Lúcia Helena, Ellen Cristine Giese, Ana Cristina Ribeiro-Duthie, and Fernando Antonio Freitas Lins. "Sustainability and the circular economy: a theoretical approach focused on e-waste urban mining." *Resources Policy* (2019): 101467.

Yadav, Vinod Kumar, Yogesh K. Chauhan, Bhim Singh, Ashish Kumar, and Deepak Berwal. "Effect of electronic waste on environmental & human health-a review." *International Journal of Applied Engineering Research* 9, no. 16 (2014): 973–4562.

Yang, Hui, Xia Huo, Taofeek Akangbe Yekeen, Qiujian Zheng, Minghao Zheng, and Xijin Xu. "Effects of lead and cadmium exposure from electronic waste on child physical growth." *Environmental Science and Pollution Research* 20, no. 7 (2013): 4441–4447.

Zeng, Xianlai, John A. Mathews, and Jinhui Li. "Urban mining of e-waste is becoming more cost-effective than virgin mining." *Environmental Science & Technology* 52, no. 8 (2018): 4835–4841.

Zhang, Tao, Jujun Ruan, Bo Zhang, Shaoyou Lu, Chuanzi Gao, Lifei Huang, Xueyuan Bai, Lei Xie, Mingwei Gui, and Rong-liang Qiu. "Heavy metals in human urine, foods and drinking water from an e-waste dismantling area: identification of exposure sources and metal-induced health risk." *Ecotoxicology and Environmental Safety* 169 (2019): 707–713.

Zhao, Xing-Ru, Zhan-Fen Qin, Zhong-Zhi Yang, Qian Zhao, Ya-Xian Zhao, Xiao-Fei Qin, Yong-Chuan Zhang, Xian-Li Ruan, Yin-Feng Zhang, and Xiao-Bai Xu. "Dual body burdens of polychlorinated biphenyls and polybrominated diphenyl ethers among local residents in an e-waste recycling region in Southeast China." *Chemosphere* 78, no. 6 (2010): 659–666.

7 Role of Media in E-waste Awareness Generation toward Social Sustainability
An Indian Scenario

Chayan Acharya

CONTENTS

7.1 INTRODUCTION

E-waste or electronic waste basically comprises electronic devices which are discarded or rejected. Due to the rapid growth of and production in electronic industry, every year a huge number of products developed by this industry becomes obsolete and eventually turns into e-waste. Electronic waste or e-waste generally means old, broken, and beyond repairable electronic items including PC, laptop, television, refrigerator, CD player, iPad, tab, smartphones, and so on. These are called e-waste

DOI: 10.1201/9781003301899-10

as they reach their end-of-life (EoL) state, and they are discarded by the consumers. Thus, e-waste is generated from comparatively luxurious and basically tough products employed for telecommunications, daily office work, data processing as well as entertainment in both personal and professional spaces. The components of e-waste are over 1,000 types. Both hazardous and non-hazardous materials are present in e-waste (ABDI 2013). Metals, plastic, rubber, glass, and ceramic are present in high volumes. Old fluorescent tubes contain mercury and phosphorus. Old circuit boards and discarded batteries contain lead, lithium, cadmium, and antimony which are highly poisonous. Mercury and its variants cause mercury poisoning and Minamata disease (Bu.edu 2020). Medical equipment such as discarded X-ray machines contain radioactive elements such as cesium which is highly carcinogenic.

If we look at the BRICS countries (Brazil, Russia, India, China, and South Africa), India produces 46,180 tons of e-waste yearly as per preliminary estimate. It is expected to exceed 800,000 tons/year soon (Wath et al. 2011). In China, 1.5 to 3.3 metric tons of waste gets illegally imported. China officially collects and treats 1.3 metric tons which is 28% of total generated wastes. Recent estimations of the situation of China indicate that e-waste generation will increase to about 15.5 metric tons and 28.4 metric tons in 2020 and 2030, respectively (Zeng et al. 2016; Baldé et al. 2017). The untreated waste electric and electronic equipment (WEEE), which is another synonym of e-waste, is still creating huge mountainous issues that are highly dependent on illegal import of e-waste that have grave impacts on the environment as well as on society (Song and Li 2014). Brazil generated around 14.12 Mt tons of e-waste in 2013–14 (UNU-IAS SCYCLE 2015). Russian Federation generates 1.478 Mt of e-waste. The Russian Federation market handles 70 million units of electronic and electrical household appliances a year (Komissarov 2012). Another faster growing economy South Africa produced 64,045 t of e-waste in 2011, of which 6,884 tons (11%) were recycled and 57,161 tons disposed into landfills (DEA 2012). Studies have revealed that, in all these nations, a lack of awareness has played a key role toward proper disposal attitudes (Ghosh et al. 2014; Baidya et al. 2020).

Global e-waste generation data in different country groups are presented in Figure 7.1. Rich countries like the United States, Canada, and European countries are also not too far behind when it comes to the amount of generation of e-waste. America produced 11.3 Mt e-waste in 2016, and only 1.9 Mt of it got collected and recycled (USEPA 2016). There is no national policy on e-waste management in Canada. Despite the absence of the national legislation, the member states have enacted regulations at a state level except the state of Yukon and Nunavut. There are different nonprofit organizations in different states, which are collecting and channelizing the e-waste for recycling. Nearly 20% of the e-waste generated in the country was recycled by these enterprises. There are some collection centers as well. The collection efficiency can be increased by creating more collection centers as well as by increasing the general awareness on e-waste in the country (Kumar and Holuszko 2016).

The huge amount of e-waste generated is not just the fault of the consumer but also the manufacturers are to blame, because they bring out new models every now and then. Also, in a competitive market where marketing policies are strategized, the users are lured in to buy new products with some minor upgradation in features

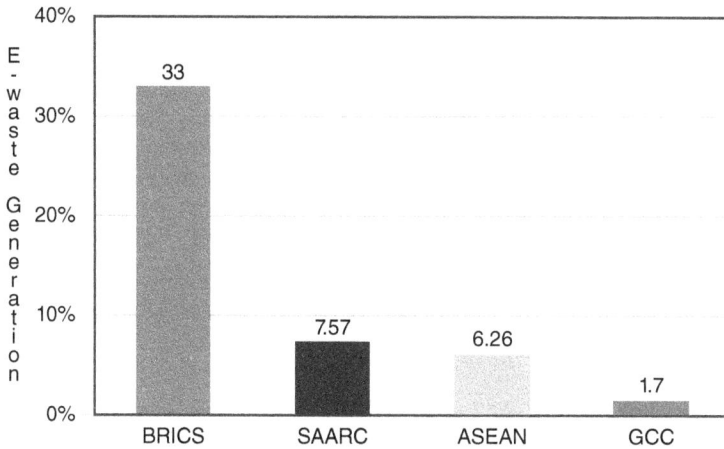

FIGURE 7.1   E-waste generation in different country groups (as of 2019) (Forti et al. 2020).

(Debnath et al. 2022). But the problem is not with buying, rather the proper disposal behavior of the consumers is the concern. It is a common scenario in developing and least developed countries where people tend to throw e-waste mixed with MSW (Debnath 2020). In some of the developed nations, people do not dispose of their e-waste in the proper bin. In country like India, Bangladesh, and Pakistan people sell their e-waste to *kabadiwallahs* in exchange of money (Das et al. 2020). But this is not intentional always. It is because of the lack of awareness. A survey-based study that employed Quality Function Deployment (QFD) shows that awareness is one of the biggest constraints toward proper e-waste management (Ghosh et al. 2014). Baidya et al. (2020) used a combined AHP–QFD method to find out issues for e-waste recycling plants based on sustainability aspects. They also found awareness to be an important criterion for social sustainability. In a recent survey-based study, it was found that 25% of respondents who were aware of e-waste chose to dispose via *kabadiwallahs* (Birawat et al. 2020). This can be attributed toward a lack of awareness. Hence, awareness plays a key role in consumer disposal attitude.

7.2   EXISTING KNOWLEDGE GAP AND MAIN OBJECTIVE OF THIS STUDY

Journalism has played a crucial role over the years when it comes to expose an environmental damage, which has a devastating effect on public health. Media coverage can attract an international attention on how to manage e-waste. But in the case of e-waste, there are very few mainstream media coverages that happened in the last 5 years which became sensational.

In the past, Indian media covered environmental disasters like Union Carbide gas leak incident (1984) (Lotha 2020) and oleum gas leak at Shriram Agro Industry (1987) (News 5th 2020). The recent case of LG Polymers Gas leak at Vizag (2020) (Sachdev 2020) also caught the attention of media, but when it comes to e-waste which is also deadly, media is silent because death comes at a slower pace. It is hard to identify the damages on public health when the Indian public health system is under so many challenges (Kasthuri 2020).

Our main objective of this study will be focusing on media presentation of e-waste over last 5 years. Media included in this study are print media and their web representations, social media, and electronic media, articles, journals written with media perspective, YouTube, and other web videos. When we started searching, we found one or two articles on this topic in the leading newspapers like *The Hindu, The Telegraph*, and *The Statesman*. The Press Information Bureau of India published only two press releases on this topic, which came from the Ministry of Environment, Forest and Climate Change in between 2017 and 2020 (PIB Delhi 2018a, 2018b). There are plenty of articles from other dedicated sources like Downtoearth.Org or Toxics link, but there is a huge overlook when it comes to the mainstream media which is more available to the common people.

The search for papers on this subject led to lots of papers sourced from Rajya Sabha Secretariat to Ministry of Environment, Forest and Climate Change. There is a plethora of papers available on online thesis reservoirs like Researchgate.net, Shodhganga, and academia.edu. But most of the papers are dealing with current e-waste scenario and management. There was hardly any paper on mainstream media's take on e-waste. This lack of media's attention leads to the lack of awareness of this growing menace among the common people, creating a knowledge gap. We'll try to discuss the future activities that mainstream media should follow to raise awareness about e-waste in the Findings and Discussions section.

7.3 METHODOLOGY

In this study, the primary method used was literature review and web search using some keywords. Direct search results of different websites of mainstream media were included in this study as well. The secondary method was forming a semi-structured questionnaire and taking semi-structured interviews of academicians, IT professionals, and e-waste recyclers as a case study. An awareness song, the video of the making of the song, and the results they produced were also reviewed for the case studies.

7.4 ROLE OF PRINT MEDIA

The articles mentioned here are on the basis of web presence of mainstream Indian newspapers. So, there are limitations when research is based on Internet data, as many open-source websites let anyone willing to contribute, alter, and rewrite. So, the result changes time to time. Search engines also prefer to showcase the articles with higher views and engagements. Reliability of these available data is totally based on observation. Tommaso Venturini warns us about use of web as a research tool. In his own words—*"Search engines are not the web; the web is not the Internet; the Internet is not the digital and the digital is not the world"* (Venturini 2012).

Search boxes of the websites of many newspapers produce results that are totally unrelated to the keyword 'E-waste'. This problem can be observable in the case of *The Hindu* of Hinduja group, though they have published some stunning articles about e-waste (Adlakha 2019). Hindu BusinessLine also has the same search-engine-archiving issues. A thorough search resulted only two articles from 3 June 2016 (Pandit 2016) to 10 February 2020 (Singh 2020). This issue is also applicable when it comes to *Indian Express*, a leading daily English newspaper. In a search result that popped out 914 articles, a scrutiny was done among them for the first 240 articles. Only 10 articles were relevant among them, dated back from 22 April 2018 (Shah 2018) to 12 April 2020 (Express Web Desk 2020).

Times of India also published a good number of e-waste-related articles, but their website's search box also shows up irrelevant articles. A search popped out a total of 16 articles, the earliest starting from 19 September 2019 (Figure 7.2).

Sometimes, 'found no results' messages show up even though a certain number of articles have been published on that newspaper. The leading Hindi newspaper *Dainik Bhaskar* showed no results, whereas *Navbharat Times* of Times of India group has a dedicated page called 'pollution janche' which provides air pollution statistics of northern India's metro cities, but no search box option for the readers.

On the other hand, popular Hindi newspaper *Dainik Jagaran* has published a surprisingly stunning number of articles on e-waste. The earliest article was published on 25 January 2013. From 2013 to July 2020, *Jagaran* has published a total of 87 articles, with the keyword 'E-waste' in their titles. Most of the articles are informative, with a neutral point of view about regional activities of different state governments about e-waste.

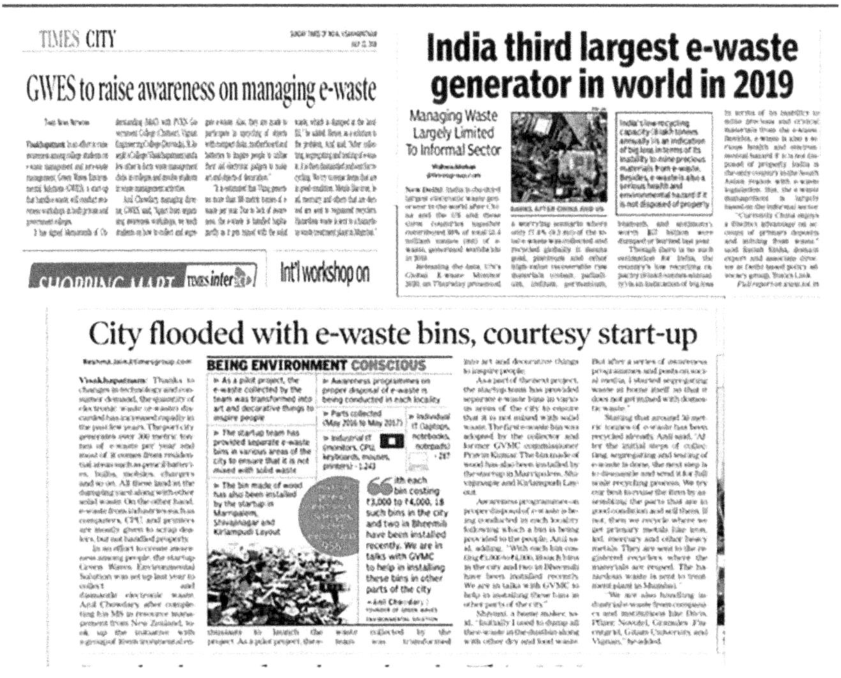
TIMES CITY

GWES to raise awareness on managing e-waste

India third largest e-waste generator in world in 2019

Managing Waste Largely Limited To Informal Sector

Int'l workshop on

City flooded with e-waste bins, courtesy start-up

BEING ENVIRONMENT CONSCIOUS

FIGURE 7.2 Initiatives taken by e-waste recyclers to create awareness in print media.

TABLE 7.1

E-waste Keyword Search Result from Leading Newspapers

Name of the Newspaper	Number of Articles Found with Keyword 'E-Waste'	Time Period
The Hindu	2	2019–20
Hindu BusinessLine	2	2016–20
Indian Express	10	2018–20
Times of India	16	2019–20
Dainik Jagaran	87	2013–20
Hindustan Times	24	2015–20
Anandabazar Patrika	6	2014–20
Total	147	2013–20

Hindustan Times also has a dedicated environmental news page. From 19 April 2015 to 18 July 2020, a total of 24 articles have been published with the titles bearing 'E-waste' keyword.

Anandabazar Patrika, a leading Bengali daily newspaper from ABP group, published a total of six articles about e-waste from 1 May 2014 to 4 March 2020.

Table 7.1 shows the data we got from 8 mainstream leading Newspapers.

These searches don't portray the whole scenario but only a small fraction of the whole. India has more than 1,05,443 newspapers according to the Registrar of Newspapers for India (RNI) as on 31 March 2015 (Pavithra and Dubbudu 2020). Therefore, information gathered from Internet is nothing but an echo of thoughts broadcasted by their networks. Venturini rightly observes

> Even if portals and search engines are constantly expanding their databases, they cannot grow as fast as the web. Everyday hundreds of thousands of new pages are created and only a fraction is reached by the search crawlers. Sometimes contents remain invisible because they are too marginal or ephemeral, sometimes because they are concealed by their authors, sometimes they are just forgotten. Even if more and more information is exchanged via the hypertext transfer protocol (http) and under the form of an xhtml page, a large slice of electronic traffic travels through other routes. E-mails, teleconferences, chats, peer-to-peers exchanges, document transfers and many other data do not transit via web protocols.
>
> (Venturini 2010)

Therefore, we hope these findings may lead to better online database systems for the web presence of newspapers, and a strong starting point for e-waste awareness through the media among the 425 million Indian readers (Malvania 2020).

7.5   ROLE OF ELECTRONIC MEDIA

There are two types of news channel broadcasts in India.

a) Private News Networks
b) Government News Networks

For the e-waste-related news, we searched the websites of leading news channels and their online streaming platforms.

DD news is a government news channel broadcasted by Prasar Bharati which has a reach over 44.9% population of India. Searching their website with the keyword 'E-waste' showed one unrelated article. Using another keyword 'Electronic Waste' brings three results, and one of them is relevant. DD news uses YouTube for online streaming. A search on their YouTube channel brought three relevant news reports.

Two top reports from DD News are:

- Campaign against E-waste, broadcasted on 14 October 2019 (DD News 2019):Jayant, a boy from Ghatkopar, Mumbai, assembled a computer out of E-waste. This news broadcasted twice—on 16 March 2017 (DD News 2017a) and on 8 May 2017 (DD News 2017b).
- DD News Kolkata also broadcasted a report. On 3 August 2017, they broadcasted a report on E-waste management in Kolkata and West Bengal (DD News Kolkata 2017). There are very less relevant news reports from DD news available on the Internet. Another government news channel Rajya Sabha TV, now known as Sansad TV showed one report titled "The pulse: E-waste Crisis-What's the solution?" on 15 September 2019 (Rajya Sabha TV 2020).

Press Information Bureau of India posted one report by Digital India on their YouTube channel 5 years ago on 22 February 2017, titled 'Digital India: E-waste management', which was more of a TV commercial than a report (PIB India 2017).

Regarding private news networks, New Delhi Television Ltd or NDTV also uses YouTube as their online streaming medium. A search on YouTube with the keyword 'NDTV E-Waste' brought six relevant results. The oldest was broadcasted 15 years ago, on 16 November 2007, titled 'E-waste poses environmental hazard' (NDTV 2007). The latest from them was broadcasted, on 21 February 2020, titled 'Scrap to sculpture: Art installation from E-waste' (NDTV 2020).

Tv Today Network Ltd (Aaj Tak), a Noida-based channel, similarly uses YouTube for streaming. A search for e-waste-related broadcasts popped zero results. Similar results can be observable in the case of BBC India. While BBC World has some stunning news reports about e-waste, their Indian branch produced no results. Another prime network TV18 Broadcast Ltd (IBN7) also showed no relevant results. The result is same in the case of Republic TV.

ABP Network Pvt Ltd produced one relevant broadcast titled "Recyclekaro e waste factory in news", published on 20 July 2016 (Gupta 2016).

Zee News from Zee Media Pvt Ltd has done a significant number of reports on e-waste. The oldest on the list was published on 16 June 2016, 6 years ago, titled "Allahabad: A student teaching drone making from trash" (Zee News 2020). A total of seven reports were broadcasted by Zee News. The latest report was broadcasted 4 years ago on 22 June 2018. The report was on Namo e-waste, a startup recycling company by Akshay Jain (Zee News 2018).

TABLE 7.2
E-waste Keyword Search Result from Streaming Platforms of News Channels

Name of the News Channel	Number of Reports Found with Keyword 'E-waste' in Streaming Platforms	Time Period
DD News	3	2017–20
Rajya Sabha TV	1	2019–20
PIB	1	2017–20
NDTV	6	2007–20
ABP News Network	1	2016–20
Zee News	7	2016–20
Aaj Tak, BBC India, Republic TV, IBN7	0	_____
Total	19	2007–20

The number of relevant broadcasts is low. Most of the channels use YouTube as their online streaming partner. Proper archiving of reports and use of proper tags are missing in most cases. But the Digital India initiatives and boom of mobile Internet users in India will definitely change the current situation. It's ironic to watch an e-waste-related report on a smartphone which will become e-waste itself sooner or later.

7.6   ROLE OF SOCIAL MEDIA

Social media platforms like Facebook, Twitter, and Instagram have a huge reach and a large number of users. People follow trends, and things can go viral in a very short span. Social media can be very much handy, spreading awareness about e-waste. There are many pages, communities, and forums for e-waste campaign. But the activities of these institutions are not always regular and cannot be mapped easily. However, a brief search was conducted which revealed the following—

i) In the social media platform of Facebook, there are thousands of pages on e-waste. A huge number of these are pages run by either recyclers or some focused groups. There are some dedicated groups which are intended for buying and selling e-scraps. Some of the groups also provide information on how to recycle e-waste or provide e-waste market updates. Very few recycler pages are active and regularly updated with posts, except some international recyclers. Most of the recyclers use this platform to showcase their activities, achievements, and collaborations.

ii) Instagram was more complex, and a brief search revealed there are 69.3k and 11.4k posts with #ewaste and #ewasterecycling hashtags, respectively. On the other hand, less than 100 posts were found with #ewasteupcycle and #ewasteupcycling. The immediate impression is that the platform is largely being used to market upcycled products and disseminate recycling programs run by recyclers, rather than general awareness generation. A few posts were found that called for appropriate e-waste disposal, but that is very limited.

iii) Twitter was explored similarly, but the results were more confusing to be deciphered. But it was clear that there are some popular hashtags such as #WEEE, #BinIT, #ewasterecycling, and #sustainability. Regular tweets from e-waste recyclers around the globe including the UK, Ireland, Australia, and India were observed with proper hashtags. This platform seemed to have more content and awareness-generating activities compared to others mentioned earlier.

7.7   CASE STUDIES AND SUCCESS STORIES

To know the opinion of the people about media, who have some awareness on the topic, a semi-structured questionnaire was formed, and a series of semi-structured interviews were conducted. A total of four interviews were taken. Two of the interviewees were e-waste experts. Among the other two interviewees, one is a licensed e-waste handler, and other one is an IT professional and refurbished electronic product user. We'll also study two YouTube videos by students of Jadavpur University, who have prepared an e-waste song to spread the awareness and a making video of that song.

7.7.1   CASE 1: SECOND-HAND ELECTRONICS USER

Interviewee A: An IT professional and refurbished electronic equipment user.
Category: Aware user

Question: Tell us something about yourself.
Answer: Hi, I have been an avid e-waste recycler. I have purchased second-hand smartphone and laptops for my personal use. And will continue to do so. My motto is if it works why not make it rework.

Question: Do you think mainstream media doesn't give any importance to topics like e-waste?
Answer: No; the mainstream media doesn't give importance to this topic because most MNCs and sponsors do not fund media regarding it as they themselves do not sponsor it.

Question: If mainstream media starts reporting more on e-waste research outputs, policy changes, and technological developments, will it enhance the e-waste formal recycling rates?
Answer: Maybe there is need for media to create awareness, but it cannot change it as it needs more public participation.

Question: What should be the other media ways to enhance awareness?
Answer: Creating online content; interactable site with solution-driven visuals can create impact.

7.7.2   CASE 2: E-WASTE RESEARCHER

Interviewee B: A doctorate degree holder in biotechnology. Expert on bioleaching of e-waste, environmental-engineering-related issues.
Category: Young professional

Question: What should be the role of media in generating e-waste awareness?
Answer: Well, media have to give a lot of importance to awareness-raising things, because you see awareness is very important to people to know what? to know how? to know whom to contact? Everything. Media is the best thing to spread something in mass; it could be awareness, could be news, anything. E-waste also requires something like that and you see like gain, interest gains, so media people should turn it into something interesting which people do listen. That is more important. See like some advertisement, you get Vidya Balan and Amitabh saying like what to do what not to do to manage solid waste and all, same way media should teach people in such a way that people understands the effects of hazardous waste. Show media plays very important role because it can actually teach people who are illiterate, to clarify effects to the people who are really aware of what electronic waste is, so it plays a major role in my opinion.
Question: Do you think, mainstream media doesn't give any importance to topics like e-waste?
Answer: Well I can't tell that way, because you see even in this lockdown period in most of the schools children were taught by the mobile phones. And in fact, in Tamil Nadu, in Coimbatore, we have an authorized recycler, green era recyclers as you might know him. So, this person what he did is directly contacted media persons to raise awareness by asking them to collect waste phones and refurbish it and give it to the poor children for their study purpose. So, media plays a very big role.
Question: Do you think, mainstream media doesn't give any importance to topics like e-waste?
Answer: I understand the question the thing is investigate the nuclear power plants; nuclear power plants are most powerful concentrated things because the impacts are known to people in an active way. Because they are highly Radioactive and once, they emit something Radioactive then it's going to cause cancer so that is something the awareness level of people. Whereas e-waste in compare to radioactive materials the awareness level is too low which was our first question that's why though it comes in first page it's least concentrated by the people, the Awareness level is rapidly proportional to this thing I guess. You see e-waste has not created something disastrous or the gap that happens between politicians, so I think it doesn't come on the first page.

7.7.3   CASE 3: RECYCLING COMPANY

Interviewee C: Founder of an e-waste recycling company
Category: Entrepreneur

Question: Tell us something about yourself.
Answer: I am founder of an authorized E-waste recycling firm in Coimbatore, Tamil Nadu. I am an enviropreneur working towards the sustainability goals and to

make an impact towards recycling, disposal and wealth from waste. At Green era recyclers we have handled 200 tons of e-waste. We are projecting for better collection and process rate in upcoming years as we have clinched reputation for safe disposal of e-waste.

Question: Do you think the awareness level of e-waste in India is low and why?

Answer: Yes. Governmental collaboration to create an awareness is poor. Like ULB and PCB can combine also with required ministry to put forth actions.

Question: What should be the role of media in generating e-waste awareness?

Answer: I believe it's the producer's part. Whenever an Ad is floated, it should end with a recycling awareness point. Which can be used by media.

Question: Do you think, mainstream media doesn't give any importance to topics like e-waste?

Answer: Exactly, just to put forth that they are diverse it's being covered. But they have to come up with provision especially for waste management.

Question: Do you think, if mainstream media starts reporting more on e-waste research outputs, policy changes, and technology developments, will it enhance the e-waste formal recycling rate?

Answer: Yes. Along with the method to disposal.

Question: Do you that proper archiving of e-waste news in national and regional level will be helpful?

Answer: Yes.

Question: What is your opinion on awareness songs, videos and short films on e-waste?

Answer: In present media world. I think it's not so effective compared to ads, newspaper and on-air radio.

Question: What should be the other media ways to enhance awareness?

Answer: Governmental platforms.

7.7.4 CASE 4: BUDDING E-WASTE EXPERT

Interviewee D: Research scholar, UK. Specialization: E-waste Supply Chain and Sustainability.
Category: Early career researcher

Question: Do you think the awareness level of e-waste in India is low and why?

Answer: Indeed the awareness level is low, simply because if you want the biggest proof, it is the informal sector and as you know, despite the number of formal e-waste recyclers have increased from 96 to 312 over the last 10 years almost, maybe in last 5 to 6 years. Despite the increase of the numbers, the recyclers are sitting idle, they are not getting enough feed, and also most of it is handled by the informal sector, that's the reason number 1. The reason number 2 is if you take any random 10 or 100 people, I doubt that there will be hardly 1 or 2 person who has the knowledge or even heard the phrase e-waste. So that is another reason to support my answer and if you say why the awareness level is low? I mean that is arguably a topic for a big debate—lack of education or lack of role of media. Because media don't want to you know, give much footage to this topic. There are other reasons which I should not talk. Recently there are changes in the environmental policies which we did notice. But there is lack of implementation in the grassroots level. I mean policy documents do exist, but they don't compliment the overall awareness spreading procedures among the common people of India. India is a very big diverse country. It is not easy to spread awareness. We are fighting against plastic waste since the beginning of 2000, but still after twenty years where are we? Still standing on the pile of plastic garbage. e-waste is comparatively new to all of us.

Question: Do you think, if mainstream media starts reporting more on e-waste research outputs, policy changes, and technology developments, will it enhance the e-waste formal recycling rate?

Answer: There are two parts of this question. One you are talking about research outputs, and the other you are talking about the policymakers. Now we need policy-makers who are interested in scientific research outputs, that is the first priority and requirement, and assuming we have that then of course the scientific research outputs should be shared in the media. You know, I won't say that research findings have not reported by print media like findings by IIT Bombay, Dr Chakraborty who has been working on soil contamination from the e-waste recycling site that was featured in print media. I think it was 2017 April. So that was featured I'm not saying print media disseminating research outputs. But there lies the responsibility of the researchers also they should come out and they should also express it to the public. I mean very few professors hold press conferences after their findings very few and those who do a part of them actually doing it for the good and part of them doing it for their own publicity. Now that's debatable issue, but the researchers must come forward and disseminate their knowledge. So, it's a two-way symbiotic model, so the researcher should come forward and these people should also come forward and lay down the platform for the researchers to disseminating their findings.

Question: What is your opinion on awareness songs, videos and short films on e-waste?

Answer: I think, you know, the songs, videos, short films, cinema all these things have a very strong message bearing capability. Now the audience must be educated. Now let's say I am making a very good film with huge star cast and big budget and I am constantly telling about recycle e-waste, recycle e-waste, recycle e-waste. People will get bored and my film will not run for more than 3 weeks, but if I do a film on the background topic of e-waste pollution, one such film was created in the South India, I can't remember the name. But it was part of a series film, the second or the third film, topic was that some person wants to import hazardous waste from Australia. And there is a police inspector who doesn't go by the laws of Physics and he jumps into the matter and saves India and saves the world. Fine, I mean those things are important in Indian cinema otherwise it won't Run, but at least the back-ground topic could have been glorified a little bit more, why it is hazardous? Even battery is hazardous, drinking too much Coca Cola is hazardous. So what hazard-ous? why it is important? they could have include some scenes like people teaching about e-waste or at least waste management, they should have done this, they should have done that, now our national anthem is played before any movie starts in the movie theatre. And also, we have been shown the visuals of people getting cancer, smoking kills, then Life Insurance and some winterwear, all these things are there but instead of some skin cream, if they show something on e-waste or recycling, that might be helpful. Now natural mentality of people is like, we have come for an entertaining movie and this Khan should do Hi voltage action, but including these ads have compelled people to buy insurance or Vico-turmeric cream in the past. So, including these messages can address the whole youth. And out of hundred if five gets a spark that's a good start. Now I cannot expect I write an awareness song and from the next day people stop throwing e-waste on the landfill or selling it to proper recyclers, I cannot think of that. My idea of the song was that I am writing a song which have not been done before, nobody have written a full length song on e-waste, so my idea was that, let me do it and see how the response is. Well technically the response is pretty slow in compared to Bollywood or any pop songs, it is expected. Even I had a very low expectation, but luckily, I have found that it somehow crossed my expectation a bit. I didn't expect that within one year and two months there will be nearly thousand views and also some good inputs from others. You know when making the song the genre of the song plays a big role, so it might vary. I mean whatever I have written I have done it in English and Bengali as well. The Bengali version is yet to be published. The English version is there. I did it in English so that it can reach the wider audience, the mass, but then again even it is in English those who understand English they have questioned why it was not done in Hindi some said you should have done it in Punjabi and some said why not an Odia version. I have already kept it almost copyright free with some of the conditions, anybody can use the same music the track, I will provide the track if they want it convert it translate it to their vernacular language keeping the meaning intact, they can change the lyrics this bit and that, but it should address e-waste. That was the basic idea. You know, many people came to me saying, I will do a Odia version, some people told me I will do a Tamil version, I will do a Kannada version, and they stopped there, cause I don't know whether they are not interested in music or they don't believe music can help, these are very powerful medium but what I lacked I could not develop a powerful video, there should have been a powerful video. I'm working on that. In near future I will do that. Though I did a making video, but it didn't populate very well. It's lagging behind reality.

7.7.5   CASE 5: E-WASTE SONG AND MAKING VIDEO

A team of talented Jadavpur University students came together and created an e-waste song (Debnath 2019a, 2021a) to generate awareness. The song was in Bengali and English and belongs to punk rock genre with a powerful lyric which gives the message of recycling the e-waste. A documentary video was also produced which describes the making of the e-waste song (Debnath 2019b, 2021b). The song got 1.2k views over the period of a year. The making video got nearly 350 views. The song and video didn't come under any commercial banner. It was a nameless production and aimed to spread awareness in a relatively small circle.

7.8 FINDINGS AND DISCUSSIONS

The basic focus of this study was on the mainstream media appearance of e-waste over the period of last 5 years. The mainstream media and its web representations are considered in this study. Though this study had a limited dataset, but a clear reflection of mainstream media trends can be observable.

The key issues that evolute from this study and interviews are—

- Numbers of articles and reports are low when it comes to e-waste in the mainstream media.
- There are no proper archiving algorithms of e-waste-related news in the websites of mainstream media.
- Awareness level is low about e-waste.
- Government should produce advertisements about e-waste awareness and methods of disposal.
- Short films, music videos, awareness songs, and animation films for children should be encouraged.
- Government should bring strict policies on e-waste handling prioritizing formal e-waste handlers.

Media is considered as the fourth pillar of the democracy. Without an active support of mainstream media, issues like e-waste will never be addressed. Low awareness level will lead people to many problems, and health hazard is one of them. Media with government collaboration can impact the current situation and raise awareness. The media stories balancing multiple and often opposing viewpoints can generate further research topics. Since media can teach the people, the further activities of media will reflect on the future of e-waste scenarios.

More and more e-waste-related articles should be published, and environment-related pages should be incorporated by the print media. Web design of leading print media and electronic media needs to be more user friendly, and web search should not pop out results that are not related with the keyword. With the government support, media should work on interactive ways to spread awareness at grassroots level. Government should develop cultural programs for rural audience, and e-waste-related advertisements must be displayed at movie theatres before any screening. Short films, music videos, and awareness songs for e-waste disposal know how must be encouraged, and there should be competitions for such cultural activities. Last but not least, government should bring legislative changes for existing laws and also new laws, ensuring more and more formal e-waste recycling instead of informal e-waste handling.

7.8.1 Social Sustainability

In this section, we present a general discussion about the impact of e-waste awareness among the society. It is hard to assume that only the awareness itself which is generated by planned campaigns of different media can bring the change in the overall thought patterns regarding e-waste. According to Western Australia Council of Social Services (WACOSS),

> Social sustainability occurs when the formal and informal processes; systems; structures; and relationships actively support the capacity of current and future generations to create healthy and liveable communities. Socially sustainable communities are equitable, diverse, connected and democratic and provide a good quality of life.

Social sustainability determination is not easy, and it varies with geography, ethnicity, and religion. Hence, the factors are more complex to be scored. Here, we present possible avenues of social sustainability through e-waste awareness generation.

i) If media, through systematic campaigns, impacts the overall point of view about e-waste, then, we can expect a change in overall e-waste handling management in India. Formal e-waste handlers will be beneficiaries because people will go to them for the disposal of their e-waste. More and more eco-friendly products will be manufactured—for example, jute bags instead of plastic bags, earthen pots, and paper cups instead of their plastic variants.

ii) We can expect a sector creating merchandise from e-waste for reusing. Example being keychains with letters from a dismantled computer keyboard.

iii) Gold extraction from e-waste can open new avenues of wealth generation.

iv) Awareness generation in hilly regions can exclusively increase social sustainability because people from hills are nature inclusive. Hence, with strict rules and proper training, eco-system destruction in the hills will reduce.

v) Government policies ensuring penalty for illegal e-waste trafficking/handling is another approach to look at.

These are some impacts which can bring social sustainability in the long run.

7.9   CONCLUSION

The menace of e-waste is vibrant, and the clarity of the global picture hints toward a sustainable and circular e-waste management system worldwide. There are country-specific issues and challenges which need to be taken care of. It is proven with several studies that the awareness generation is of higher importance in the waste management scenario. In developing countries, the issues are quite similar, and replicability of a model with minor adjustment is certainly a way out. In this chapter, we delved deeper in understanding the role of print media, electronic media, and awareness songs and videos toward developing awareness of e-waste in the Indian scenario. Four interviews and one success story have been presented, which provide the opinion of experts on awareness activities. A structured questionnaire was used which has been presented almost verbatim. The findings suggest that, currently, e-waste awareness is quite low in India and the media doesn't take issues such as e-waste seriously. Also, in online media, the search algorithms are not very much efficient. Some of the events are covered in print media, but they are never a front pager and often in special columns. The story of the awareness song suggests that the reach is not very wide as compared to popular songs, but it is quite accepted. It appears from the interview of the experts that specific steps toward developing strict policy, their implementation, and policy will be helpful in deciding the level of awareness in the future of this country. While it is a clear case that studies such as these are scant in reported literature, which also suggests the lack of interest in the subject itself, more studies of such style and findings will be helpful to the policymakers and other relevant stakeholders.

ACKNOWLEDGMENT

The author would like to acknowledge the help received from the Department of Film Studies, Jadavpur University, India. Additionally, inputs received from Prof. Amar Chandra Das, Ananda Mohan College (now retired), is acknowledged.

REFERENCES

ABDI (Agência Brasileira de Desenvolvimento Industrial). 2013. *Logística Reversa de Equipamentos Eletroeletrônicos—Análise de Viabilidade Técnica e Econômica*. www. mdic.gov.br/arquivos/dwnl_1362058667.pdf

Adlakha, N. 2019. "Where's Our E-Waste Going?" *The Hindu*. www.thehindu.com/sci-tech/ energy-and-environment/global-e-waste-production-is-all-set-to-reach-120-million-tonnes-per-year-by-2050/article29259488.ece. Accessed 26 August 2019.

Baidya, Rahul, Biswajit Debnath, Sadhan Kumar Ghosh, and Seung-Whee Rhee. 2020. "Supply Chain Analysis of E-Waste Processing Plants in Developing Countries." *Waste Management & Research* 38, no. 2: 173–183.

Baldé, C. P., V. Forti, V. Gray, R. Kuehr, and P. Stegmann. 2017. *The Global E-Waste Monitor–2017*. United Nations University (UNU), International Telecommunication Union (ITU) & International Solid Waste Association (ISWA), Bonn/Geneva/Vienna, pp. 978–992. ISBN Electronic Version.

Birawat, Khushbu K., Biswajit Debnath, Shushmitha L. Gowda, and Sadhan Kumar Ghosh. 2020. "Willingness of Students and Academicians to Participate in E-Waste Management Programmes—A Case Study of Bangalore." In *Urban Mining and Sustainable Waste Management*. Springer, Singapore, pp. 249–261.

Bu.edu. 2020. *Minamata Disease, Sustainability*. Boston University. www.bu.edu/sustainability/ minamata-disease/. Accessed 29 June 2020.

Das, Ankita, Biswajit Debnath, Nipu Modak, Abhijit Das, and Debasish De. 2020. "E-Waste Inventorisation for Sustainable Smart Cities in India: A Cloud-Based Framework." In *2020 IEEE International Women in Engineering (WIE) Conference on Electrical and Computer Engineering (WIECON-ECE)*. IEEE, Bhubaneswar, India, pp. 332–335.

DD News. 2017a. *Jayant Makes Computers Out of Junk or E-Waste*. Video. https://youtu.be/ cHSJhBbW0Hw. Accessed 4 August 2020.

DD News. 2017b. *Good News: 16-Year-Old Mumbai Boy Assembles Computer From E-Waste*. Video. https://youtu.be/W27yLgwc3Qs. Accessed 4 August 2020.

DD News. 2019. *Campaign Against E-Waste . . . High Time to Erase the E-Waste*. https:// youtu.be/0MBN2ezTUrU. Accessed 4 August 2020.

DD News Kolkata. 2017. *E Waste Management at Kolkata and West Bengal Area Special News Story Telecast at Dd Bangla*. Video. www.youtu.be/2Q4zvW_xNDo. Accessed 4 August 2020.

Debnath, B. 2019a. *Making of the E-Waste Song*. www.youtube.com/watch?v=sXblQjXiwKQ. Accessed 31 August 2020.

Debnath, B. 2019b. *E-Waste Song [English Version].* www.youtube.com/watch?v=04-xu 49f0hw. Accessed 20 April 2020.

Debnath, B. 2020. "Towards Sustainable E-Waste Management Through Industrial Symbiosis: A Supply Chain Perspective." In *Industrial Symbiosis for the Circular Economy.* Springer, Cham, pp. 87–102.

Debnath, B. 2021a. *The E-Waste Song: A Brief Background.* doi:10.13140/RG.2.2.33935.10408

Debnath, B. 2021b. *Making of the E-Waste Song: An Overview.* doi:10.13140/RG.2.2.24235.77600

Debnath, B., Das, A., & Das, A. (2022). Towards circular economy in e-waste management in India: Issues, challenges, and solutions. In *Circular Economy and Sustainability* (pp. 523–543). Elsevier.

Department of Environmental Affairs (DEA). 2012. *National Waste Information Baseline Report.* Department of Environmental Affairs, Pretoria, South Africa, 22pp.

Express Web Desk. 2020. "This Lockdown, Go on a Treasure Hunt Inside Your Homes." *The Indian Express.* https://indianexpress.com/article/world/lockdown-e-waste-gold-silver-old-phones-6358963/. Accessed 12 April 2020.

Forti, V., C. P. Baldé, R. Kuehr, and G. Bel. 2020. *The Global E-Waste Monitor 2020 Quantities, Flows, and the Circular Economy Potential.* United Nations University (UNU)/ United Nations Institute for Training and Research (UNITAR)—Co-Hosted SCYCLE Programme, International Telecommunication Union (ITU) & International Solid Waste Association (ISWA), Bonn/Geneva/Rotterdam. ISBN Digital: 978-92-808-9114-0

Ghosh, S. K., R. Baidya, B. Debnath, N. T. Biswas, and Lokeswari, M. De D. 2014. "E-Waste Supply Chain Issues and Challenges in India using QFD as Analytical Tool." In *Proceedings of International Conference on Computing, Communication and Manufacturing, ICCCM.* Curran Associates, Inc, pp. 287–291.

Gupta, R. 2016. *Recyclekaro's Ewaste Factory in News.* Video. https://youtu.be/LHXst6tWTtY. Accessed 4 August 2020.

Kasthuri, A. 2020. "Challenges to Healthcare in India—The Five A's". *Pubmed Central (PMC).* www.ncbi.nlm.nih.gov/pmc/articles/PMC6166510/.

Komissarov, V. A. 2012. *WEEE is the Most Growing Waste Flow in the World. Situation with WEEE Management in Russia and Other Countries, Russia, Moscow.* http://ac.gov.ru/files/content/2535/komissarov-v-a-pdf.pdf

Kumar, A., and Holuszko, M. 2016. Electronic Waste and Existing Processing Routes: A Canadian Perspective. *Resources* 5: 35.

Lotha, G. 2020. *Bhopal Disaster | Causes, Effects, Facts, & History.* Encyclopedia Britannica. www.britannica.com/event/Bhopal-disaster.

Malvania, U. 2020. "Print Readership in India Jumps 4.4% to 425 Million in Two Years: Report". *Business-Standard.Com.* www.business-standard.com/article/current-affairs/print-readership-in-india-jumps-4-4-to-425-million-in-two-years-report-19042700079_1.html.

NDTV. 2007. *E-Waste Poses Environmental Hazards.* Video. https://youtu.be/Us9iXxUStlY. Accessed 4 August 2020.

NDTV. 2020. *Scrap to Sculpture: Art Installations from Electronic Waste.* Video. https://youtu.be/Ezjm5uCAyKA. Accessed 4 August 2020.

News, 5th. 2020. "Case Summary-M.C. Mehta v/s Union of India | 5Th Voice News". *5th Voice.News.* https://5thvoice.news/legalnews/NzQzMw==/CASE-SUMMARY-MC-MEHTA-VS-UNION-OF-INDIA.

Pandit, V. 2016. "India's E-Waste Growing at 30% Annually." *@Businessline.* www.thehindubusinessline.com/info-tech/indias-ewaste-growing-at-30-annually/article8686442.ece. Accessed 3 June 2016.

Pavithra, M., and Dubbudu, R. 2020. "More than A Lakh Newspapers & Periodicals are Registered in the Country". *FACTLY.* https://factly.in/indian-newspapers-more-than-one-lakh-newspapers-periodicals-registered-in-the-country/.

PIB Delhi. 2018a. *"E-Waste Management Rules Amended for Effective Management of E-Waste in The Country": Union Environment Minister.* https://pib.gov.in/PressReleseDetail. aspx?PRID=1526177.

PIB Delhi. 2018b. *Management of E-Waste.* https://pib.gov.in/PressReleseDetail.aspx?PRID= 1519141.

PIB India. 2017. *Digital India: E-Waste Management.* Video. https://youtu.be/CmP67zq5hfo. Accessed 4 August 2020.

Rajya Sabha TV. 2020. *The Pulse: E-Waste Crisis—What's The Solution?* Video. https://youtu. be/bjuk983LhKs. Accessed 4 August 2020.

Sachdev, V. 2020. "Vizag Gas Leak: LG Polymers has 'Absolute Liability' Under Law". *Thequint.* www.thequint.com/news/law/vizag-gas-leak-legal-responsibility-lg-polymers- absolute-liability-supreme-court-oleum-bhopal-gas-cases.

Shah, N. 2018. "Digital Native: The E-Wasteland of Our Times." *The Indian Express.* https:// indianexpress.com/article/technology/tech-news-technology/digital-native-the-e-waste land-of-our-times-5146406/. Accessed 22 April 2018.

Singh, J. 2020. "E-Waste Recycling with Zero Waste Concept." *@Businessline.* www.the hindubusinessline.com/news/science/e-waste-recycling-with-zero-waste-concept/ article30783816.ece. Accessed 10 February 2020.

Song, Q., and Li, J. 2014. "A Systematic Review of The Human Body Burden of E-Waste Exposure in China". *Environ. Int.* 68: 82–93. doi:10.1016/j.envint.2014.03.018.

UNU-IAS SCYCLE. 2015. *Step E-Waste World Map. Database Available from STEP—Solving the E-Waste Problem 2015.* http://www.step-initiative.org/Overview_India.html

U.S. Environmental Protection Agency. 2016. *Electronic Products Generation and Recycling Methodology Review.* www.epa.gov/sites/production/files/2016-12/documents/electronic_ products_generation_and_recycling_methodology_review_508.pdf

Venturini, T. 2010. "Building on Faults: How to Represent Controversies with Digital Methods". *Public Understanding of Science* 21, no. 7: 796–812. doi:10.1177/0963 662510387558.

Venturini, T. 2012. Building on Faults. How to Represent Controversies with Digital Methods. *Public Underst. Sci.* 21.

Wath, Sushant B., P. S. Dutt, and Tapan Chakrabarti. 2011. "E-Waste Scenario in India, Its Management and Implications." *Environ. Monit. Assess.* 172, no. 1–4: 249–262.

Zee News. 2018. *E-Waste Recycling Startup Aims at Clean and Green India.* Video. https:// youtu.be/O8f5toT9NCU. Accessed 4 August 2020.

Zee News. 2020. *Allahabad: A Student Teaching Drone Making from Trash.* Video. https:// youtu.be/e6H6SkNjD5A. Accessed 4 August 2020.

Zeng, X., Gong, R., Chen, W. Q., and Li, J. 2016. Uncovering the Recycling Potential of "New" WEEE in China." *Environ. Sci. Technol.* 50, no. 3: 1347–1358.

8 Economic Potential of Resource Recovery from E-waste

Rohit Panchal, Hema Diwan and Anju Singh

CONTENTS

8.1 INTRODUCTION

Electronic waste (e-waste) is one waste stream that is having huge implications on waste management initiatives. Due to technological innovations, however, this challenge has turned into an opportunity as e-waste is considered as a resource due to huge recovery potential associated with electronic products (Widmer et al., 2005; Robinson, 2009). Because of this mindset, it is also referred to as a misplaced resource which has economic value if tapped efficiently. The waste products like mobile phones, computers, television sets, and refrigerators are the sources of secondary resources in the form of materials like metals, plastics, and other miscellaneous materials.

Across the globe, the e-waste is witnessing a surge due to the rising standard of living and innovations in the electronic industry. It has been reported that the consumption of electronic products worldwide escalates up to annually by 2.5 million metric tons (Mt) (Forti et al., 2020). Global e-waste Monitor 2020 cites that e-waste quantum will reach up to 74.7 Mt by 2030 (Forti et al., 2020). The electronic products will eventually get discarded in the environment once the products reach their end-of-the-life stage (Thiebaud et al., 2018a).

National economies across the globe are witnessing a growing need to tap wealth from waste through business models that unleash waste management into

DOI: 10.1201/9781003301899-11

145

an opportunity for economic returns as e-waste is one of the fastest-growing waste streams in the world (Panchal et al., 2021). India generates about 2000,000 tons of e-waste annually and is the fifth largest e-waste generator, after the United States, China, Japan, and Germany. This scenario requires scientific handling and management of discarded products to avoid environmental impacts as most of the electronic waste still land in a landfill without any scientific treatment. Till date, the major quantum of India's e-waste (95% of India's waste electrical and electronic equipment (WEEE)) is recycled in an unscientific manner due to the sector being unregulated and informally controlled. So, there is a growing concern for quantification of e-waste generation so that planning can be done to regulate the flow of waste in the supply chain and consequently plan its disposal strategies to optimise the resource recovery from the discarded products. Thus, national economies call for their efficient management (Singh et al., 2020). This also calls for conducive regulations and policy reforms to streamline the management efforts and formalise the waste supply chain to tap the secondary raw materials through reuse, recycling, and recovery.

The business of resource recovery from e-waste is a promising approach, but it requires economic feasibility. The secondary resources available in e-waste can be quantified at a national/regional, product, and element level (Islam and Huda, 2019a). In the current context, it is imperative to understand the metabolism of anthroposphere of metal recovery by estimating the flows and stocks in the system. Material flow analysis and substance flow analysis can be employed for the quantification of secondary resources (Wang et al., 2018). The material flow analysis involves quantification of the stocks and flows of critical metals in the waste streams at a systemic level with due consideration on the disposal channels to the environment. The various functions that lead to material flow analysis are inflow data of EEE, service lifetime and storage time, and transfer coefficients for the specific flows. Prediction models are used to assess and quantify the flows using the probabilistic approach (Dwivedy and Mittal, 2010). Ikhlayel (2016) and Tran et al. (2018) have done e-waste generation studies through modelling different scenarios for predicting future outflows of the end-of-the-life products. Similarly, multiple methodologies for estimation of lifespans like leaching model (Wang et al., 2013), consumer surveys, and Weibull function have been proposed (Habuer et al., 2014; Golev et al., 2016; Oguchi et al., 2010; Huang et al., 2020). The resource recovery from e-waste is driven by precious and critical raw materials (Sethurajan, 2019). In this context, an economic assessment is crucial to estimate the recovery value of the e-waste stream (Cucchiella et al., 2015). This, in turn, will entail more thrust on efficient recycling systems, thereby creating a reverse supply chain of EEE.

There is limited data on e-waste generation, flows in various waste streams, and future obsolete steams. However, this data is very crucial for formalising and regulating the e-waste management and consequent tapping of the potential of economic recovery from the e-waste streams. Hence, the present study gives an overview of research techniques on quantifying resource embedded in e-waste and further estimating their economic potential. Figure 8.1 depicts the flow of recovery economic potential of common, precious, and critical raw materials from the geographical boundary of India.

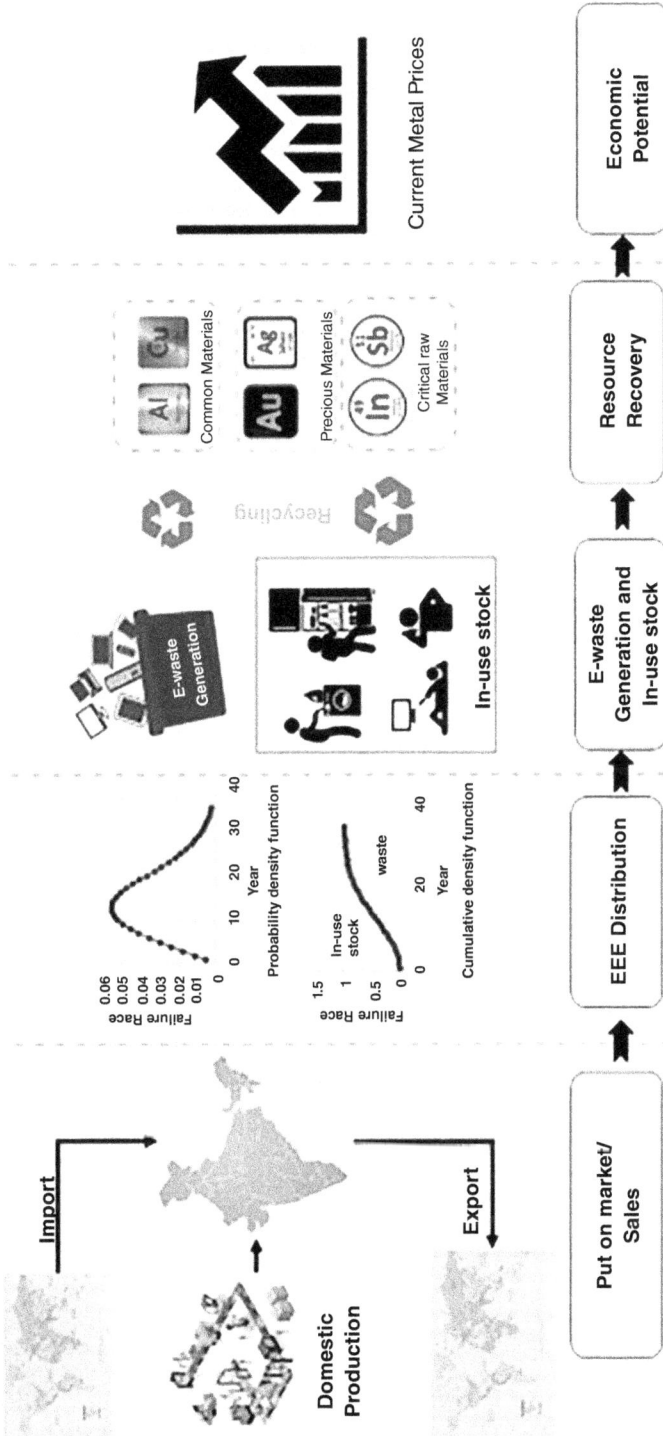

FIGURE 8.1 Procedure for estimating recovery economic potential from e-waste at the national level.

8.2 STUDY METHODOLOGY

The narrative review/literature review examines the scientific literature of the given field using published materials (Grant et al., 2009). The research articles were retrieved from Scopus and Web of Science (WoS) database using the keywords "E-waste", "electronic waste", "waste electrical and electronic equipment", "waste electrical and electronic equipment", and "dynamic material flow analysis" to answer the research question. Only those articles and review papers are included which are published in the English language. The papers in line with the resource recovery of estimated e-waste generation, in-use stock, or storage stock are included in the study. The papers which exclusively talk about estimating e-waste generation but not on resource recovery are excluded from the study. The technical reports are also considered in the review, such as reports developed by United Nations University. The articles are classified into (a) data source and method used for estimating put-on market/EEE sales and (b) the methodology used for forecasting and back casting the put-on market data. Furthermore, these are classified into lifespan (fixed or dynamic) and their distribution, country, research technique, and e-waste classification/products considered. The articles are further categorised into studies of resource recovery from e-waste generation, in-use stock and storage stock, and the materials recovered (common, precious, and critical raw materials). The literature review process is given in Figure 8.2.

8.3 RESULTS AND DISCUSSIONS

The resource recovery from e-waste in a geographical boundary depends on different variables such as put-on market (PoM), lifespan, collection, e-waste generation, and in-use stock. The most sensitive aspect in quantifying the e-waste generation is the put-on market data which has more influence than the lifespan of a product (Forti et al., 2018). The resource recovery from a country depends on the sales of EEE. Articles focusing on the resource recovery from e-waste generation, in-use stock, or storage stock is given in Table 8.1.

8.3.1 E-WASTE CLASSIFICATION

A uniform EEE classification is essential for international reporting of e-waste generation and to compare various country-level e-waste statistics. Till now, only United Nations University (UNU) has outlined such a classification based on Harmonized Commodity Description and Coding System (HS) code and European waste electrical and electronic equipment (WEEE) directive (Wang et al., 2012). Nine hundred product types have been classified into 54 UNU-Keys using approximately 270 HS code. The country-level e-waste classification differs widely. For example in India, according to e-waste management rules 2016, the EEE is classified into two types, namely information technology and telecommunication equipment and consumer electronics (CPCB, 2016). The e-waste classification for India is given in Table 8.2.

Material Comprehensive Research

To identify the research gaps and future directions in economic potential of resource recovery from E-waste literature

Definition of Search String: "E-waste", "electronic waste", "waste electrical and electronic equipment", "dynamic material flow analysis", AND "waste electrical and electronic equipment" in Topic (Title, Author Keyword, Abstract and Keyword-Plus)

Filter's adoption:

• Only peer-reviewed articles and review papers published in English language journals were considered for the study.
• The techinical reports are also considered in the review, such as reports developed by United Nations University.
• The articles were selected from both Scopus and Web of Science database.

Selection of Papers

First Criteria (Inclusion):
The papers in line with the resource recovery of estimated e-waste generation, in-use stock or storage stock are included in the study.

Second Criteria (Exclusion):
The papers which exclusively talk about estimating e-waste generation but not on resource recovery are excluded from the study.

Content Analysis

Classification of research articles based on
(a) data source and method used for estimating put-on market/EEE sales
(b) methodology used for forecasting and back casting the put-on market data
(c) lifespan (fixed or dynamic) and the probability distribution
(d) country
(e) research technique
(f) E-waste classification/products considered
(g) resource recovery from e-waste generation, in-use stock, and storage stock
(h) materials recovered (common, precious, and critical raw materials)

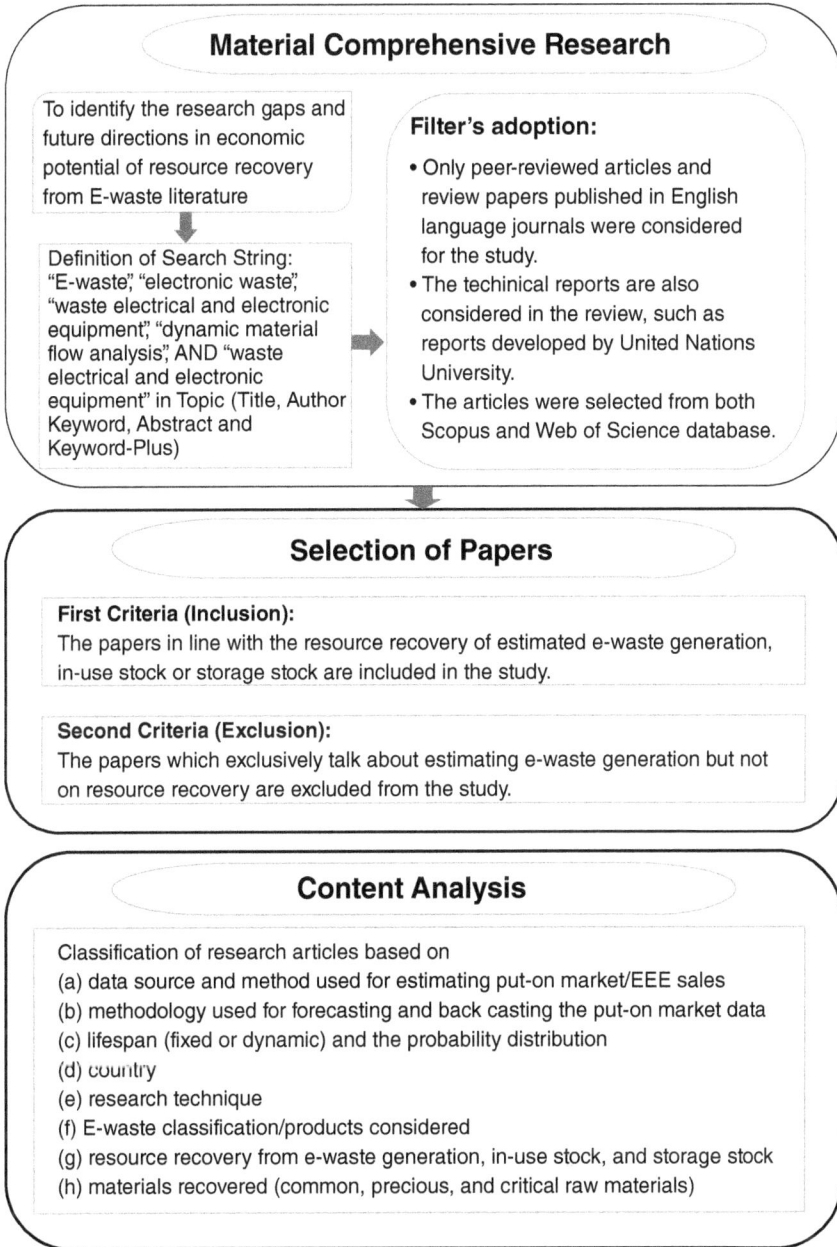

FIGURE 8.2 Literature review process.

TABLE 8.1
Key Studies on Estimating Recovery Economic Potential from E-waste

References	Put-On Market (POM) and Data Source	Backcast/ Forecast/ Both	Lifespan/ Distribution	Country	Research Technique (Inclusion of Variables Among Sales, Stock, and Lifespan)	Resource Recovery from E-waste Generation/ In-use Stock/ Storage Stock/ All	Products	Resource Recovery				Sensitivity and Uncertainty Analysis
								Common Materials	Precious Metals	Rare-earth/ Critical Raw Materials	Others	
Islam and Huda, 2020	Euromonitor International (Apparent consumption method)	Both—linear regression	Dynamic, Weibull distribution	Australia	Weibull distribution-based sales-stock-lifespan model	E-waste Generation and In-use Stock (2010–2030)	Dishwashers, home laundry appliances, large cooking appliances, microwaves, refrigeration appliances, air treatment products, food preparation appliances, heating appliances, personal care appliances, vacuum cleaners, small cooking appliances, portable players, imaging devices, home audio and cinema, video players, video games hardware	Fe, Cu, Al, plastic, Pb, Sn, Zn,	Au, Ag, Pd	Nd, Ta, Dy	-	Sensitivity—variable material price Uncertainty—weight distribution (Monte Carlo simulation)
Islam and Huda, 2019b	UNCOMTRADE (Apparent consumption method)	Holt's double exponential smoothing	Dynamic, Weibull distribution	Australia	Weibull distribution-based sales-stock-lifespan model	E-waste generation and in-use stock (2000–2047)	51 UNU-Keys	Fe, Cu, Al, plastic, Pb, Sn, Zn, Ni	Au, Ag, Pd, and Pt	Co, Ga, Ta, Nd, Pr	Cd	Sensitivity—variable material price Uncertainty—weight distribution (Monte Carlo simulation)

Reference	Data source	Regression	Distribution	Country	Method	Output (period)	Product types	Base metals / plastics	Precious metals	Other / critical metals	Hg	Uncertainty
Parajuly et al., 2017	Euromonitor International	Both—linear regression	Dynamic, Weibull distribution	Denmark	Dynamic material flow analysis, (sales-stock-lifespan model)	E-waste generation (1990–2025)	61 product types divided into Large Household Appliances, Small Household Appliances, IT & telecommunication equipment and Consumer electronics.	Plastics, Al, Cu, Fe, Pb, Sn, Zn	Ag, Au, Pd	Ba, Bi, Co, Ga, Sr, Ta	–	Uncertainty and Sensitivity Analysis—Monte Carlo Simulation
Thiebaud et al., 2017	ICT market report, Swiss Consumer Electronics Association, and GfK	Estimating past and current stock	Dynamic	Switzerland	Stock-driven dynamic material flow analysis	In-use stock and storage stock	Mobile phones, desktop, laptop computers, monitors, cathode ray tube, and flat-panel television, DVD players, and headphones	-	Au	Nd, In	–	
Wang et al., 2018	National Bureau of Statistics of the People's Republic of China and General Administration of Custom of the People's Republic of China (Apparent consumption method)	Logistic function	Dynamic, Weibull distribution	China	Stock-driven dynamic material flow analysis	E-waste generation (1992–2040)	TV sets	Cu, Fe, Al, Zn, Sn, Ni, plastic, Pb	Au, Ag, Pd	Ba, Sn	Hg	Uncertainty— average lifespan, weight, substance concentration, market share, maximum level of average possession of TV sets
Zeng et al., 2016	Apparent consumption method	Both—linear regression	Dynamic, Weibull distribution	China	Sales-stock-lifespan	E-waste generation and in-use stock	Refrigerator, washing machine, air conditioner, TV, desktop PC, laptop PC, mobile phone, single-machine telephone, fax machine, copier, printer, monitor, rangehood, electric water heater, gas water heater	Al, Cu, Fe	Au, Pd, Ag	Co, In, Nd	–	Uncertainty— variable market price of metals (Monte Carlo Simulation)

(Continued)

TABLE 8.1 (Continued)

Key Studies on Estimating Recovery Economic Potential from E-waste

References	Put-On Market (POM) and Data Source	Backcast/Forecast/Both	Lifespan/Distribution	Country	Research Technique (Inclusion of Variables Among Sales, Stock, and Lifespan)	Resource Recovery from E-waste Generation/In-use Stock/Storage Stock/All	Products	Common Materials	Precious Metals	Rare-earth/Critical Raw Materials	Others	Sensitivity and Uncertainty Analysis
Van Eygen et al., 2016	Eurostat	-	Dynamic, Weibull distribution	Belgium	Sales-stock-lifespan	E-waste generation	Desktop and laptop computers	Steel, Fe, Al, Cu, Ni, MnO2, plastics, organics, minerals	Ag, Au, Pd	Sb, Sb$_2$O$_3$, Sn, Bi, Co, Si, Ba, Li	Pb, Mg, Cr, Zn, Hg	-
Kalmykova et al., 2015	Combined nomenclature classification (apparent consumption method)	-	Dynamic, Weibull distribution	Sweden	Sales-stock-lifespan	E-waste generation	TVs and monitors	Plastics, Cu, steel/Fe,	Au, Pd, Ag	Sn	Pb	-
Habuer et al., 2014	Domestic production—NBSC Import and export—GAC	Logistic regression	Variable (Weibull distribution)	China	Sales-lifespan (dynamic product or substance flow analysis)	E-waste generation (1995–2030)	Refrigerator, washing machine, air conditioner, TV set, PC	Fe, Cu, Al, Sn, Zn, Ni, Pb	Ag, Au, Pd	Ba, Bi, Co, Sb, Sr	Hg	Sensitivity analysis—lifespan distribution and future market share ratio

Reference	Method	Model	Country	Lifespan	Output	Products	Common Materials	Precious Metals	Critical Raw Materials	Others	Uncertainty analysis
Tran et al., 2018	Calculation pathway with a stock-based approach	Dynamic, Weibull distribution	Vietnam	Stock-lifespan	E-waste generation, in-use stock	TV	Cu, Al, Fe, Steel, plastics, glass	Au, Ag, Pd	-	-	Uncertainty analysis—Monte Carlo Analysis
Althaf et al., 2019	Logistic forecasting	Dynamic, Weibull distribution	The United States	Sales-lifespan	E-waste generation	CRT TVs, audio-visual media technologies - VCRs, desktop CPU, CRT monitor, printer, telephone answering devices, digital camcorders, satellite set-top boxes, basic mobile phones, laptops, DVD player, digital cameras, MP3 player, LCD monitor, LCD TV, portable navigation devices, plasma TV, cable set-top boxes, smart phone, VoIP player, IPTV, Blu-ray player, digital photo frames, tablet, LED TV, LED monitor			Co, In	Pb	

Note:
Symbols:
Common Materials—*Iron (Fe), copper (Cu), Aluminum (Al), Tin (Sn), Zinc (Zn), Nickel (Ni), Manganese Oxide (MnO_2)*
Precious Metals—*Gold (Au), silver (Ag), Palladium (Pd), and Platinum (Pt)*
Critical Raw Materials—*Neodymium (Nd), Tantalum (Ta) and Dysprosium (Dy), Cobalt (Co), Gallium (Ga), Praseodymium (Pr), Indium (In), Barium (Ba), Bismuth (Bi), Strontium (Sr), Silicon (Si), Lithium (Li)*
Others—*Lead (Pb), Cadmium (Cd), Mercury (Hg), Chromium (Cr), Manganese (Mg)*

TABLE 8.2

Categorisation of Electrical and Electronic Equipment and their Weight and Lifespan

Sr. No	Categories of Electrical and Electronic Equipment (EEE)	EEE Code	Average Weight (kg/Unit)	Average Life (Year)
	Information Technology and Telecommunication Equipment			
1.	Centralised data processing: Mainframe (MF), minicomputer (MC)	ITEW1	MF = 8.33 MC = 1.26	10 5
2.	Personal computing: Personal computers (central processing unit with input and output devices)	ITEW2	8.77	6
3.	Personal Computing: Laptop computers (central processing unit with input and output devices)	ITEW3	1.26	5
4.	Personal computing: Notebook computers	ITEW4	1.26	5
5.	Personal computing: Notepad computers	ITEW5	1.26	5
6.	Printers including cartridges	ITEW6	10.32	10
7.	Copying equipment	ITEW7	40	8
8.	Electrical and electronic typewriters	ITEW8	0.4	5
9.	User terminals and systems	ITEW9	22	6
10.	Facsimile	ITEW10	10.32	10
11.	Telex	ITEW11	0.23	5
12.	Telephones	ITEW12	0.45	9
13.	Pay telephones	ITEW13	0.45	9
14.	Cordless telephones	ITEW14	0.45	9
15.	Cellular telephones (feature phones and smart phones)	ITEW15	0.09	7 5
16.	Answering systems	ITEW16	0.45	5
	Consumer Electrical and Electronics:			
17.	Television sets (including sets based on liquid crystal display and light emitting diode technology)	CEEW1	10.2	9
18.	Refrigerator	CEEW2	40.79	10
19.	Washing machine	CEEW3	72.54	9
20.	Air-conditioners excluding centralised air conditioning plants	CEEW4	26.7	10
21.	Fluorescent and other mercury-containing lamps	CEEW5	0.08	2

Source: (Forti et al., 2018; CPCB, 2016)

8.3.2 PUT-ON MARKET

The term "put-on market" refers to the amount of EEE consumed in the given time frame and geographical boundary. The data sources on the EEE put-on market in developed countries, especially the European Union (EU), is quite notable such as

Eurostat. However, the data sources on the EEE put-on market or EEE sales are inadequate in developing countries. Baldé et al. (2017) suggested an apparent consumption method using UNCOMTRADE data with Harmonized Commodity Description and Coding System (HS) code of products as given in Eq (8.1).

$$S(t) = P(t) + I(t) + E(t) \tag{8.1}$$

where
$S(t)$ = Put-on market/EEE sales at time t
$P(t)$ = Domestic production in the country at time t
$I(t)$ = Import of EEE at time t
$E(t)$ = Export of EEE at time t

Some studies which have used apparent consumption method to estimate put-on market are Parajuly et al., 2017, Islam and Huda (2020), Islam and Huda (2019b), and Kuong et al. (2019). The main limitation of using apparent consumption method is the lack of availability of the domestic production data. The developing countries do not have data on production, which influences the e-waste generation estimation. The data on put-on market can influence the e-waste estimation greater than the lifespan of the EEE products (Forti et al., 2018). Therefore, the databases used to estimate e-waste generation are UNCOMTRADE, Euromonitor International, and the country-level statistics developed under extended producer responsibility.

8.3.3 Lifetime of Products

The lifetime of the EEE refers to the amount of time the EEE is in stock (e.g. household) (Balde et al., 2015). With the advancement of technology, the lifespan of EEE products decreases over time (Gutierrez et al., 2011). The fixed lifespan of products used in MFA studies such as Araujo et al. (2012) and Robinson (2009) has limitations since it may underestimate or overestimate the e-waste generation and stocks (Wang et al., 2013; Islam and Huda, 2020). The fixed lifespan of products develops a static model which gives information about the flows in one year but not in the time series. Moreover, it does not take into consideration the whole lifespan of product which starts from the initial year in which the products are put into the market to the year when the market will saturate for the product. The lifetime of products is determined by the consumer survey in many studies such as Thiebaud et al. (2018b) and Polak and Drapalova (2012). It was established from studies such as Wang et al., (2013), Balde et al., (2015), Baldé et al. (2017), and Forti et al. (2018) that the lifetime of products follows a two-parameter Weibull distribution. Shape (α) and scale (β) are two parameters of Weibull distribution for lifetime estimation. The parameters of Weibull distribution for all 54 UNU-keys are available in Balde et al. (2015). The studies of Zeng et al. (2016), Islam and Huda (2019b), and Islam and Huda (2020) have used lifetime Weibull distribution parameters developed by Balde et al. (2015). The equation for probability density function (PDF) and cumulative density function (CDF) of Weibull distribution is given in Eq. (8.2) and Eq. (8.3), respectively. The

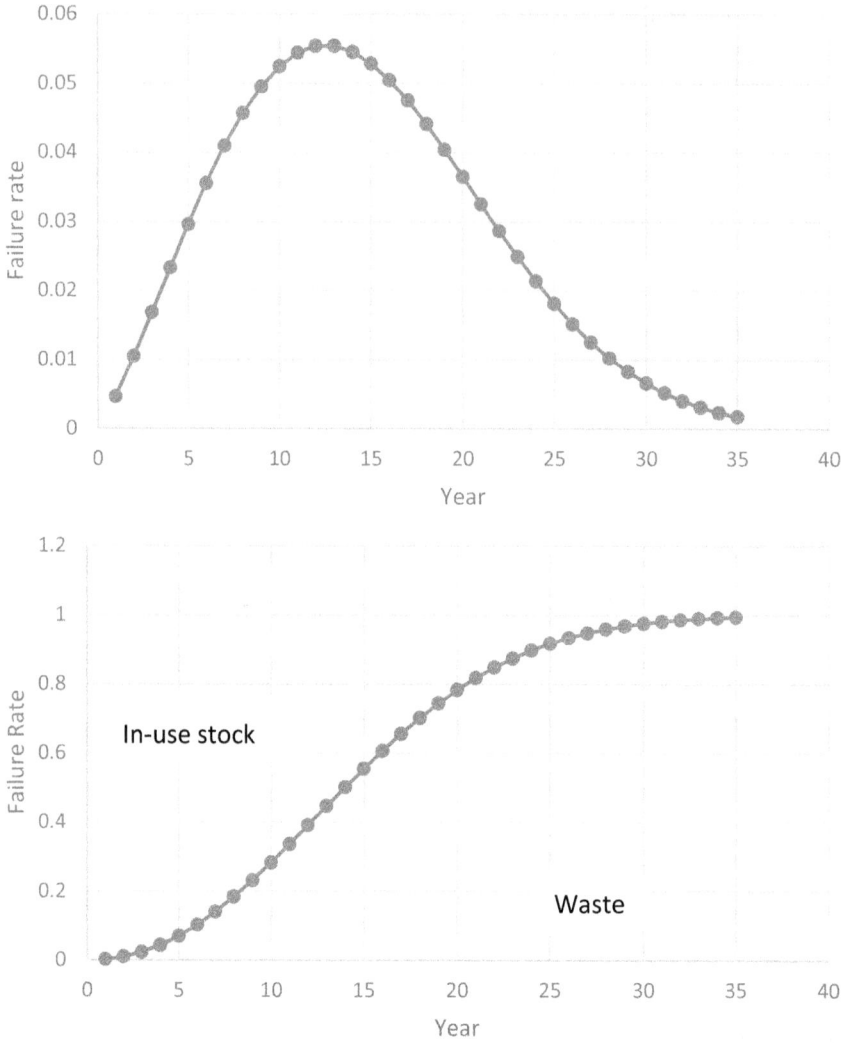

FIGURE 8.3 (a) Probability Density Function of refrigeration appliances and (b) Cumulative Density Function (CDF) of refrigeration appliances.

PDF and CDF of the Weibull distribution for refrigerators with shape (α) and scale (β) parameters are 2.2 and 16.5 as given in Figure 8.3 (Balde et al., 2015).

$$f(x) = \begin{cases} \dfrac{\beta}{\alpha}\left(\dfrac{x}{\alpha}\right)^{\beta-1} e^{-\left(\frac{x}{\alpha}\right)^{\beta}}, & x \geq 0 \\[3mm] 0, & x < 0 \end{cases} \tag{8.2}$$

$$F(x) = 1 - e^{-\left(\frac{x}{\alpha}\right)^{\beta}} \tag{8.3}$$

Huang et al. (2020) concluded from his study that there is a significant difference in the lifespan of EEE products between consumer survey and formal recycling plants. Even the authors further elaborate that the lifetime of products determined from the survey of the formal recycling plant is more than that determined by the consumer survey (Huang et al., 2020).

8.3.4   E-WASTE GENERATION

Estimation of WEEE generation is essential to assess the opportunity for WEEE operators, especially recyclers. There are various models to approximate WEEE generation such as market supply method, consumption use method, econometric analysis, questionnaire-based survey, time step, Carnegie Mellon method, batch leaching, and material flow analysis method (Singh et al., 2020; Zeng et al., 2016). The selection of method depends on the reliability and robustness of the model (Wang et al., 2013). Yedla (2016) suggested the integration of methods for estimating e-waste generation. In recent studies, the use of the sales-stock-lifespan model is evident in studies, for instance Parajuly et al. (2017); Islam and Huda (2019b), and Islam and Huda (2020). The estimation of e-waste generation is given by Eq. (8.4).

$$D(x) = \int_0^n f(x) P(x) \, dx \qquad (8.4)$$

where
x indicates year
$D(x)$ indicates e-waste generation in year x (in tons)
n indicates total number of products in the e-waste category
$f(x)$ —obsolescence rate in year x given in Eq. (8.2).
$P(x)$—net weight in year x

8.3.5   STOCK

The EEE at the end of life is a reservoir of common, precious, and critical raw materials which are known as urban mine. These materials are available as stock in households, businesses, etc. The stock is of two types, namely in-use stock and storage stock (Muller, 2014). The EEE which are presently utilised by consumers are considered as in-use stock while the products which are stored after their end of life form storage stock (Thiebaud et al., 2018b). The dynamic material flow analysis studies have mainly focused on in-use stock but not on storage stock as observed in Table 8.1. The in-use stock is given by Eq. (8.5).

$$S_j = \sum_i^n F_i(x) \times c_{ij} \qquad (8.5)$$

where
S_j is resource stock with the e-waste (in tons)
$F_i(x)$ is cumulative generated weight of the e-waste in the year x (in tons)
c_{ij} is the content of the jth resource in the ith category of the e-waste

8.3.6 RESOURCE RECOVERY

There are 69 elements found in EEE which contain common, precious, and critical raw materials (Forti et al., 2020). The secondary raw materials (SRMs) are considered as "critical" on the basis of economic importance and supply chain risk (Blengini et al., 2017). For example the critical raw materials for India are indium, gallium, neodymium, yttrium, europium, lanthanum, praseodymium, cerium, gadolinium, terbium, dysprosium, yttrium, tantalum, palladium, platinum, and antimony (Gupta et al., 2016). However, the list of critical raw materials for the EU includes antimony, baryte, bauxite, beryllium, bismuth, borates, cobalt, coking coal, fluorspar, gallium, germanium, hafnium, heavy rare earth elements (HREEs), indium, lithium, low rare earth elements (LREEs), magnesium, natural graphite, natural rubber, niobium, PGMs, phosphate rock, phosphorous, scandium, silicon metal, strontium, tantalum, titanium, tungsten, and vanadium (CRMALLIANCE, 2020). Hence, the list of critical raw materials differs with that of the country. The quantity of SRMs available for recycling in the given year can be determined by Eq. (8.6) as given here.

$$SRM_n = \sum_m N_m W_{mn} \qquad \forall\, n \tag{8.6}$$

where

SRM_n = Quantity of secondary raw material n available for recycling in the given year (g)

N_m = Number of units of WEEE, m generated in the given year

W_{mn} = Average weight of SRM, n of given WEEE, m (g/unit)

The material recovery from e-waste is a function of market prices (D'Adamo et al., 2019). The economic potential of resource recovery from e-waste is determined from Eq. (8.7).

$$REP_n = SRM_n R_n P_n \qquad \forall n \tag{8.7}$$

where

REP_n = Recovery economic potential (USD) of metal, n

SRM_n = Quantity of secondary raw material n available for recycling in the given year (g)

R_n = Recycling rate (%) of metal, n

P_n = Average price of metal, n in (USD/g)

Sensitivity analysis is an important aspect in determining the recovery economic potential. Market prices is one of the major fluctuating variables in estimating e-waste generation and their corresponding resource recovery (Zeng et al., 2016). Another sensitive variable is the material content embedded in various EEE. Cucchiella et al. (2015), Oguchi et al. (2011), and Reuter et al. (2013) have made available the material composition for various EEE. The dismantling of WEEE can be done by manual dismantling, semi-automated dismantling, and automated dismantling. Since the automatic disassembly causes loss of precious and critical raw materials in shredding, therefore, manual dismantling is preferred to enhance resource recovery from e-waste (Fontana et al., 2021; Buchert et al., 2012). The WEEE is highly heterogeneous, and thus one recycling technology cannot be generalised for resource recovery (Sethurajan, 2019). Pyrometallurgical technology, mild extracting technology, biometallurgical technology, electrochemical technology, supercritical technology, vacuum metallurgical technology, and other metallurgical technologies (ultrasonical, mechanochemical technology, etc.) can recover metals from WEEE (Zhang and Xu, 2016). The industrial applications of pyrometallurgy and hydrometallurgical technologies have improved in recent years. However, hydrometallurgy is preferred over pyrometallurgy due to multiple factors like reduced gas emission, no slag generation, and higher recovery rate (Sethurajan, 2019).

8.4 CONCLUSION

The resources embedded in e-waste attracts policymakers, businesses, and recycling system managers. Therefore, it is of utmost importance that suitable mathematical models and tools are developed to evaluate e-waste generation, in-use stock, and storage stock at the country level. Different models are developed considering one or multiple variables among sales, stock, and lifespan. Although there is no single accepted model to capture dynamics of e-waste generation and stock, but recent studies have considered the sales–stock–lifespan model to be more robust and accurate. However, it estimates only in-use stock neglecting the storage stock. There is a considerable difference between the lifetime of EEE products estimated using consumer survey and formal recycling plants. Moreover, the product lifetime differs with a country or the geographical boundary. Hence, more studies are required to determine the lifetime distributions of EEE in various countries considering both consumer survey and the survey of formal recycling plants. The drawback of the present study is that it is based on narrative review instead of systematic literature review, which can be conducted for further future directions.

REFERENCES

Althaf, S., C. W. Babbitt, and R. Chen. 2019. "Forecasting electronic waste flows for effective circular economy planning." *Resour. Conserv. Recycl.* 151:10. doi:10.1016/j.resconrec.2019.05.038.

Araujo, M. G., A. Magrini, C. F. Mahler, and B. Bilitewski. 2012. "A model for estimation of potential generation of waste electrical and electronic equipment in Brazil." *Waste Manag.* 32 (2):335–342. doi:10.1016/j.wasman.2011.09.020.

Baldé, C. P., Vanessa Forti, Vanessa Gray, Ruediger Kuehr, and Paul Stegmann. 2017. *The global e-waste monitor*. United Nations University (UNU), International Telecommunication Union (ITU) & International Solid Waste Association (ISWA), Bonn/Geneva/Vienna. Accessed 4 September 2020. https://collections.unu.edu/eserv/UNU:6341/Global-E-waste_Monitor_2017__electronic_single_pages_.pdf

Balde, C. P., R. Kuehr, K. Blumenthal, S. Fondeur Gill, M. Kern, P. Micheli, E. Magpantay, J. Huisman. 2015. *E-waste statistics: guidelines on classifications, reporting and indicators*. United Nations University, IAS—SCYCLE, Bonn, Germany. https://i.unu.edu/media/ias.unu.edu-en/project/2238/E-waste-Guidelines_Partnership_2015.pdf

Blengini, G. A., P. Nuss, J. Dewulf, V. Nita, L. T. Peiro, B. Vidal-Legaz, C. Latunussa, L. Mancini, D. Blagoeva, D. Pennington, M. Pellegrini, A. Van Maercke, S. Solar, M. Grohol, and C. Ciupagea. 2017. "EU methodology for critical raw materials assessment: policy needs and proposed solutions for incremental improvements." *Resour. Pol.* 53:12–19. doi:10.1016/j.resourpol.2017.05.008.

Buchert, Matthias, Andreas Manhart, Daniel Bleher, and Detlef Pingel. 2012. *Recycling critical raw materials from waste electronic equipment.* Öko-Institut eV, Freiburg, 49, February 24. Accessed 7 September 2020. www.oeko.de/oekodoc/1375/2012-010-en.pdf

CPCB (Central Pollution Control Board). 2016. *Government of India ministry of environment, forest and climate change notification, March 23.* Accessed 3 September 2020. https://cpcb.nic.in/displaypdf.php?id=RS1XYXN0ZS9FLVdhc3RlTV9SdWxlc18yMDE2LnBBkZg==

CRMALLIANCE. 2020. *List of critical raw materials.* Accessed 8 September 2020. www.crmalliance.eu/critical-raw-materials

Cucchiella, F., I. D'Adamo, S. C. L. Koh, and P. Rosa. 2015. "Recycling of WEEEs: an economic assessment of present and future e-waste streams." *Renew. Sustain. Energy Rev.* 51:263–272. doi:10.1016/j.rser.2015.06.010.

D'Adamo, I., F. Ferella, M. Gastaldi, F. Maggiore, P. Rosa, and S. Terzi. 2019. "Towards sustainable recycling processes: wasted printed circuit boards as a source of economic opportunities." *Resour. Conserv. Recycl.* 149:455–467. doi:10.1016/j.resconrec.2019.06.012.

Dwivedy, M., and R. K. Mittal. 2010. "Estimation of future outflows of e-waste in India." *Waste Manag.* 30 (3):483–491. doi:10.1016/j.wasman.2009.09.024.

Fontana, D., F. Forte, M. Pietrantonio, and S. Pucciarmati. 2021."Recent developments on recycling end-of-life flat panel displays: a comprehensive review focused on indium." *Crit Rev Environ Sci Technol.* 28. doi:10.1080/10643389.2020.1729073.

Forti, Vanessa, Cornelis P. Balde, Ruediger Kuehr, and Garam Bel. 2020. *The global e-waste monitor 2020: quantities, flows and the circular economy potential.* United Nations University (UNU), International Telecommunication Union (ITU) & International Solid Waste Association (ISWA), Bonn/Geneva/Vienna. Accessed 6 September 2020. www.itu.int/en/ITU-D/Environment/Documents/Toolbox/GEM_2020_def.pdf

Forti, Vanessa, Kees Baldé, and Ruediger Kuehr. 2018. *E-waste statistics: guidelines on classifications, reporting and indicators.* United Nations University (UNU), International Telecommunication Union (ITU) & International Solid Waste Association (ISWA), Bonn/Geneva/Vienna. Accessed 7 September 2020. https://collections.unu.edu/eserv/UNU:6477/RZ_EWaste_Guidelines_LoRes.pdf

Golev, A., D. R. Schmeda-Lopez, S. K. Smart, G. D. Corder, and E. W. McFarland. 2016. "Where next on e-waste in Australia?" *Waste Manag.* 58:348–358. doi:10.1016/j.wasman.2016.09.025.

Grant, M. J., and A. Booth. 2009. "A typology of reviews: an analysis of 14 review types and associated methodologies." *Health Info. Libr. J.* 26 (2):91–108. doi:10.1111/j.1471-1842.2009.00848.x.

Gupta, Vaibhav, Tirtha Biswas, and Karthik Ganesan. 2016. *Critical non-fuel mineral resources for India's manufacturing sector, a vision for 2030.* Council on Energy, Environment and Water (CEEW) and National Science and Technology Management Information System (NSTMIS), Dept. of Sci. and Technology, Govt. of India, CEEW Report, July. https://dst.gov.in/sites/default/files/CEEW_0.pdf

Gutierrez, E., B. Adenso-Diaz, S. Lozano, and P. Gonzalez-Torre. 2011. "Lifetime of household appliances: empirical evidence of users behaviour." *Waste Manag. Res.* 29 (6):622–633. doi:10.1177/0734242x10377914.

Habuer, J. Nakatani, and Y. Moriguchi. 2014. "Time-series product and substance flow analyses of end-of-life electrical and electronic equipment in China." *Waste Manag.* 34 (2):489–497. doi:10.1016/j.wasman.2013.11.004.

Huang, H. T., X. Tong, Y. Cai, and H. Tian. 2020. "Gap between discarding and recycling: estimate lifespan of electronic products by survey in formal recycling plants in China." *Resour. Conserv. Recycl.* 156:7. doi:10.1016/j.resconrec.2020.104700.

Ikhlayel, M. 2016. "Differences of methods to estimate generation of waste electrical and electronic equipment for developing countries: Jordan as a case study." *Resour. Conserv. Recycl.* 108:134–139. doi:10.1016/j.resconrec.2016.01.015.

Islam, M. T., and N. Huda. 2019a. "Material flow analysis (MFA) as a strategic tool in e-waste management: applications, trends and future directions." *J. Environ. Manag.* 244:344–361. doi:10.1016/j.jenvman.2019.05.062.

Islam, M. T., and N. Huda. 2019b. "E-waste in Australia: generation estimation and untapped material recovery and revenue potential." *J. Clean. Prod.* 237:21. doi:10.1016/j.jclepro.2019.117787.

Islam, M. T., and N. Huda. 2020. "Assessing the recycling potential of "unregulated" e-waste in Australia." *Resour. Conserv. Recycl.* 152:15. doi:10.1016/j.resconrec.2019.104526.

Kalmykova, Y., J. Patricio, L. Rosado, and P. E. Berg. 2015. "Out with the old, out with the new—the effect of transitions in TVs and monitors technology on consumption and WEEE generation in Sweden 1996-2014." *Waste Manag.* 46:511–522. doi:10.1016/j.wasman.2015.08.034.

Kuong, I. H., J. H. Li, J. Zhang, and X. L. Zeng. 2019. "Estimating the evolution of urban mining resources in Hong Kong, up to the year 2050." *Environ. Sci. Technol.* 53 (3):1394–1403. doi:10.1021/acs.est.8b04063.

Muller, E., L. M. Hilty, R. Widmer, M. Schluep, and M. Faulstich. 2014. "Modeling metal stocks and flows: a review of dynamic material flow analysis methods." *Environ. Sci. Technol.* 48 (4):2102–2113. doi:10.1021/es403506a.

Oguchi, M., S. Murakami, H. Sakanakura, A. Kida, and T. Kameya. 2011. "A preliminary categorisation of end-of-life electrical and electronic equipment as secondary metal resources." *Waste Manag.* 31 (9–10):2150–2160. doi:10.1016/j.wasman.2011.05.009.

Oguchi, M., S. Murakami, T. Tasaki, I. Daigo, and S. Hashimoto. 2010. "Lifespan of commodities, part II." *J. Ind. Ecol.* 14 (4):613–626. doi:10.1111/j.1530-9290.2010.00251.x.

Panchal, R., Singh, A. and Diwan, H. 2021. "Economic potential of recycling e-waste in India and its impact on import of materials". *Resour. Pol.* 74:102264. doi.org/10.1016/j.resourpol.2021.102264.

Parajuly, K., K. Habib, and G. Liu. 2017. "Waste electrical and electronic equipment (WEEE) in Denmark: flows, quantities and management." *Resour. Conserv. Recycl.* 123:85–92. doi:10.1016/j.resconrec.2016.08.004.

Polak, M., and L. Drapalova. 2012. "Estimation of end of life mobile phones generation: the case study of the Czech Republic." *Waste Manag.* 32 (8):1583–1591. doi:10.1016/j.wasman.2012.03.028.

Reuter, M. A., Hudson, C., Van Schaik, A., Heiskanen, K., Meskers, C., and Hagelüken, C. 2013. *Metal recycling: opportunities, limits, infrastructure.* A report of the working group on the global metal flows to the international resource panel. Accessed 1 September 2020. www.wrforum.org/wp-content/uploads/2015/03/Metal-Recycling-Opportunities-Limits-Infrastructure 2013Metal_recycling.pdf

Robinson, B. H. 2009. "E-waste: an assessment of global production and environmental impacts." *Sci. Total Environ.* 408 (2):183–191. doi:10.1016/j.scitotenv.2009.09.044.

Sethurajan, M., E. D. van Hullebusch, D. Fontana, A. Akcil, H. Deveci, B. Batinic, J. P. Leal, T. A. Gasche, M. A. Kucuker, K. Kuchta, I. F. F. Neto, Hmvm Soares, and A. Chmielarz. 2019. "Recent advances on hydrometallurgical recovery of critical and precious elements from end of life electronic wastes-a review." *Crit. Rev. Environ. Sci. Technol.* 49 (3):212–275. doi:10.1080/10643389.2018.1540760.

Singh, A., R. Panchal, and M. Naik. 2020. "Circular economy potential of e-waste collectors, dismantlers, and recyclers of Maharashtra: a case study." *Environ. Sci. Pollut. Res.* 19. doi:10.1007/s11356-020-08320-3.

Thiebaud, E., L. M. Hilty, M. Schluep, H. W. Boni, and M. Faulstich. 2018a. "Where do our resources go? Indium, neodymium, and gold flows connected to the use of electronic equipment in Switzerland." *Sustainability* 10 (8):17. doi:10.3390/su10082658.

Thiebaud, E., L. M. Hilty, M. Schluep, and M. Faulstich. 2017. "Use, storage, and disposal of electronic equipment in Switzerland." *Environ. Sci. Technol.* 51 (8):4494–4502. doi:10.1021/acs.est.6b06336.

Thiebaud, E., L. M. Hilty, M. Schluep, R. Widmer, and M. Faulstich. 2018b. "Service lifetime, storage time, and disposal pathways of electronic equipment a Swiss case study." *J. Ind. Ecol.* 22 (1):196–208. doi:10.1111/jiec.12551.

Tran, H. P., T. Schaubroeck, D. Q. Nguyen, V. H. Ha, T. H. Huynh, and J. Dewulf. 2018. "Material flow analysis for management of waste TVs from households in urban areas of Vietnam." *Resour. Conserv. Recycl.* 139:78–89. doi:10.1016/j.resconrec.2018.07.031.

Van Eygen, E., S. De Meester, H. P. Tran, and J. Dewulf. 2016. "Resource savings by urban mining: the case of desktop and laptop computers in Belgium." *Resour. Conserv. Recycl.* 107:53–64. doi:10.1016/j.resconrec.2015.10.032.

Wang, F., J. Huisman, Kees Baldé, and Ab Stevels. 2012. *A systematic and compatible classification of WEEE.* Paper presented at the annual meeting for the Society of Electronics Goes Green 2012+, Berlin, September 9–12.

Wang, F., J. Huisman, A. Stevels, and C. P. Balde. 2013. "Enhancing e-waste estimates: improving data quality by multivariate input-output analysis." *Waste Manag.* 33 (11):2397–2407. doi:10.1016/j.wasman.2013.07.005.

Wang, M. X., X. L. You, X. Li, and G. Liu. 2018. "Watch more, waste more? A stock-driven dynamic material flow analysis of metals and plastics in TV sets in China." *J. Clean. Prod.* 187:730–739. doi:10.1016/j.jclepro.2018.03.243.

Widmer, R., H. Oswald-Krapf, D. Sinha-Khetriwal, M. Schnellmann, and H. Boni. 2005. "Global perspectives on e-waste." *Environ. Impact Assess Rev.* 25 (5):436–458. doi:10.1016/j.eiar.2005.04.001.

Yedla, S. 2016. "Development of a methodology for electronic waste estimation: a material flow analysis-based SYE-waste model." *Waste Manag. Res.* 34 (1):81–86. doi:10.1177/0734242x15610421.

Zeng, X. L., R. Y. Gong, W. Q. Chen, and J. H. Li. 2016. "Uncovering the recycling potential of "new" WEEE in China." *Environ. Sci. Technol.* 50 (3):1347–1358. doi:10.1021/acs.est.5b05446.

Zhang, L. G., and Z. M. Xu. 2016. "A review of current progress of recycling technologies for metals from waste electrical and electronic equipment." *J. Clean. Prod.* 127:19–36. doi:10.1016/j.jclepro.2016.04.004.

Part IV

Critical Aspects of Sustainable E-waste Management

Focuses on Relatively Unexplored but Critical Area of E-waste Management

9 A Review of Security Threats from E-waste

Issues, Challenges, and Sustainability

*Sanchari Das, A.K.M. Salman
Hosain and Biswajit Debnath*

CONTENTS

DOI: 10.1201/9781003301899-13

9.1 INTRODUCTION

In parallel with the unbridled development of technology, such as Internet of Things (IoT) devices (Das et al., 2020a; Hadan et al., 2019; Gopavaram et al., 2019) and Augmented and Virtual Reality (AR/VR) tools (Joshi et al., 2020; Duzgun et al., 2020), we witness a digital evolution in our daily lives. Infiltration of technological advancement even in the form of mobile applications (Dev et al., 2018; Momenzadeh et al., 2020; Das et al., 2019a, b; Dev et al., 2019) and social media (Das et al., 2017; Das et al., 2018a, b; Noman et al., 2019; Das et al., 2019) has created a digital dependency on electronic gadgets (Debnath et al., 2020a; Debnath et al., 2021). With an increase in consumer electronics usage, there is an iterative increase in the electronic waste formed by discarded electronic devices (Widmer et al., 2005). When any electric or electronic equipment ceases to be of any value to the consumer, they are referred as electronic waste (e-waste) (Widmer et al., 2005). Global generation of e-waste reached a record of 53.9 million metric tons by 2019 with a growth rate of 21% in just 5 years (Forti et al., 2020). It was estimated that the GCC (Gulf Coast Countries) countries alone would produce 120 million tons of e-waste annually by 2020 (Alghazo and Ouda, 2016). The BRICS (Brazil, Russia, India, China, South Africa) nations generated nearly 33% of the total e-waste generation in 2019 (Forti et al., 2020).

People may choose to dispose of or stop using a device for various reasons. For example, with the increased smartphone usage, their lifespan is getting shortened as Khramova and Martinez mentions that smartphones are replaced in just about 2 years now (Khramova and Martinez, 2018). Researchers have also shown that mobile phones' average lifespan has reduced from 10 years to 2 years now (Debnath 2019). Reasons for replacing or disposing of a device may include the emergence of new and updated technology, device getting obsolete for more extended periods of use (Szewczyk et al., 2018), malfunction of appliance, intelligent marketing strategies (Debnath, 2020), lucrative features, and technological advancement (Debnath et al., 2019a, b), or the wish to try out new electronic devices. Despite such reasons that contribute toward e-waste generation, each disposed-of electronic item by a consumer adds a block to the never-ending pile of e-waste.

This enormous amount of e-waste will unequivocally have a deleterious impact on the environment, ecology, and human health because of the toxic materials it contains (Widmer et al., 2005). But another concern of electronic waste—the security threat arising from disposed-of electronics—often goes less noticed (Alghazo et al., 2018). Many electronic devices, such as laptops, mobile phones, and USB sticks have storage capabilities. When these devices with storage capabilities are sold or disposed of without sanitizing the data stored in them previously, they create a massive possibility of a personal or organizational data breach if they somehow end up in the wrong hands from dumpsites' second-hand market (Gutmann and Warner, 2019). It is reported that used computers from second-hand auction sites have become primary targets of cybercriminals to extract and use confidential data for financial gain (Arthur, 2009). Often data is not removed correctly before reselling or disposing of. A study on USB storage devices by Jones et al. identified that only 46% of the USB devices were deleted or formatted prior to reselling (Jones et al., 2009). If the buyer of a second-hand storage device finds any private and sensitive information of the seller in the device, the seller's privacy can be compromised as well (Conacher et al., 2020). Most consumers are found to be perplexed regarding the effectiveness of deletion methods (Diesburg et al., 2016), exposing themselves to various security threats like identity theft, phishing (Unchit et al., 2020), data theft, exposure of credit card passwords, address from remnant data in their resold or disposed-of devices (Das et al., 2020b, c, d, e, f). Even if consumers are aware of the threat from their remnant data and conduct the usual deletion process provided with the operating systems, it does not guarantee complete and safe data removal. For example, the factory reset function in smartphones may not always ensure complete deletion of data (Khramova and Martinez, 2018).

Such important aspect of e-waste is rarely discussed in popular literature. Some of our previous works focus on device-specific security issues from a supply chain perspective. However, looking at it from a system approach, i.e., analyzing e-waste management as a whole system and detailed discussion of the security aspects are scant. Hence, to understand the current research landscape in e-waste security, this chapter discusses an in-depth systematic literature review on security threats from electronic waste and end-of-life (EoL) electronics. We report on how e-waste creates security threats, what research is done so far in the domain, which devices have been investigated, the methodologies of other research conducted on e-waste and its security threat, and the kinds of solutions provided so far, and so on. The rest of the chapter is organized as follows: We start by detailing some background research on e-waste security and systematic review, which has motivated our work in Section 9.2. After that, we provide the methods and analysis of our study in Section 9.3. Our analysis results are detailed in Section 9.4 and discussion on our findings provided in Section 9.5. Finally, we conclude the systematic literature review in Section 9.6 by giving directions for future works to better the overall e-waste scenario.

9.2 RELATED WORK

Security threats arising from e-waste is often a neglected issue compared to other topics on e-waste (Debnath et al., 2019). When an electronic or electrical device has lost its functionality or value to the current user, that device is called electronic waste or e-waste (Widmer et al., 2005). When a device stops functioning and ends up as waste, it is known as Waste Electrical and Electronic Equipment (WEEE) (Debnath et al., 2019). WEEE and e-waste are generally used synonymously (Widmer et al., 2005). E-waste, WEEE, or EoL electronics denotes electronic and electrical equipment, including all components, sub-assemblies, and consumables, deemed obsolete or unwanted by a user (Bhuie et al., 2004). A confluence of different definitions are available in Table 9.1.

TABLE 9.1

Description and Definitions of the Various Terminologies Used in the E-Waste Domain by Prior Researchers

Author	Definition
EU WEEE Directive (Widmer et al., 2005)	Electrical or electronic equipment, which is waste, including all components, sub-assemblies, and consumables, which are part of the product at the time of discarding. Directive 75/442/EEC, Article 1(a) defines waste as any substance or object that the holder disposes of or is required to dispose of pursuant to national law provisions in force.
Basel Action Network (Puckett and Smith, 2002)	E-waste encompasses a broad and growing range of electronic devices ranging from large household devices such as refrigerators, air conditioners, cell phones, personal stereos, and consumer electronics to computers which have been discarded by their users.
OECD (Widmer et al., 2005)	Any appliance using an electric power supply that has reached its end of life.
StEP (Widmer et al., 2005)	E-waste refers to the reverse supply chain which collects products no longer desired by a given consumer and refurbishes for other consumers, recycles, or otherwise processes wastes.
Sinha (Sinha, 2004)	An electrically powered appliance that no longer satisfies the current owner for its original purpose.
Debnath (Debnath, 2019)	Any EoL electrical and/or electronic equipment and their accessories that have a power source or require a power source for functioning can be defined as e-waste.

With the increment of electronic device usage, the amount of e-waste increases both regionally and globally. The e-waste industry has burgeoned to an industry of $7 billion (Grant and Oteng-Ababio, 2012), and the cybercrime industry has financially depleted more than $450 billion annually worldwide (Shiloh and Fassassi 2016; Graham, 2017). These two sectors can contribute to each other when proper disposal methods are not adapted either by the consumers or by the recycling organizations (Debnath et al., 2019).

Personal and sensitive data can be recovered from the remnant data in disposed storage device (Conacher et al., 2020), which can later be misused to commit cybercrime. Personal and sensitive data can be tracked from the remnant data in the device (Conacher et al., 2020). Personal data including login credentials (Das et al., 2018a), contact list (Das et al., 2018b), messages (Dev et al., 2019), and photos (Noman et al., 2019) can be retrieved (Debnath et al., 2019). Studies on data remnants in e-waste have revealed that sensitive and confidential personal and organizational data can be recovered with data recovery tools. Suppose these data fall in the wrong hand. In that case, they can be exploited to commit severe cybercrime such as in-person tracking down and harassment, access to the previous owners' online accounts, identity theft, and illegally tampering commercially sensitive information (Conacher et al., 2020). E-waste is an expression used to include all EEE that has been discarded by consumers, not for reuse (Roychowdhury et al., 2019). Recovered remnant data from storage e-waste devices can be exploited to breach individual and organizational privacy, resulting in devastating security threats (Sansurooah and Szewczyk, 2012).

These recovered data can be misused to gain unauthorized access to the individual's private data (Debnath et al., 2020a), in-person tracking down and harassment, access to the previous owners' online accounts, identity theft, illegally tampering commercially sensitive information (Conacher et al., 2020), exposure to personal information (Jones et al., 2019), economic and political data exposure of organizations, reverse engineering of electronic devices to produce clone or regeneration (Alghazo et al., 2018), disclosure of confidential state data (Agayev et al., 2016), personally identifiable information (Szewczyk, 2011), confidential organizational and military data theft (Homaidi, 2009), confidential business information exposure (Jones, 2005), confidential organizational information exposure (Valli and Jones, 2005), etc. Research on e-waste has recovered sensitive and confidential data from storage devices in the second-hand market and e-waste dumpsites. Sansurooah and Szewczyk conducted a study on remnant data found in second-hand USB devices. They reported that 95% of the readable USB storage devices contained recoverable data. Thirty percent USB drives contained sufficient information to identify the source organizations, and 53% exposed individual identifying information (Sansurooah and Szewczyk, 2012).

9.2.1 REMANENCE OF E-WASTE

Data can remain in storage devices for various reasons. For example, Homaidi mentioned two types of data remanence in SSDs: System-level remanence and machine-level remanence (Homaidi, 2009). The file system keeps a record of all files and folders' existence and location and folders written to the storage device. Although file's record is deleted when a file deletion is performed, the file's contents usually remain undeleted and retrievable (Hughes and Coughlin, 2006; Gutmann and Warner, 2019). When disposing of storage devices such as USB sticks, most devices fail to effectively remove the previous data (Sansurooah and Szewczyk, 2012). These remaining data can be recovered easily using open source and paid data recovery tools. Often, the default data removal option provided in the device does not efficiently remove data from the device. Khramova and Martinez reported that the default erasure process in mobile phones fails to irretrievably erase the data, which allows them to retrieve the user data directly from the NAND flash bypassing the controller (Khramova and Martinez, 2018).

Jones et al. conducted a series of research on remnant data in second-hand storage devices in different years: computer disks (Jones et al., 2005), mobile phones, Personal Digital Assistants (PDAs) and Blackberry (RIM) devices (Jones et al., 2008), disposed-of organizational hard disks (Jones, 2005), second-hand hard disks analysis in 2009 (Jones et al., 2010), second-hand USB storage device in 2009 (Jones et al., 2009), second-hand hard disk analysis of UAE in 2016 (Jones et al., 2016), and second-hand hard disk analysis of the UK in 2016 (Jones et al., 2017). In each research, they could recover a considerable amount of personal and organizational sensitive remnant data in second-hand storage devices.

Gutmann and Warner mentioned that consumers usually get confused between "*erasing data*" and "*deleting data*". These terms might be similar in regular use, but, from a technical perspective, they are different. Erasing of data means making it irretrievable by overwriting with other data, while deleting refers to Operating System (OS) forgetting specific data and marked as available for overwrite which allows new data to be stored in its place when required. This process makes data often retrievable until it has been overwritten (Gutmann and Warner, 2019). Confusion regarding deleting and erasing, identification of inconsistencies in User Interface (UI), and confusing and occasionally misguidance in two widely used Os was investigated by them. They identified a need for guidelines and best practices on the General Data Protection Regulation (GDPR) compliant erasure and presented a set of implications for training suitable to improve the consistency between UIs and data protection legislation (Gutmann and Warner, 2019).

Diesburg et al. also found consumers being confused about deleting and erasing procedure and were able to recover data from 83.3% of the second-hand USB sticks (Diesburg et al., 2016). Medlin and Cazier conducted a research on 50 second-hand and donated hard disk drives and found 300,000 files containing personally identifiable information (Medlin and Cazier, 2010). Robins et al. reported remnant personal and organizational sensitive data in used USB sticks collected through online auction sites (Robins et al., 2016). Valli et al. conducted a series of research on remnant data on used hard disks (Valli and Jones, 2005), second-hand hard disks of Australia in 2007 (Valli and Woodward, 2007) and 2008 (Valli and Woodward, 2012), and organizational hard disks (Valli, 2004) and found confidential personal and organizational information which can lead to data breach and cybercrime.

9.2.2 SYSTEMIZATION OF KNOWLEDGE

A systematic literature review or systematization of knowledge research works provides a holistic overview and a basis for the research undertaken. Such analysis helps to direct future works in respective fields and helps grasp the overall scenario in a considerably less amount of time (Majam and Theron, 2006). Systematic reviews are a type of scientific research that aims to integrate objectively and systematically the results of empirical studies on a particular research problem to determine the state of the question in its field of study (Ferreras-Fernandez et al., 2016).

For example, Das et al. in their "*All About Phishing Exploring User Research through a Systematic Literature Review*" revealed that although researchers and practitioners often provided technical solutions to solve phishing-related issues, only 13.9% of relevant published papers from 2004 to 2018 included any user-focused study—primarily focusing on usability or testing of tools developed by the researchers rather than exploring the ways different kinds of users approach and make sense of phishing attempts. The absence of crucial details about study participants was also reported in the study (Das et al., 2019e).

Another systematic review by Das et al. presented that most research works on Multi-Factor Authentication (MFA) focused on new authentication technologies but lacked risk perception analysis—summing up to only 9.1% papers with any user evaluation research. Their meta-analysis of user-focused studies showed that researchers found a lower adoption rate inevitable for MFAs, while avoidance was pervasive among mandatory use (Das et al., 2019c, d). Actionable recommendations to pave future research scope, primarily aiming to include more diverse population for user study evaluations, were provided in this systematic review which aligns with prior works on multi-factor authentication (Das et al., 2019d, 2020f; Das, 2020b).

Aforementioned works provide us a comprehensive notion of the importance of a systematic review. The literature review's importance is unequivocal in any science allowing the development of scientific knowledge on the basis of the current knowledge and considers the schools of thought and scientific paradigms to build and report new knowledge (Saur-Amaral, 2011). Thus, this chapter's goal was to detail the systematic review of electronic waste security to understand the landscape of electronic waste security research. Through the detailed analysis, we evaluated security threats; reviewed research works on e-waste and the security threats arising from it, reasons behind the security issues, and solutions proposed to minimize the risks (Debnath et al. 2019a; Debnath et al. 2020b; Das, 2020b); and provided directions for future works in this field of study.

9.3 METHODOLOGY

9.3.1 DATA COLLECTION

We collected data from scholarly articles published in conferences as full papers to conduct the systematic literature review. To execute the data extraction process, a keyword bank was generated, including a comprehensive set of terminologies related to our research and alternative terminologies pertaining to the broad area of e-waste security threat. The keywords were screened to eliminate any possible duplication. The final set of keywords was used to explore different search engines (detailed here) to generate the list of peer-reviewed papers we analyzed. Wherever available, the advanced search options were used to constrain the search. All sources accessed and used are referenced in this chapter.

9.3.2 SEARCH ENGINES AND SEARCH KEYWORDS

The following ten databases were explored for the collection of research articles. They are listed here—IEEE, ResearchGate, ACM Digital Library, Springer, ProQuest, ScienceDirect, Google Scholar, Publish or Perish, Scopus, and SSRN. The number of papers extracted from these databases are shown in Table 9.2.

The search strings used for article search were modulated by relevant terminologies in the existing literature. We also searched for existing alternative terminologies used in the area of e-waste security threat. The Boolean operator "AND" was used for linking the different keywords used in the search. The keywords used are presented in Table 9.3.

TABLE 9.2

The Number of Papers Collected and Analyzed in the Final Section from Different Digital Libraries

Database Source	Total Articles Found	Abstract Screening	Full Text Screening
IEEE	4,108	154	24
ResearchGate	5,398	223	32
ACM Digital Library	468	79	14
Springer	2,138	189	9
ProQuest	52,812	75	3
ScienceDirect	5,769	456	7
Google Scholar	129,300	10,509	41
Scopus	272	57	3
SSRN	3,527	364	5

TABLE 9.3

List of Database Search Keywords Used to Evaluate the Different Categories in the Prior Literature

Sl. No.	Category	Keywords
1	E-waste management and peripherals	E-waste, WEEE, ICT waste, IOT waste, second-hand memory, used USB storage, second-hand device, e-waste reuse, second-hand market, second-hand memory card, second-hand hard disks, second-hand SSD, used storage device, used phone, reuse phone
2	Data security	Cybersecurity, data security, data privacy, security threats, residual data, privacy, sharing mobile, remnant data
3	Methodology and techniques	Data erasure, data deletion, data sanitization, media sanitization, erasing data, data sharing, data deletion, secure deletion, disk sanitization, SoK, data removal

9.3.3 DATABASE SEARCH

This section summarizes the findings of the papers collected from different digital libraries through different stages, namely—initial total collection of the articles, articles selected after abstract screening, and the final list of articles after full-text screening.

9.3.4 SCREENING METHOD

After the data collection, a screening process was carried out by going through the titles, abstracts, keywords, and the full text's quick lookup wherever necessary. Publications related to specific security threats of e-waste were included in the final set of publications. Articles published in peer-reviewed journals were given first preference, followed by book chapters with ISBN and reports published by verified national and international organizations. Publications were excluded if they were on other topics (unrelated to our research), about e-waste but not on its security threats, in languages other than English, or had an updated version of the paper from the same author. It is to be noted that there was substantially less research on e-waste and its security threat juxtaposed to e-waste related to other topics, i.e., environmental effects, recycling, the flow of e-waste, etc. The publication time frame was not considered for the screening-out process for a fewer number of papers. The papers cited and referred by the initially collected publications were also explored. They were included in the data set if they were related to the topic, followed by the extraction of relevant keywords for further database probing. This process was further repeated for the secondary cited and referred papers as well. Additionally, we also investigated the profiles of the authors and co-authors who had worked on e-waste security threats. Their work was included in our dataset and in the event of our search missing any relevant works.

9.4   RESULTS

9.4.1   Devices Studied

The devices considered in the research works were either storage devices or devices with storage functionalities. Mobile phones accounted for 30.76% of the devices (Debnath et al., 2019; Debnath et al., 2020b; Khramova and Martinez, 2018; Ahmed et al., 2017). The next two largely researched devices were computer hard disk drives (HDDs), solid-state devices (SSDs) (46.15%), and USB storage devices (23.08%) (Jones et al., 2016; Lim et al., 2014; Sansurooah and Szewczyk, 2012). Disposed or second-hand memory cards were considered in 9.62%. Some less frequently inspected devices were scanners, printers, floppy disks, personal digital assistance devices, ADSL routers, and so on (Szewczyk, 2011; Krumay, 2016; Agayev et al., 2016).

9.4.2   Method of Studies

More than half of the research articles (32 out of 58 papers) followed the forensic analysis methodology of remnant data (Jones et al., 2009; Lim et al., 2014; Jones et al., 2010). In this procedure, disposed or second-hand devices were collected from the second-hand market or dumpsites, and they were forensically analyzed if any data could be recovered. As presented in most of the articles, the authors created a forensic image and stored the device in a secure place. The analysis was conducted on the secured image (Jones et al., 2009; Valli, 2004). First, they were analyzed if any data was not deleted. As mentioned before, a simple data deletion process does not always ensure complete data removal from the storage device—demonstrated in the factory reset study of mobile phones (Khramova and Martinez, 2018). If the storage device was found to have undergone any deletion procedure, the data recovery process was carried out to recover the deleted data. Afterward, the recovered remnant data were analyzed for sensitivity and volume (Szewczyk and Sansurooah, 2012; Robins et al., 2016).

Fourteen out of 58 papers have been found to conduct user studies—either online or in-person interviews and surveys—to gather data on security threats arising from e-waste and to explore the underlying reason for not considering experiencing any data deletion procedure before the disposition of used devices (Homaidi, 2009; Karlson et al., 2009; Debnath et al., 2020b). Online surveys were mostly conducted through Google Forms and through social media. In-person interviews included group discussion (Von Zezschwitz and Hang, 2012), site visit (Debnath et al., 2019a, b), questionnaire (Yeboah-Boateng, 2012), ethnographic analysis (Guha et al., 2017), etc.

Four out of 58 papers are found to have conducted a literature survey. The remaining 16 papers adopted divergent methodologies that include a comparison of e-waste flow and its security threats in different years of the same region as well as for other regions in the same year (Alghazo and Ouda, 2016), ethnographic study (Guha et al., 2017), showcasing how remnant data can be used in cyber-crime by recovering remnant data (Doyon-Martin, 2015), presenting an overview of different policies related to e-waste management (Gumbo and Kalegele, 2015), efficiency of a built-in factory reset of mobile phones (Khramova and Martinez, 2018), device-specific data deletion methods (Lee et al., 2011), conducting a focus group and analyzing the discussion of the group (Von Zezschwitz and Hang, 2012), comparing different deletion methods and showcasing their efficacy (Wang et al., 2020), comparing deletion methods of other Operating Systems and their User Interfaces (Gutmann and Warner, 2019), discussing why different data sanitation methods fail (Shu et al., 2017), vulnerabilities of data deletion from flash drives (Garg et al., 2020), and so on.

9.4.3 User Studies

As mentioned earlier, 35 out of 58 papers instituted diverse user studies, including online surveys, in-person questionnaires, interviews, and data surveys. The most frequent user study was collecting EoL storage devices from consumers and analyzing them for remnant data (Szewczyk et al., 2018; Jones et al., 2019; Conacher et al., 2020). Online surveys were conducted mostly in Google Forms from various social media platforms (Guha et al., 2017; Debnath et al., 2020a). In-person questionnaires and interviews of selected group of people were also conducted to better comprehend the reasons contributing to data breaches from electronic waste (Debnath et al., 2019; Debnath et al., 2020a, b). Ethnographic studies in device repair shops were conducted by Guha et al. (2017).

9.4.4 Sources of E-waste Used

Second-hand storage devices have been reported to constitute the sample e-waste in 35 out of 58 papers mostly collected from various online auction sites (Conacher et al., 2020) (eBay, Amazon, etc.), auction shops, and second-hand markets (Szewczyk et al., 2018). Nine of the papers reported that they collected the devices from dumpsites or disposal sites (Alghazo and Ouda, 2016; Agayev et al., 2016; Zingerle and Kronman, 2019; Forgor et al., 2019). Some of the studies conducted research on e-waste from e-waste repair shops (Guha et al., 2017), donations (Medlin and Cazier, 2010), and unauthorized recyclers (Debnath et al., 2020b).

9.4.5   SECOND-HAND MARKET AND REPAIR SHOPS

Second-hand markets were one of the vital aspects or entities in the supply chain network of e-waste. Comprehensive discussion on the impact of second-hand markets is prominent in recent literature (Debnath, 2019; Debnath et al., 2019; Debnath et al., 2020b). The second-hand market—being a stochastic and vibrant area—is standing on the policy loopholes. Second-hand markets and repair shops contribute to the increment of the electronic items' lifespan by repairing, refurbishing, or reselling used electronic devices, thereby constituting themselves as crucial factors of a circular economic policy framework. A comprehensive case study on this notion has been provided by Debnath (Debnath, 2019). The impact of repair shops is predominantly observed in the technical aspect, whereas the economic or business perspective perceives the second-hand market's influence. Despite all the pros, an extensive discussion on the issue of security threats arising from these entities is less highlighted in the literature. A brief overview of security threats from the second-hand market and repair shops is given in some literature (Collard, 2017; Debnath et al., 2019; Debnath et al., 2020a).

9.4.6   SECURITY THREAT

Consumers have been reported to be inclined to delete their data using the default operating system deletion program before disposing of or reselling their used electrical storage devices. These programs seldom fail to remove the device's data due to system-level and machine level remanence (Homaidi, 2009). It was reported by several research articles that unremoved data of EoL electronics could be recovered using various tools, including open-sourced and paid data recovery tools and forensic tools, which, in turn, could lead to severe cybercrimes if they somehow end up being in possession of cybercriminals or hackers (Diesburg et al., 2016). It was reported that the 67.31% of the sensitive data recovered were in the form of images, videos, and documents previously stored in those devices (Jones et al., 2019).

Another form of security threat can arise via a reverse engineering of the EoL electronics (Roychowdhury et al., 2019). Any e-waste requires a systematic dismantling for the recovery of several parts, including ECs, which can pose security threats from the hardware perspective, i.e., via Reverse Engineering (RE) (Debnath et al., 2016). Copious amounts of specialized equipment have been reported to be available for the execution of RE, e.g., optical high-resolution digital microscope, Transmission Electron Microscope (TEM), Scanning Electron Microscope (SEM), probe stations, and logic analyzers (Quadir et al., 2016). RE can occur in—a) system level which involves the extraction of firmware from ROM and bitstreams from FPGA (Quadir et al., 2015); b) Printed Circuit Board (PCB) level that includes identification of ECs, datasheet mining followed by non-destructive imaging of PCB layers (Grand, 2014; Quadir et al., 2015); and c) chip-level RE which has decapsulation, delayering, cross-sectional imaging, and post-analysis (Quadir et al., 2015). These circumstances can result in dupe products, leading to business volatility.

A comprehensive discussion on unwanted exposure of sensitive personal data and confidential organizational data resulting from the recovery of remnant data present in second-hand or disposed-of devices was prominent in 71.32% of the articles. Unwanted personal and corporate data breach can result in severe cybercrime, including the threat of acquiring profile details of social networking sites, emails, private messages, photos, identity theft, data tampering (Debnath et al., 2020b), in-person tracking down and harassment (Conacher et al., 2020), access to the seller's online accounts (Conacher et al., 2020), illegal tampering of commercially sensitive information (Conacher et al., 2020), exposing password of cloud storage (Debnath et al., 2019), political and military data exposure (Alghazo et al., 2018; Homaidi, 2009), confidential state data (Agayev et al., 2016), loss of privacy and reputation in the corporate sector resulting from the data breach (Raman and Pramod, 2013), etc. Such cybercrimes can have far-reaching impact on the lives of several including vulnerable population, such as children (Streiff et al., 2018; Streiff et al., 2019; Das, 2020a), teenagers (Karmakar and Das, 2021; Das et al., 2020; Karmakar and Das, 2020), or even the elderly (Das et al., 2019, 2020). Moreover, cybercrime can happen at a system-level via a reverse engineering of EoL devices and supply chain levels by mixing ECs recovered from e-waste with brand new ones, leading to faulty device production (Roychowdhury et al., 2019).

We found that 24 out of 58 papers analyzed briefly discussed the data collected from the discarded technology and mentioned potential security threats from such unintentional data repository. Therefore, those papers were included in this chapter. However, a detailed discussion on the security threats and what type of vulnerability is exploited is not carried out, indicating a need for a thorough research on this extremely critical topic.

9.5   DISCUSSIONS

In this chapter, we have collected and analyzed relevant papers on an understudied concept of e-waste security. The devices studied in these peer-reviewed articles were primarily with storage capabilities. These studied devices were mostly collected from second-hand markets or auction sites—to recover and analyze the remnant data stored. Both online and offline interviews and questionnaire studies were conducted in some articles to portray the security issues' consumer-side aspect. Diverse security threats that can arise from the remnant data in these disposed or resold storage devices were depicted by the researchers, as mentioned in Section 9.4. In this discussion section, we extract the data from our results and analysis to see the researchers' potential solutions and the sustainability of these devices, given the discarded nature of them. Among the proposed solutions to assuage the security threats, we found raising awareness among consumers about the remnant data to be the cardinal one.

9.5.1 Solutions Proposed

E-waste has become a global concern in recent years. The additional appendage is the inclusion of security threats. Several issues are giving rise to security threats, but there are solutions as well. The following subsections describe the proposed solutions in detail.

9.5.1.1 User Awareness

The most suggested solution for diminishing security threat from e-waste was to create awareness among consumers (81.58% of the articles) about the remnant data in their used or disposed-of electronic storage devices (Jones et al., 2017; Alghazo et al., 2018; Debnath et al., 2019; Forgor et al., 2019; Debnath et al., 2020a). It was observed from the user studies that consumers are not aware of the remnant data in disposed-of storage devices (Jones et al., 2005). Disposed storage devices were found to contain retrievable deleted data and even undeleted data (Raman and Pramod, 2013). Dumpsites and second-hand markets—the two significant sources of e-waste—appear to be the primary targets for spying and cybercrime as confidential personal, military, and state data can be easily recovered and misused from the disposed or used storage devices if not properly sanitized. Creating consumer consciousness on the potential threats that may arise from improperly sanitized storage devices prior to disposition and ensuring ubiquitous accessibility of data erasure tools were supported thoroughly. Organizations were suggested to conduct a routine risk assessment of devices (Jones et al., 2017) and audit any possible future demand of data present in the devices before disposal, resell, or donation. Organizations can be susceptible to confidential data breach and data exposure through unsanitized storage devices, which can be insidious to the organization's reputation and safety (Debnath et al., 2020a).

9.5.1.2 Using Data Deletion Tools

The default data deletion process does not ensure complete data deletion, thus, leaving users susceptible to confidential personal data exposure. Hence, the adoption of legally approved data erasure tools was strongly suggested by 76.52% of the articles. Open-source and commercially available erasure tools such as HDDerase (Yeboah-Boateng, 2012), USRM (Zhang et al., 2014), Darik's Boot, and Nuke (Medlin and Cazier, 2010) were suggested to erase the remnant data efficiently. Combining two or more data erasure tools to erase data before disposing or reselling can be a method with high efficacy to ensure a complete removal of remnant data (Fundo et al., 2014). Ensuring data erasure conforming to the U.S. Department of Defense (DoD) standards was strongly suggested (Bennison and Lasher, 2004). Providing built-in data erasure tools (Medlin and Cazier, 2010) or providing a manual switch to storage devices by the manufacturers for complete data erasure, setting robust deletion methods, and utilizing external block overwriting software (Medlin and Cazier, 2010) were also found to be strong suggestions for data erasure.

9.5.1.3 Physical Destruction Storage Devices

Destroying used storage devices to a data-unretrievable state before disposing of is regarded as the Physical Destruction method, which, if done accurately, can be the most effective procedure for eliminating security issues, be it mechanical destruction or using a magnetic field to destroy the stored data. Among diverse physical destruction procedures, degaussing was mentioned in 46.78% of the articles. Degaussing means exposing the storage device to a strong magnetic field to destroy the storage device to an unretrievable state (Medlin and Cazier, 2010; Yeboah-Boateng, 2012; Forgor et al., 2019). Mechanical destruction of storage devices was also rendered feasible at a personal level if destruction to an unretrievable state can be achieved.

In formal recycling units, the storage devices were first exposed to a high magnetic field, which erases the data, followed by mechanical shredding, aka physical destruction (Services, 2013). This process has been rendered to be the most feasible procedure in several research works (Jones et al., 2016; Diesburg et al., 2016; Zhang et al., 2014; Yeboah-Boateng, 2012; Storer et al., 2010; Bennison and Lasher, 2004).

9.5.1.4 E-waste Regulation Policy

Ensuring stringent e-waste regulation policies can better the management and flow of e-waste, which, in turn, can mitigate security threats considerably. Hence, the necessity of revision of rules is evident to ensure better management of e-waste. Used, resold second-hand storage devices can be acquired easily from second-hand markets (Robins et al., 2016; Diesburg et al., 2016; Jones et al., 2005; Valli and Jones, 2005; Jones et al., 2008), repair shops (Guha et al., 2017), and dumpsites (Alghazo and Ouda, 2016; Agayev et al., 2016; Zingerle and Kronman, 2019; Forgor et al., 2019) which contributes to cybercrime as data stored previously in those devices can be recovered easily if proper sanitization beforehand is not ensured. Therefore, creating strict regulation policies for these sources and management of e-waste flow are paramount (Alghazo et al., 2018; Doyon-Martin, 2015; Yeboah-Boateng, 2012). We also found a severe lack of proper training and grooming among the resellers and repairers of second-hand devices. Providing sufficient training (Debnath et al., 2019; Glisson et al., 2016; Lim et al., 2014) about data safety to the recyclers, repairers, and second-hand market sellers and making them aware of the security threat from e-waste can play a significant role in attenuating security threats from e-waste.

9.5.1.5 Other Suggested Solutions

Besides aforementioned procedures, several other methods have also been suggested by the researchers. These approaches include ISO-based framework for secure e-waste management (Debnath et al., 2020a), efficient data encryption (Forgor et al., 2019; Jones et al., 2017), supply chain framework (Baidya et al., 2020), life cycle-based framework (Debnath et al., 2020b), comprehensive frameworks for extended producer responsibility (Alghazo et al., 2018), providing incentives for proper data recycling (Alghazo et al., 2018), and tax rebate to encourage data sanitization practice, and so on.

9.5.2 Sustainability Aspects

The security threat of e-waste is an emerging aspect of e-waste management. While the whole world is striving to deal with the wrath of e-waste both in terms of physical disposal and digital disposal, it is imperative to investigate the sustainability aspects. One must keep in mind that sustainability aspects do not necessitate the discussion to be sustainable. Instead, they discuss critical elements that can be useful in identifying paths toward sustainability. Following the footsteps of contemporary literature (Ghosh et al., 2018; Baidya et al., 2020; Debnath and Ghosh, 2019), we are presenting a qualitative discussion from the perspective of sustainability. As prescribed in the Bruntland Report (Imperatives, 1987), sustainability has three pillars—environmental, economic, and social. However, the boundaries of sustainability are not explicitly defined. Since the e-waste management system's effectiveness is mostly dependent on the supply chain network (SCN), we choose the fourth pillar to be demand (Chattopadhyay et al., 2020; Debnath et al., 2022). In other words, our further discussion is centered around four pillars of sustainability—environmental, economic, demand, and social.

9.5.2.1 Environmental Sustainability

Environmental sustainability is arguably the most important from the environmentalists' perspective. It is essential to understand the possible ecological impacts of critical analysis. Life Cycle Assessment (LCA) is the best method for this purpose. However, this is beyond the scope of this chapter. Hence, we analyze the situation following the principles of LCA, which reveals the following points:

1) E-waste requires systematic dismantling for further RE processing. Usually, the recycling units follow atypical mechanical processing lineup, including stripping, shredding, milling, etc. These machines consume high energy. As a result, the carbon dioxide emission will be higher, impacting the category of global warming potential. Compared to that, skilled manual dismantling with protective gear will have a positive environmental impact.
2) Though manual dismantling can lead to more security issues in all three levels, i.e., system-level, PCB-level, and chip-level, in a proper recycling unit. However, it allows one to rip off another benefit. As described by Roychowdhury et al. (2019), the ECs can be recovered and can be reused on the basis of their functionality after testing. Reusing ECs, wherever possible, will lead to resource conservation, reducing virgin mining, thereby reducing environmental impacts. The potential impact categories include—marine aquatic toxicity, freshwater aquatic toxicity, climate change, abiotic depletion, eutrophication, acidification, global warming, and so on (Baidya et al., 2020).
3) Proper policy implementation is key to a sustainable e-waste management system. As proposed by (Debnath et al., 2019), several policy tools and strategies can help regularize and reorganize the informal sector and the repairing market. This will significantly reduce informal recycling activities and mitigate potential security threats due to unauthorized repairing activities. There will be positive environmental impacts having an effect

on the following categories—acidification, climate change, eutrophication, water depletion, aquatic toxicity, abiotic depletion, and so on.

4) If a circular economy framework is implemented (Debnath et al., 2018), it will enhance resource circulation and reduce environmental burden. Hence, EC recycling (Debnath et al., 2016) through a proper channel will enhance CE establishment possibilities.

9.5.2.2 Economic Sustainability

The most important aspect of sustainability is the economic aspect of the business perspective. The economics of e-waste management is a debatable issue. The primary revenue is generated via metal recycling. The electronic items are becoming lighter in recent times as the metal portion decreases and the polymer portion increases (Debnath, 2020). As a result, the market volatility has increased, and the focus has been on other parts of e-waste. We present a generalized analysis from an economic aspect, and the significant points are listed here:

1) The security threats arising from e-waste are executed by people of different ages with a singular motive of earning money. The overall benefit of e-waste as a resource is that it contains different types of metals and materials. On the one hand, the RE techniques can pose security threats if it ends up in the wrong hands; on the other hand, RE can increase the repairability by redesigning ICs.

2) Reuse of ECs and ICs, wherever possible, will lead to circular economy establishment. A CE model can ensure resource circulation, provided the SCN is quite efficient.

3) With proper policy implementation, the unauthorized repairing shops and other relevant stakeholders can be licensed. This will reduce security issues without disrupting the economics and livelihood of the people concerned.

4) As presented in the literature (Debnath, Roychowdhury et al., 2016; Debnath, Chowdhury et al., 2018), the CE ensures that ECs are tested and reused wherever possible. The testing can be either in-house, or it could be outsourced. This can give rise to specialized services that will only carry out high-precision testing, which will be good from an economic perspective.

5) RE can be considered a potential technological solution for the future, proliferating differently depending on the situation.

9.5.2.3 Demand Sustainability

Demand is an essential driver of any SCN that fuels the driving force for business. Demand is considered as the fourth pillar of sustainability. In this case, it is imperative as the security threats arise from different parts of the supply chain. A detailed analysis of security threats from the supply chain perspective is presented in recent literature (Debnath et al., 2019; Debnath et al., 2020b). Our analysis of demand sustainability reveals the following points:

i) A key aspect is the EC recycling supply chain, a subnetwork of the overall complex network. The demand sustainability will be highly dependent on the efficiency of the EC recycling supply chain.

ii) There is a demand for reused ECs for low-cost toy manufacturing in countries like China. This ignites the SCN, which includes several unscrupulous activities mentioned in literature (Debnath et al., 2016; Roychowdhury et al., 2019).

iii) New industries such as testing industries are expected to be mushrooming. Additionally, companies in the SCN that supply parts to the manufacturers and refurbishers will also be under the demand game's umbrella.

9.5.2.4 Social Sustainability

Social sustainability is a very complex thing as it includes several aspects such as religion, mentality, and behavior. These are also variables based on geographical location and local beliefs. Since social LCA is beyond the scope of this chapter, we present a generalized discussion on the basis of social indicators described by United Nations (DiSano, 2002)—

i) Awareness related to e-waste is a significant issue worldwide, particularly in developing nations. Security threat associated with e-waste is a new and emerging issue that requires immediate attention and can only be achieved through proper awareness campaigns.

ii) A better and circular e-waste management system ensures higher job opportunities. As discussed earlier, the proliferation of new industries such as high-precision testing, parts supplier, and RE will provide job opportunities.

iii) Utilization of RE and EC testing as a service will help increase repairability. This will be an intangible benefit for society.

iv) As mentioned before, for environmental and economic sustainability, policy implementation is essential. Exemplary policy implementation will ensure a better opportunity for the people associated with repair and refurbish business to authorize their business. Such policy implementations will enhance their social status and legalize their business, providing economic proliferation opportunities.

9.6 CONCLUSION

Approximately 53.6 million metric tons (Mt) of e-waste have been produced worldwide, contributing to 7.3 kg per capita in the year 2019 alone. The global generation of e-waste is projected to grow to 74.7 Mt by 2030—almost doubling in only 16 years. In addition to the rising e-waste, concerns about these discarded technologies' security threats are a concern (Jones et al., 2017; Debnath, 2019; Diesburg et al., 2016), yet less explored. To understand the current state-of-the-art research in the field of e-waste security, we conducted a detailed systematic review, which is reported in this chapter. This systematic literature review evaluated and analyzed the research trends on often neglected security risks arising from the remnant data in e-waste. We found that security threats stemming from the disposed or resold electronic device— the foremost reason being a lack of awareness among consumers— is often a neglected notion compared to environmental or other issues ascribing to e-waste. Our study on several research works revealed that confidential and sensitive data could be recovered from disposed-of electronic storage devices if not appropriately removed (Lim et al., 2014; Jones et al., 2010, 2009, 2008). The majority of the recommendations made by the researchers were to properly sanitize storage devices before disposing or reselling and growing consumer awareness (Jones et al., 2016; Valli, 2004; Szewczyk et al., 2018). Still, we found fewer usability studies on the data removal tools that were suggested for device sanitization. It is essential to understand and evaluate the user perspective and the feasibility of such technological solutions. Future work can be done to develop a data removal tool with a high concentration on User Interface, as Gutmann and Warner suggested (Gutmann and Warner, 2019).

ACKNOWLEDGMENT

We would also like to acknowledge the research institutions and labs of the researchers involved with this chapter—Secure and Privacy Research in New-Age Technology (SPRINT) Lab, University of Denver, Colorado, USA; Chemical Engineering Department, Jadavpur University, India; and Department of Mathematics, Aston University, UK. Any opinions, findings, and conclusions or recommendations expressed in this material are solely those of the author(s).

REFERENCES

Agayev, B., S. Mehdiyev, and T. Aliyev. (2016). Electronic information carrier as an object of information security. *Problems of Information Society 7*(1), 41–49.

Ahmed, S. I., M. R. Haque, J. Chen, and N. Dell. (2017). Digital privacy challenges with shared mobile phone use in Bangladesh. *Proceedings of the ACM on Human-Computer Interaction 1*(CSCW), 1–20.

Alghazo, J., and O. K. Ouda. (2016). Electronic waste management and security in GCC countries: a growing challenge. In *2nd International Conference on Integrated Environmental Management for Sustainable Development (ICIEM)*.

Alghazo, J., O. K. Ouda, and A. El Hassan. (2018). E-waste environmental and information security threat: GCC countries vulnerabilities. *Euro-Mediterranean Journal for Environmental Integration 3*(1), 13.

Arthur, C. (2009). Before you sell your computer, smash the hard drive, says which. *The Guardian.* www.theguardian.com/technology/2009/jan/08/hard-drive-security-which.

Baidya, R., B. Debnath, S. K. Ghosh, and S.-W. Rhee. (2020). Supply chain analysis of e-waste processing plants in developing countries. *Waste Management & Research 38*(2), 173–183.

Bennison, P. F., and P. J. Lasher. (2004). Data security issues relating to end of life equipment. In *IEEE International Symposium on Electronics and the Environment, 2004. Conference Record, 2004*, pp. 317–320. IEEE.

Bhuie, A., O. Ogunseitan, J.-D. Saphores, and A. Shapiro. (2004). Environmental and economic trade-offs in consumer electronic products recycling: a case study of cell phones and computers. In *IEEE International Symposium on Electronics and the Environment, 2004. Conference Record, 2004*, pp. 74–79. IEEE.

Chattopadhyay, A. K., B. Debnath, R. El-Hassani, S. K. Ghosh, and R. Baidya. (2020). *Cleaner Production in Optimized Multivariate Networks: Operations Management through a Roll of Dice.* arXiv preprint arXiv:2003.00884.

Collard, S. (2017, February). *Is Buying a Second-Hand Phone Worth the Risk?* www.abc.net.au/news/2017-02-21/is-it-worth-investing-in-a-second-hand-smartphone/890044 (accessed January 28, 2021).

Conacher, J., K. Renaud, and J. Ophoff. (2020). *Caveat Venditor, Used USB Drive Owner.* arXiv:2006.11354v1.

Das, S. (2020a). Eyes in your child's bedroom: Exploiting child data risks with smart toys. In *USENIX Conference.* www.usenix.org/conference/enigma2020/presentation/das (accessed January 28, 2021)

Das, S. (2020b). *A Risk-Reduction-Based Incentivization Model for Human-Centered Multi-Factor Authentication.* Ph.D. Thesis, Indiana University.

Das, S., J. Abbott, S. Gopavaram, J. Blythe, and L. J. Camp. (2020a). User-centered risk communication for safer browsing. In *International Conference on Financial Cryptography and Data Security*, pp. 18–35. Springer.

Das, S., J. Dev, and L. J. Camp. (2019a). *Privacy Preserving Policy Framework: User-Aware and User-Driven.* Available at SSRN 3445942.

Das, S., J. Dev, and K. Srinivasan. (2018a). Modularity is the key a new approach to social media privacy policies. In *Proceedings of the 7th Mexican Conference on Human-Computer Interaction*, pp. 13. ACM.

Das, S., A. Dingman, and L. J. Camp. (2018b). Why johnny doesn't use two factor a two-phase usability study of the fido u2f security key. In *2018 International Conference on Financial Cryptography and Data Security (FC).*

Das, S., J. Goard, and D. Murray. (2017). How celebrities feed tweeples with personal and promotional tweets: celebrity twitter use and audience engagement. In *Proceedings of the 8th International Conference on Social Media & Society*, pp. 1–5.

Das, S., A. Kim, and L. J. Camp. (2021). Organizational security: implementing a risk-reduction-based incentivization model for MFA adoption. In *Proceedings of the International Conference on Financial Cryptography and Data Security.*

Das, S., A. Kim, B. Jelen, L. Huber, and L. J. Camp. (2020b). Non-inclusive online security: older adults' experience with two-factor authentication. In *Proceedings of the 54th Hawaii International Conference on System Sciences.*

Das, S., A. Kim, B. Jelen, J. Streiff, L. J. Camp, and L. Huber. (2020c). *Why Don't Older Adults Adopt Two-Factor Authentication?* In *Proceedings of the 2020 SIGCHI Workshop on Designing Interactions for the Ageing Populations-Addressing Global Challenges.*

Das, S., A. Kim, and S. Karmakar. (2020d). *Change-Point Analysis of Cyberbullying-Related Twitter Discussions During Covid-19.* arXiv preprint arXiv:2008.13613.

Das, S., A. Kim, Z. Tingle, and C. Nippert-Eng. (2019b). *All About Phishing: Exploring User Research through a Systematic Literature Review.* arXiv preprint arXiv:1908.05897.

Das, S., S. Mare, and L. J. Camp. (2020e). Smart storytelling: video and text risk communication to increase MFA acceptability. In *2020 IEEE Humans and Cyber Security Workshop (HACS 2020) in Association with the 6th IEEE International Conference on Collaboration and Internet Computing (CIC).* IEEE.

Das, S., J. Streiff, L. L. Huber, and L. J. Camp. (2019c). Why don't elders adopt two-factor authentication? Because they are excluded by design. *Innovation in Aging 3*(Suppl 1), S325.

Das, S., B. Wang, and L. J. Camp. (2019d). MFA is a waste of time! understanding negative connotation towards MFA applications via user generated content. In *Proceedings of the Thirteenth International Symposium on Human Aspects of Information Security & Assurance (HAISA 2019).*

Das, S., B. Wang, A. Kim, and L. J. Camp. (2020f). MFA is a necessary chore!: exploring user mental models of multi-factor authentication technologies. In *Proceedings of the 53rd Hawaii International Conference on System Sciences.*

Das, S., B. Wang, Z. Tingle, and L. J. Camp. (2019e). *Evaluating User Perception of Multi-Factor Authentication: A Systematic Review.* arXiv preprint arXiv:1908.05901.

Debnath, B. (2019). Sustainability of WEEE recycling in India. In *Re-Use and Recycling of Materials Solid Waste Management and Water Treatment*, pp. 15–32. River Publishers.

Debnath, B. (2020). Towards sustainable e-waste management through industrial symbiosis: a supply chain perspective. In *Industrial Symbiosis for the Circular Economy*, pp. 87–102. Springer.

Debnath, B., J. M. Alghazo, G. Latif, R. Roychoudhuri, and S. K. Ghosh. (2020a). An analysis of data security and potential threat from IT assets for middle card players, institutions and individuals. In *Sustainable Waste Management: Policies and Case Studies*, pp. 403–419. Springer.

Debnath, B., R. Chowdhury, and S. Ghosh. (2018). Towards circular economy in e-waste recycling via metal recovery from e-waste (MREW) facilities. In *Proceedings of ISWA 2018 World Congress*, pp. 22–24. ISWA.

Debnath, B., A. Das, and A. Das. (2021). Towards circular economy in e-waste management in India: issues, challenges and solutions. In *Circular Economy and Sustainability*, vol. II. Elsevier (In Press).

Debnath, B., A. Das, S. Das, and A. Das. (2020b). Studies on security threats in waste mobile phone recycling supply chain in India. In *2020 IEEE Calcutta Conference (CALCON)*, pp. 431–434. IEEE.

Debnath, B., S. Das, and A. Das. (2019a). Study exploring security threats in waste phones a life cycle based approach. In *2019 IEEE Smart World, Ubiquitous Intelligence Computing, Advanced Trusted Computed, Scalable Computing Communications, Cloud Big Data Computing, Internet of People and Smart City Innovation*. IEEE.

Debnath, B., S. Das, A. Das, and A. Das. (2019b). Smart devices for smart cities in India exploring security vulnerabilities of e-waste management. In *1st international Workshop on Security Issues in the World of Smart Cities (WSIWSC 2019)*. Institute of Electrical and Electronics Engineers Inc.

Debnath, B., R. El-Hassani, A. K. Chattopadhyay, T. K. Kumar, S. K. Ghosh, and R. Baidya. (2022). Time evolution of a Supply Chain Network: kinetic modeling. *Physica A: Statistical Mechanics and its Applications 607*, 128085.

Debnath, B., and S. K. Ghosh. (2019). Hydrogen generation from biorefinery waste: recent advancements and sustainability perspectives. In *Waste Valorisation and Recycling*, pp. 557–572. Springer.

Debnath, B., P. Roychowdhury, and R. Kundu. (2016). Electronic components (EC) reuse and recycling–a new approach towards WEEE management. *Procedia Environmental Sciences 35*, 656–668.

Dev, J., S. Das, and L. J. Camp. (2018). Privacy practices, preferences, and compunctions: WhatsApp users in India. In *HAISA*, pp. 135–146. University of Plymouth.

Dev, J., S. Das, Y. Rashidi, and L. J. Camp. (2019). *Personalized WhatsApp Privacy: Demographic and Cultural Influences on Indian and Saudi Users*. Available at SSRN.

Diesburg, S., C. Feldhaus, M. A. Fardan, J. Schlicht, and N. Ploof. (2016). Is your data gone? measuring user perceptions of deletion. In *Proceedings of the 6th Workshop on Socio-Technical Aspects in Security and Trust*, pp. 47–59. doi:10.1145/3046055.3046057

DiSano, J. (2002). *Indicators of Sustainable Development: Guidelines and Methodologies*. United Nations Department of Economic and Social Affairs. www.un.org/esa/sustdev/publications/indisd-mg2001.pdf.

Doyon-Martin, J. (2015). Cybercrime in west Africa as a result of transboundary e-waste. *Journal of Applied Security Research 10*(2), 207–220.

Duzgun, R., P. Mayer, S. Das, and M. Volkamer. (2020). Towards secure and usable authentication for augmented and virtual reality head-mounted displays. In *Who Are You?! Adventures in Authentication Workshop (WAY), Co-Located with 16th Symposium on Usable Privacy and Security (SOUPS 2020), August 7-11, 2020*.

Ferreras-Fernandez, T., H. Martın-Rodero, F. J. Garcıa-Penalvo, and J. A. Merlo-Vega. (2016). The systematic review of literature in LIS: an approach. In *Proceedings of the Fourth International Conference on Technological Ecosystems for Enhancing Multiculturality*, pp. 291–296.

Forgor, L., W. Brown-Acquaye, J. K. Arthur, and S. Owoo. (2019). Security of data on e-waste equipment to Africa: the case of Ghana. In *2019 International Conference on Communications, Signal Processing and Networks (ICCSPN)*, pp. 1–5. IEEE.

Forti, V., C. P. Baldé, R. Kuehr, and G. Bel. (2020). *The Global E-waste Monitor 2020: Quantities, Flows and the Circular Economy Potential.* United Nations University (UNU)/United Nations Institute for Training and Research (UNITAR)—Co-Hosted SCYCLE Programme, International Telecommunication Union (ITU) & International Solid Waste Association (ISWA).

Fundo, A., A. Hysi, and I. Tafa. (2014). Secure deletion of data from SSD. *International Journal of Advanced Computer Science and Applications 5*(8), 131–134. doi:10.14569/IJACSA.2014.050820

Garg, A., S. Chakraborty, M. Malik, D. Kumar, S. Singh, and M. Suri. (2020). *Investigation of Data Deletion Vulnerabilities in NAND Flash Memory-Based Storage.* arXiv preprint arXiv:2001.07424.

Ghosh, A., B. Debnath, S. K. Ghosh, B. Das, and J. P. Sarkar. (2018). Sustainability analysis of organic fraction of municipal solid waste conversion techniques for efficient resource recovery in India through case studies. *Journal of Material Cycles and Waste Management 20*(4), 1969–1985.

Glisson, W. B., T. Storer, A. Blyth, G. Grispos, and M. Campbell. (2016). *In the Wild Residual Data Research and Privacy.* arXiv preprint arXiv:1610.03229.

Gopavaram, S. R., J. Dev, S. Das, and J. Camp. (2019). *IoT Marketplace: Informing Purchase Decisions With Risk Communication.*

Graham, L. (2017). Cybercrime costs the global economy $450 billion: CEO. *CNBC.* www.cnbc.com/2017/02/07/cybercrime-costs-the-global-economy-450-billion-ceo.html (accessed January 28, 2021).

Grand, J. (2014). Printed circuit board deconstruction techniques. In *WOOT 2014.* www.usenix.org/system/files/conference/woot14/woot14-grand.pdf (accessed January 28, 2021).

Grant, R., and M. Oteng-Ababio. (2012). Mapping the invisible and real "African" economy: urban e-waste circuitry. *Urban Geography 33*(1), 1–21.

Guha, S., M. R. Rifat, F. H. Shezan, N. Dell, et al. (2017). Privacy vulnerabilities in the practices of repairing broken digital artifacts in Bangladesh. *Information Technologies and International Development 13*(Special Section), 186–199.

Gumbo, M., and K. Kalegele. (2015). E-waste management: awareness, strategies, facilities, sources and treatment in Tanzania. In *Information and Knowledge Management*, vol. 5, pp. 7–28. Citeseer.

Gutmann, A., and M. Warner. (2019). Fight to be forgotten: exploring the efficacy of data erasure in popular operating systems. In *Annual Privacy Forum*, pp. 45–58. Springer.

Hadan, H., N. Serrano, S. Das, and L. J. Camp. (2019). Making IoT worthy of human trust. *SSRN Electronic Journal 47.*

Homaidi, O. A. (2009). *Data Remanence: Secure Deletion of Data in SSDs.* Master Thesis. School of Computing Blekinge Institute of Technology. www.diva-portal.org/smash/get/diva2:832529/FULLTEXT01.pdf (accessed January 28, 2021)

Hughes, G., and T. Coughlin. (2006). *Tutorial on Disk Drive Data Sanitization.* cmrr.ucsd.edu/people/Hughes/DataSanitizationTutorial.pdf.

Imperatives, S. (1987). *Report of the World Commission on Environment and Development: Our Common Future.* (accessed February 10).

Jones, A. (2005). How much information do organizations throw away? *Computer Fraud & Security 2005*(3), 4–9.

Jones, A., O. Angelopoulou, and L. Noriega. (2019). Survey of data remaining on second hand memory cards in the UK. *Computers & Security 84*, 239–243.

Jones, A., O. Angelopoulou, S. Vidalis, and H. Janicke. (2017). The 2016 hard disk study on information available on the second-hand market in the UK. In *European Conference on Cyber Warfare and Security*, pp. 193–199. Academic Conferences International Limited.

Jones, A., T. Martin, and M. Alzaabi. (2016). The 2016 analysis of information remaining on computer hard disks offered for sale on the second hand market in the UAE. *Journal of Digital Forensics, Security and Law 11*(4), 6.

Jones, A., V. Mee, C. Meyler, and J. Gooch. (2005). Analysis of data recovered from computer disks released for resale by organisations. *Journal of Information Warfare 4*(2), 45–53.

Jones, A., C. Valli, and G. Dabibi. (2009). The 2009 analysis of information remaining on USB storage devices offered for sale on the second-hand market. In *7th Australian Digital Forensics Conference*. Edith Cowan University. doi:10.4225/75/57b2836b40cca

Jones, A., C. Valli, G. S. Dardick, I. Sutherland, G. Dabibi, and G. Davies. (2010). The 2009 analysis of information remaining on disks offered for sale on the second hand market. *Journal of Digital Forensics, Security and Law 5*(4), 3.

Jones, A., C. Valli, and I. Sutherland. (2008). Analysis of information remaining on hand held devices offered for sale on the second hand market. *The Journal of Digital Forensics, Security and Law: JDFSL 3*(2), 55.

Joshi, S., K. Stavrianakis, and S. Das. (2020). Substituting restorative benefits of being outdoors through interactive augmented spatial soundscapes. In *The 22nd International ACM SIGACCESS Conference on Computers and Accessibility*, pp. 1–4. ACM. doi:10.1145/3373625.3418029

Karlson, A. K., A. B. Brush, and S. Schechter. (2009). Can I borrow your phone? understanding concerns when sharing mobile phones. In *Proceedings of the SIGCHI Conference on Human Factors in Computing Systems*, pp. 1647–1650. ACM. doi:10.1145/1518701.1518953

Karmakar, S., and S. Das. (2020). Evaluating the impact of covid-19 on cyberbullying through Bayesian trend analysis. In *Proceedings of the European Interdisciplinary Cybersecurity Conference*, pp. 1–6. ACM. doi:10.1145/3424954.3424960

Karmakar, S., and S. Das. (2021). Understanding the rise of twitter-based cyberbullying due to covid-19 through comprehensive statistical evaluation. In *Proceedings of the 54th Hawaii International Conference on System Sciences*.

Khramova, M., and S. Martinez. (2018). Analysis of data remanence after factory reset, and sophisticated attacks on memory chips. In *Smartphones: Repair, Remanufacturing and Reuse of Components at the 2018 CARE Innovation*.

Krumay, B. (2016). The e-waste-privacy challenge. In *Annual Privacy Forum*, pp. 48–68. Springer.

Lee, B., K. Son, D. Won, and S. Kim. (2011). Secure data deletion for USB flash memory. *J. Inf. Sci. Eng. 27*(3), 933–952.

Lim, C., I. Firdausi, and A. Bresnev. (2014). Forensics analysis of corporate and personal information remaining on hard disk drives sold on the secondhand market in Indonesia. *Advanced Science Letters 20*(2), 522–525.

Majam, T., and F. Theron. (2006). The purpose and relevance of a scientific literature review: a holistic approach to research. *Journal of Public Administration 41*(3), 603–615.

Medlin, B. D., and J. A. Cazier. (2010). A study of hard drive forensics on consumers' PCs: data recovery and exploitation. *Journal of Management Policy and Practice 12*(1), 27–35.

Momenzadeh, B., S. Gopavaram, S. Das, and L. J. Camp. (2020). Bayesian evaluation of user app choices in the presence of risk communication on android devices. In *International Symposium on Human Aspects of Information Security and Assurance*, pp. 211–223. Springer.

Noman, A. S. M., S. Das, and S. Patil. (2019). Techies against Facebook: understanding negative sentiment toward Facebook via user generated content. In *Proceedings of the 2019 CHI Conference on Human Factors in Computing Systems*, pp. 468. ACM.

Puckett, J., and T. Smith. (2002). Exporting harm: the high-tech trashing of Asia the Basel action network. In *Silicon Valley Toxics Coalition*. Basel Action Network.

Quadir, E. S., J. Chen, D. Forte, N. Asadizanjani, S. Shahbazhohmadi, L. Wang, J. Chandy, and M. Tehranipoor. (2016). A survey on chip to system reverse engineering. *ACM Journal on Emerging Technologies in Computing Systems (JETC) 13*(1), 1–34.

Raman, R., and D. Pramod. (2013). A study on data privacy, protection & sanitization practices during disk disposal by Indian educational institutes. *International Journal of Computer Science Issues (IJCSI) 10*(2 Part 1), 53.

Robins, N., P. A. Williams, and K. Sansurooah. (2016). I know what you did last summer . . . an investigation into remnant data on USB storage devices sold in Australia in 2015. In *Proceedings of the Australasian Computer Science Week Multi Conference*, pp. 1–8. ACM. doi:10.1145/2843043.2843356

Roychowdhury, P., J. Alghazo, B. Debnath, S. Chatterjee, and O. Ouda. (2019). Security threat analysis and prevention techniques in electronic waste. In *Waste Management and Resource Efficiency*, pp. 853–866. Springer.

Sansurooah, K., and P. Szewczyk. (2012). A study of remnant data found on USB storage devices offered for sale on the Australian second-hand market in 2011. In *10th Australian Information Security Management Conference*, vol. 82. Citeseer.

Saur-Amaral, I. (2011). Towards a methodology for literature reviews in social sciences. *investigaçao e Intervençâo em Recursos Humanos* (3), 1–10.

Services., S. L. (2013). *IT Asset Disposal, Version 2.0.* (accessed February 2, 2020).

Shiloh, J., and A. Fassassi. (2016). *Cybercrime in Africa: Facts and Figures*. www.scidev. net/sub-saharan-africa/features/cybercrime-africa-facts-figures/ (accessed January 28, 2021).

Shu, J., Y. Zhang, J. Li, B. Li, and D. Gu. (2017). Why data deletion fails? A study on deletion flaws and data remanence in android systems. *ACM Transactions on Embedded Computing Systems (TECS) 16*(2), 1–22.

Sinha, D. (2004). *The Management of Electronic Waste: A Comparative Study on India and Switzerland*. St. Gallen7 University of St. Gallen.

Storer, T., W. B. Glisson, and G. Grispos. (2010). Investigating information recovered from re-sold mobile devices. In *Privacy and Usability Methods Pow-wow (PUMP) Workshop*, p. 2. ACM, University of Abertay.

Streiff, J., S. Das, and J. Cannon. (2019). Overpowered and underprotected toys empowering parents with tools to protect their children. In *2019 IEEE 5th International Conference on Collaboration and Internet Computing (CIC)*, pp. 322–329. IEEE.

Streiff, J., O. Kenny, S. Das, A. Leeth, and L. J. Camp. (2018). Who's watching your child? exploring home security risks with smart toy bears. In *2018 IEEE/ACM Third International Conference on Internet-of-Things Design and Implementation (IoTDI)*, pp. 285–286. IEEE.

Szewczyk, P. (2011). Analysis of data remaining on secondhand ADSL routers. *Journal of Digital Forensics, Security and Law 6*(3), 3.

Szewczyk, P., and K. Sansurooah. (2012). The 2012 investigation into remnant data on second-hand memory cards sold in Australia. In *10th Australian Digital Forensics Conference, Novotel Langley Hotel, Perth, Western Australia, 3-5 December, 2012*. doi:10.4225/75/57b3d443fb870

Szewczyk, P., K. Sansurooah, and P. A. Williams. (2018). An Australian longitudinal study into remnant data recovered from second-hand memory cards. *International Journal of Information Security and Privacy (IJISP) 12*(4), 82–97.

Unchit, P., S. Das, A. Kim, and L. J. Camp. (2020). Quantifying susceptibility to spear phishing in a high school environment using signal detection theory. In *International Symposium on Human Aspects of Information Security and Assurance*, pp. 109–120. Springer.

Valli, C. (2004). Throwing out the enterprise with the hard disk. In *Australian Computer, Network & Information Forensics Conference*, pp. 124–129.

Valli, C., and A. Jones. (2005). A UK and Australian study of hard disk disposal. In *Proceedings of 3rd Australian Computer, Network & Information Forensics Conference*, pp. 74–78. https://ro.ecu.edu.au/cgi/viewcontent.cgi?article=3762&context=ecuworks (accessed January 28, 2021).

Valli, C., and A. Woodward. (2007). Oops they did it again: the 2007 Australian study of remnant data contained on 2nd hand hard disks. In *5th Australian Digital Forensics Conference*. Edith Cowan University. doi:10.4225/75/57ad5aac7ff30

Valli, C., and A. Woodward. (2012). *The 2008 Australian Study of Remnant Data Contained on 2nd Hand Hard Disks: The Saga Continues.* arXiv preprint arXiv:1207.6025.

Von Zezschwitz, E., and A. Hang. (2012). Towards privacy-aware mobile device sharing. In *4th International Workshop on Security and Privacy in Spontaneous Interaction and Mobile Phone Use (IWSSI/SPMU)*.

Wang, D., J. Tang, M. Jia, Z. Xu, and H. Han. (2020). Review of NAND flash information erasure based on overwrite technology. In *2020 39th Chinese Control Conference (CCC)*, pp. 1150–1155. IEEE.

Widmer, R., H. Oswald-Krapf, D. Sinha-Khetriwal, M. Schnellmann, and H. Böni. (2005). Global perspectives on e-waste. *Environmental Impact Assessment Review 25*(5), 436–458.

Yeboah-Boateng, E. O. (2012). Using fuzzy cognitive maps (FCMs) to evaluate the vulnerabilities with ICT assets disposal policies. *International Journal of Electrical & Computer Sciences IJECS-IJENS 12*(5), 20–31.

Zhang, P., W. Zhang, S.-Z. Niu, and Z.-P. Huang. (2014). User level secure deletion for USB flash disks. In *The 2014 2nd International Conference on Systems and Informatics (ICSAI 2014)*, pp. 1072–1077. IEEE.

Zingerle, A., and L. Kronman. (2019). Information diving on an e-waste dump in west Africa–artistic remixing of a global data breach. In *Conference on Computation, Communication, Aesthetics X*.

10 Electronic Component Recycling
Solutions toward Green Development

Indrashis Saha, Rayan Kundu and Saswati Gharami

CONTENTS

10.1 INTRODUCTION

E-waste typically includes a wide variety of electronic products from extensive range of domestic equipment like refrigerators, air conditioners, cell telephones, stereos, and consumer goods. The e-waste law e-waste. The guidelines complied with the guidelines laid down in the Guideline on Electrical and Technological Waste for all EEE in one of ten category groups (Spitczok von Brisinski et al. 2014).

DOI: 10.1201/9781003301899-14

E-waste consists of over 1,000 compounds, all harmful and contaminating when recycled, but this is not well documented. When e-waste is landfilled or incinerated, there are significant radiation risks. Drainage leach chemicals and incinerators release harmful substances, including dioxins, into the setting. Moreover, e-waste recycling has a significant effect on work and the atmosphere, primarily when the recycling industry is dedicated to majorly optimizing benefits and does take much interest to preserve the environment and the well-being of its employees (Niu et al. 2017a). An e-waste recycling organization needs to be established not just because e-waste impacts people's lives adversely but also because it is a future business proposition with positive viewpoints. It can be easily found that e-waste includes rare metals and noble metals if metals are associated with e-waste. This means that e-waste is a rich, rare metal resource, and its careful treatment can help protect the climate and guarantee that social and financial benefits continue to grow.

The manufacturing of PCBs for virtually all EEE is the backbone of the electronics industry. Over the last few years, overall global PCB output in South-East Asia and mainland China has increased by 10.8% and 8.7%, respectively. (14.4 percent). Currently, China is extracting Rare Earth Metals (REMs) because of the possible low production output of them in future, and recycling is banned because of their unusual physical and chemical properties. Several dangerous products including heavy metals, PVCs, and flame retardants (BFR) can be safely identified in the typical PCBs. So proper recycling of Rare Earth Elements (REEs) from pre-consumers is the only way to build a green, low-carbon economy parallel to meeting the demands of the growing market.

Chemical and mechanical processes are two conventional waste PCB disposal methods (Kaya 2016). Kaya (2016) presented a comprehensive review of various physico-chemical and metallurgical processes of recycling e-wastes and their advantages and disadvantages for a transition to green e-waste utilization, with a special focus on the extraction of both metallic and non-metallic substances. Pyrolysis, oxidation, hydration, and electrolysis are commonly used chemical processes. Because of their low costs and simple operations, most waste PCBs are usually processed in the backyards or small workshops through primary processes such as open burning or acid washing. Dioxin and furan releases can also contribute to coordinated combustion and pyrometallurgy. During the recycling process, significant amounts of liquid waste acid would be generated and carefully should be removed for hydration and electrolysis. A significant number of researchers have applied different mechanical methods for extracting metals from PCBs, including isolation of the structure, jigging, and density separation. However, these tests take time, waste, and air pollution (Chen et al. 2020). Several literatures on the recycling of PCBs as a whole were reported (Luda 2011; Kaya 2016; Kaya and Sözeri 2019), but only few cases of recovery of precious metals from these electronic components of PCBs are executed and only few literatures are available on the recovery of these precious metals from all the individual components in a PCB (Ilyas et al. 2021; Debnath et al. 2016a, b).

The largest obstacle for biodiversity is the practice of informal recycling (Widemar et al. 2005; Dutta and Goel 2021; Ohajinwa et al. 2017). The e-waste export standard for individual collectors facilitates the growth of the broad informal recycling industry. Informal recycling operations can occur in small sheds, or even in the backyards, through manual removal and free combustion. Many of the times, in order to recover metals, tablets are also soaked in acid baths in order to remove gold or other metals. Acid baths, if dumped into seawater, have significant impact on infants, in the form of elevated amounts of lead in blood, with dirt and heavy metals in the water. These acts endanger not only the persons engaging in this activity, but also other residents and the environment (Charles et al. 2017).

Simple mechanisms for retrieving electrical components from circuit boards such as decommissioning, disassembling, and desoldering methods may be used to recover electronics from the household equipment/appliances explored in this study. The present literature presents a new-age review of proper dismantling/desoldering of ECs from PCBs and of proper treatment and recycling techniques of ECs to recover the precious metals. This report analyzed the types of waste in the countryside and specifically described its characteristics in terms of removal and disposal and to quantify the sum of electric components that can be recovered from e-waste. The e-waste shape, volume, and chemical composition are reviewed corresponding to the treatment technologies implemented, and the usability/functionality of the reclaimed goods will be determined (Kaya 2016).

10.2 METHODOLOGY

The study is an exhaustive review of electronic component recycling. Literature review followed by a critical analysis was adopted for the purpose. Since there are different types of electronic components (ECs) mounted on PCBs such as capacitors, resistors, diodes, and transistors, scientific databases were explored in combination with aforementioned names and keywords such as "e-waste", "recycling", "recovery", "green solvent", "green chemistry", "circular economy", etc. Thereafter, industry aspects were also explored for EC recycling. Finally, possible solutions were proposed along with future directions.

10.3 ELECTRONIC PART RETRIEVAL AND MONITORING

The dismantling and disassembling of plastic tubes, metals, wires, and electrical components such as diodes, resistors, condensers, and ICs can be obtained from the equipment and electronic devices. Components of electronics should be separated from the PCBs by desoldering each unit (Barnwal and Dhawan 2020). Later on, the computer components obtained can be checked to figure out if they are still good or faulty. Only the optical multimeter is used to measure the recovered electronic components, and the assessment of whether the part is adequate or faulty is decided by focusing solely on capacitance, inductivity, resistance, open circuits, or short circuits between component conduits. Some other factors such as estimated gain and the evaluation do not involve linearity, voltage breakdown, and other electrical parameters.

10.3.1 Classification of Various Electronic Components

The production sector in electronics is one of the world's fastest developing markets. Parallelly, the aggressive publicity combined with the frequent lure of new designs or technologies in the electronics sector allows certain consumer electronics to become obsolescent (Barnwal and Dhawan 2020). Global e-waste crossed more than 53.6 million tons in 2019 and develops around 4% to 5% year-on-year. E-waste has thus been a significant social and environmental issue for many countries today (Chen et al. 2020).

10.3.1.1   Integrated Circuits

The integrated circuits (ICs) are miniaturized electronic circuits assembled on the surface of a thin substratum of semiconductors used as central electronic components installed on PCBs. Due to the advantages of low cost, compact size, and durability, ICs were commonly used in all electronic products, and they revolutionized the electronic environment. According to the Chinese National Statistical Bureau, the number of ICs transported into China in 2018 amounted to 417.6 billion, up by 10.8% year-on-year. Not unexpectedly, significant volumes of waste ICs are frequently removed from the e-waste stream by removing electrical equipment (Barnwal and Dhawan 2020). The global market value of transportation of ICs was $43,240 million in 2017 and is expected to get doubled by 2025 (Ajiboye et al. 2020). As a semiconductor of the second generation, gallium arsenide has electron mobility five to six times higher than silicone. GaAs-based ICs are commonly use in certain places where there are no productive alternatives because of their unique characteristics. According to the United States Geological Study, the United States' GaAs intake in 2017 is grossly 800 tons, and ICs accounted for 70% of US domestic intake of gallium (Zhang et al. 2019). Successful isolation of plastics and metals before hydrometallurgical or pyrometallurgical processing aids in reducing acid usage significantly and the production of leach residuals and slag and the number of several processing phases.

10.3.1.2   RAM

To detect historical temporal patterns in module participants emerging from shifts in development processes, linear regression analysis of the compositional data is organized in compliance with sample chronology and predict future trends to use them to determine future recycling capacity more accurately. It has been demonstrated that dynamic random access memory (DRAM) is "high quality" with steady amounts of gold and silver over time, an 80% decline in palladium in 1991–2008, and a rise in the copper content of 0.23 g/module/year, with a 75% increase expected in 2008 by 2020. Due to possible shifts in future equipment participation, the quality of future recycling potential predictions for e-waste using existing methods centered on standardized compositional data from current devices is doubtful. The effect on recycling of waste laptops, smartphones, cell phones, and tablets from Europe 2020, a 75% rise in copper content, on current estimates utilizing static compositional results, is considered (Charles et al. 2017). In recent years, there has been an increasing demand for precious rare earth metal (REM) because of the anticipation of lack of precious metals in future, leading to concerns over the security of resources globally. Data on time-dependent trends in precious metal and copper content is an important perspective to analyze and verify these concerns in a global platform. Charles et al. (2017) presented a review of data through a comprehensive analysis of recovery of REMs, with the rise of copper content on DRAM.

10.3.1.3 Transistor

Recent demand for low-cost multifunctional electronic equipment has stimulated the curiosity in organic thin-film transistors or OTFTs. The usage of OTFTs as chemical sensors has been promising in producing electronic noses and the identification of nerve agents (Elkington et al. 2014). The long-term reliability and credibility of the device in environmental working conditions are critical problems in the overall development of OTFT. Pentacene OTFTs earned considerable exposure to the instability of environmental components such as oxygen and humidity between the tiny molecular-based OTFTs. Several processes have been suggested to justify this volatility in pentacene OTFTs, including the adsorption of water on grain borders and impurities induced by oxygen (Park et al. 2009). The surface morphology of pentacene films has had a significant impact on air pollution during these processes. For thiophene-based OTFTs, similar correlations were recorded between surface morphology and sensor response. Metal phthalocyanines (MPcs), another well-studied category of organic semiconductors, have been applied for modern OTFT-related sensors.

Nevertheless, a small number of OTFT degradation experiments have been published, where MPcs are used as the active organic substrate. An approach to isolating the cause of system deterioration "aging" in Cu Pc OTFTs may provide insight into MPc films' experiences with atmospheric active ingredients (Park et al. 2009).

10.3.1.4 Semiconductors, Resistors, and Diodes

A benchmark study on waste metallic film resistor recycling was carried out by Ruan et al. (2017a). In China, from an e-waste recovery plant with license, these researchers found waste PCBs of televisions from which the resistors were assembled. They used metallic film resistors which were molded from a ceramic core engulfed by a metal film and two nickel pins. These metallic film resistors had characteristics outstanding than the other resistors such as carbon film resistors and winding resistors. Thus, metallic film resistors were the only resistors which were used for the fabrication of printed circuit boards. They comprised a ceramic matrix, metallic (conductive) film, end-cap, paint, color ring, and metallic pin (Ruan et al. 2017a). Ruan et al. (2017b) presented the recovery technology to extract nickel and ceramic elements from waste-film resistors. Nickel is a significant material in the manufacturing of alloys, stainless steel, and electroplates. It is, therefore, a toxic substance to people and the world in the interim and is highly capable of interacting with proteins and nucleic acid to affect the human health in a negative way, often likely to build up in the kidneys, spleen, and liver to develop cancer. Hence, waste metal film resistor recovery technologies were implemented to eliminate the leakage of nickel into the atmosphere. The weight of the ceramic content was roughly 36.2 wt%. The proportions of metallic film and organic content, respectively, were 1.3 and 2.7 wt%, respectively.

Generally, electrostatic separations are extensively used to recover the mixtures containing semiconductors, conductors, and non-conductors. Xue et al. (2012) in their study presented a novel approach to recover conductors, semiconductors, and nonconductors, i.e., copper, extrinsic silicon, woven glass-reinforced resin, respectively, by electrostatic separation. In their study, they used a roll-type corona electrostatic separator, and they varied the input variables such as voltage, roll speed, and size grade to observe the variation in removal efficiency for a combination of two from the mixture of three, namely semiconductors, conductors, and nonconductors.

10.3.1.5 Waste Capacitor

Tantalum waste condensers are an exciting resource for tantalum recovery as about 40% of Ta supply worldwide is procured through condensers. There are several recovery processes from off-specification lots of the output of condensers (Lessard et al. 2015). The emphasis is on extracting pure tantalum powder directly from the scrap. It is known from historical evidence that each German discards around 0.6 g of tantalum per year, which is mostly fed to waste incinerators into the residual waste stream. Besides, extra quantities are entered into the total waste stream by separate storage mechanisms such as electronic scrap. This so-called post-consumer scrap is also a good source of tantalum. However, tantalum condensers have not been retrieved from PCBs, which represent their key field of use. Copper and noble metals such as gold and platinum are also the primary yields from scrap PCB (Spitczok von Brisinski et al. 2014). Many efforts have been made to isolate the condensers from the remaining sections of the scrap PCB. Niu et al. (2017a) proposed a modern solution to the recycling of tantalum condenser metals. Instead of mechanical milling methods, tantalum is extracted chemically with an ionic solvent only from the other sections of the condenser. The anode of tantalum is not damaged and is thus fully isolated. Some metals such as manganese, tin, and silver are dissolved in ionic liquids, and the latter two can be extracted by electrodeposition from ionic liquids (Niu et al. 2017a).

Wang and Xu (2017) in their paper proposed an eco-friendly technique to recycle waste capacitors (aluminum electrolytic capacitors, AECs) from WPCBs. In their experimental study, they investigated the effect of heating temperatures and holding time on weight loss rate. To determine the suitable heating temperature for the thermal decomposition, they used gravimetric analyzer to analyze the components present, namely, non-metallic components, electrolytic paper, and sealing rubber plug after collection and separation. They separated aluminum through crushing, sieving, and magnetic separation. They varied the heating temperature at 500°C, 520 °C, and 460°C with holding time of 60 min, and, with the results obtained for the weight loss rate, they found that 500°C was the optimized desired heating temperature. On the basis of results, they performed a scaled-up thermal decomposition and found that the weight loss of aluminum electrolytic capacitor was 35.79%, which was near to the result of heating treatment tests (36.64%). Through the integrated process of crushing, they separated aluminum and iron (48.277 g) and residue (10.6426 g). The aluminum and iron mixture was further separated by magnetic separation to recover aluminum at a recovery rate of 96.52% (35.5990 g). Niu et al. (2017a, b) in their study implemented an integrated process containing pyrolysis, mechanical operations, and chloride metallurgy to recover tantalum oxide from waste capacitor. In addition, Ni–Fe terminals were also recovered by mechanical–physical separation.

10.4 RECYCLING AND TREATMENT TECHNOLOGY

There are numerous methods for recovering metals from nature rather than natural sources, some are currently under research, and some are ready for commercial usage as "packages".

10.4.1 PHYSICAL/MECHANICAL RECYCLING PROCESS

When various metals contained in e-wastes are separated from the feedstock by crushing and shredding processes, later, in order to recover the valuable REEs, physical treatment processes are exploited. The processes include PCB assembly desoldering for the extraction of electric components followed by size reduction of the coarser materials by primary crusher, followed by secondary crusher and tertiary crusher like fluid energy mill or hammer mill to liberate ultrafine materials (1 to 50 μm particle size). Then in the upgrading stage, a subjective process (many to one mapping) or a combined route of physical and extraction process is exploited to liberate the valuable metals. This combined route ensures that the selected samples are roasted to determine the ash content followed by granulation tests and screening and magnetic separation—pulverizing.

10.4.2 AUTOMATED AND SEMIAUTOMATED DISMANTLING

Dismantling is a systematic removal of components, groups, or parts from e-waste. Currently, informal dismantling is common in developing countries and is quite a hindrance in the transition to a circular economy in sustainable smart cities. Mechanical dismantling is common in developed countries (Charles et al. 2017). In the process of informal dismantling, the usage of chisels and cutting tools is common to open solder joints and separate various types of metals in PCB. Usually, PCBs are cooked over a coal bath plate and melted to resell the chips, and the recovered components are soaked in acid strippers for further processing (Huang et al. 2009). Through this informal dismantling process, bad smell and black fumes are generated.

Mechanical dismantling process can be segregated into semi-automatic and automatic dismantling processes. The ECs on the PCBs are removed by a combined process of applying heat above melting point of the solder and the application of impactful external forces such as crushing and shearing. Ninety-four percent maximum disassembly ratio was obtained at a feed and entered at a speed of 0.33 cm/s and a heating temperature of 250°C with the help of an apparatus developed by Park et al. (2015) for di-assembling ECs from e-waste. Usually infrared heaters are used for soldering as a pretreatment with solder baths and liquids as the medium to supply heat. Apart from dismantling, shredding, crushing, and milling processes are executed which are exploited by machines like jaw crushers, gyratory crusher, crashing rolls, Bradford Breaker, roller mill, cage mill, hammer mills, impactor, vertical shaft impactor, ball mill, and attrition mill/pulverizer. Afterwards, the isolation of metal from the nonmetals is attained. Distinct methods such as magnetic separation (Niu et al. 2017; Wang and Xu 2017), eddy current separation, and density separation are used. Shearing process followed by magnetic separation to obtain ceramic and nickel–chromium alloy particles and ball milling process to get nano Al_2O_3 were implemented by (Ruan et al. 2017a, b). There are also reported literatures on the possible recovery of metal fractions which carry more than 50% of copper, 24% of tin, and 8% of lead by the implementation of an amalgamation of electrostatic and magnetic segregation which separates the metal part from the non-metal ones (Veit et al. 2005). Further methods include corona discharge method (Li et al. 2007), density-based separation (Peng et al. 2004), froth floatation (Vidyadhar and Das 2012), etc.

10.4.3 Metallurgical Processes

REE metals from e-waste may typically be extracted utilizing hydrometallurgical, pyrometallurgical, and biometallurgical techniques (Cui and Zhang 2008; Debnath et al. 2018).

10.4.3.1 Hydrometallurgical Processes

Hydrometallurgical processes include the previously prepared heavy e-waste acid assault to process the acid solution from dissolved metals. Hydrometallurgical process for the extraction of metals from e-waste is the advanced form of the traditional hydrometallurgical process. Leaching is done via acid, alkali, or other alcoholic solvents which are used to extricate metals in form of soluble salts. Impurities are omitted with the residues, and the recovery of isolated metals from the solution is accomplished by physico-chemical processes such as adsorption, coagulation, micro-filtration, solvent extraction, and hyper-filtration. An electrolytic leaching process may accompany the method to obtain those metals in the specific position to get the precipitated metal salts. These procedures are beneficial but need a high amount of acid solutions; also, they are ideal only for recovery and not for cleaning of infected sludges. Leaching processes are basically of four types, namely—cyanide leaching, thiourea leaching, halide leaching, and thiosulfate leaching. Precious metals as well as REEs are also recovered through leaching. Aqua regia was utilized as a leaching solvent for the recovery of gold from PCBs (Sheng and Etsell 2007). Alongside sulfuric acid and nitric acid, muriatic acid-based solutions were also used as primary leaching solvents for the recovery of precious metals from e-waste. Metals such as nickel (Mecucci and Scott 2002), tin (Gibson et al. 2003), copper (Veit et al. 2006), and silver (Petter et al. 2014) are reported to have been extracted through this process. It is easier to control the rate of the reaction, creating less environmental hazards for which slow kinetics on the other hand is one such drawback. Usually metals are recovered from pretreated PCBs (desol-dered) through leaching which is a mass transfer phenomenon commonly known as solid–liquid extraction. The common base metals are usually recovered through this process because it has a significant impact on the economics due to the bulk availability of waste PCBs. It requires low capital cost compared to other metal-lurgical processes and is used for the recovery of base metals. There are different classifications of hydrometallurgical processes depending on the composition of the substrate used (Li et al. 2004).

10.4.3.2 Pyrometallurgical Processes

They are implemented to separate metals which are present in complex matrix and when physical recycling process fails. PCBs are generally easy to recycle through the pyrometallurgical process. It is focused on the separation of e-waste metals by alloying other molten metals. This technique has achieved decent success rates, but, currently, the recovery of individual components can hardly be done. The biggest downside is that operations take place at elevated temperatures. Pyrometallurgical technique includes smelting process in a blast furnace or Cu smelter to recover mainly nonmetallic substances (Kaya and Sözeri 2009; Kaya 2016). Since this technique can accept any form of scrap, so e-waste which is highly hazardous can efficiently be used as a raw material for the recovery of Cu, Ag, and Au (Sum 1991) but has difficulty in recovery of metals such as Fe and Al (Cui and Zhang 2008). For the extraction of Bi, Pb, and Sb and other metals with high vapor pressure, vacuum metallurgical separation is exploited. Pyrometallurgical treatment has its disadvantage that it emits fumes of heavy metals (metals with low melting points such as Hg, Cd, and Pb). Additionally, if the raw material contains PVC or plastics with BFRs, pyrometallurgical treatment can generate fumes of halogenated dioxins which are hazardous.

10.4.4 THERMOCHEMICAL TECHNOLOGIES

Pyrolysis is a thermal degradation process which takes place in the absence of air. There are different types of pyrolysis process reported, which are implemented for the thermal degradation of the targeted material, namely vacuum pyrolysis (Qiu et al. 2009), catalytic pyrolysis (Hall et al. 2008), microwave-assisted pyrolysis (Ruidian et al. 2007; Sun et al. 2012), and pyrolysis with biomass (Liu et al. 2013). Nowadays, plasma technology has gained importance and is exploited heavily for the treatment of e-waste, which operates at high temperature and is also eco-friendly. In the report of Ruj and Chang in 2013, the treatment of mobile phone waste using plasma was reported, and it proved that the procedure aids the recovery of metals. High-enthalpy plasma jet (Mitrasinovic et al. 2013) and plasma reactor (Rath et al. 2012) have been used and are reported in literature for the recovery of metals from e-waste.

10.4.5 BIOACCUMULATION

Biometallurgical methods are focused on bio-dissolution, bio-absorption, or bacterial or other species. This technology is particularly appropriate for recovering metals from water-based solutions or from low-to-medium condensing solids—metals that can be removed from diluted solutions and distilled in economic quantities (Manhart). The primarily used organisms are currently environmentally friendly microbes, which can remove metals from overly acidic contaminated areas (e.g., mining dumps). These methods lead to acid leachate, which consists of metals that must be refined further before being disposed of as waste (Cecchi et al. 2017; Gahan et al. 2013; Hennebel et al. 2015). Debnath et al. (2016a) presented a brief review on gold bioleaching from e-waste with a focus on sustainability. Fungi have a huge metal tolerance and strong resistance mechanisms like phytochelatin proteins which can adsorb and deactivate the toxic metals from the metal binding sites (Gadd 2007; Onofri et al. 2011; Qu and Lian 2013). Some fungi have an enormous capacity of deactivating contaminants and bioaccumulation (Zotti et al. 2014; Ceci et al. 2012). Di Piazza et al. (2017) in their study proposed an innovative procedure of using *P. expansum* isolated from a metal-contaminated site to recover metals from e-waste. They proposed a dedicated protocol to extract REEs by exploiting fungi from fungal biomass by acid digestion. Through this protocol, they collected a huge concentration of REEs for which acid digestion by the use of fungi is an effective mechanism of recycling using metal bioaccumulation techniques. Hence, the biological solution is deemed more cost-effective, smoother, and more environmentally sustainable than its chemical equivalent relative to traditional technology. Furthermore, biometallurgy has commonly been viewed as a far greener solution. This technology currently needs far lower temperatures, and the corresponding processes run under atmospheric pressure (Di Piazza et al. 2017).

10.5 DISCUSSIONS AND ANALYSIS

10.5.1 ROLE OF INDUSTRY

Electronic components play a very essential role in the development of the industry. There are numerous electronic components found which are initializing on the way to miniaturization and chips. Classification is a very vital basic venture among the fabrication, application, scientific study, and recycling of electronic components. As a result, it is of immense practical importance to create an automatic recognition scheme for electronic components, which can function in real time (Li-xiang et al. 2008). The recycling of electrical and electronic components has become a chief concern as it could affect the safety and dependability of a broad assortment of electronic systems quite possibly. The detection of recycled ICs has become extremely challenging as in most of the cases, these ICs are used for a very little phase of time as the method alterations could outpace the dreadful conditions caused by ageing. Recycled ICs are recovered or reclaimed from used systems and often misinterpreted as new components formed by an original maker. Owing to the properties of ageing throughout their earlier utilization and exploitation in the course of the recycling process, these ICs have shorter lifespan and exhibit lower performance, compared to the new and authentic ones (Alam et al. 2018).

Digital signatures are widely used to ensure the integrity of a message and its end-point validation. The verification of message integrity is essential for finding whether the message has been altered and the end-point authentication guarantees the basis of the message. To generate digital signatures, public key cryptographic approaches are among the widely used methods. In the case of the exclusion of method alterations during manufacturing, the on-chip sensors that use ring oscillators can be perfect for the recognition of recycled ICs. The frequency of the ring oscillator is measured and combined with a fixed length hash from the frequency, with a secure hash algorithm (SHA) and accumulated in the non-volatile memory (NVM) of the chip with a digital signature (Guin et al. 2015). This could be considered as the first step to work on the traceability of the recycled ICs, and with an SHA, any unethical use of recycled ICs will be prevented as SHAs generate a unique hash for each and every component. The similar application of digital signatures on programmable ROMs will be helpful in terms of both traceability and data-security (Lee and Chang 1999; Islam et al. 2018).

From a bigger perspective, using blockchain technology ensures a better result not just in terms of traceability of the recycled electronic components, but also provides a transparent system to all the stakeholders to prevent the use of any recycled component in a new device (Islam et al. 2018). The entire supply-chain network of electronic components can be transformed into a blockchain network which will allow the manufacturers to provide the best service or product to their consumers (Xu et al. 2019). Recycling of these components will have a separate branch of the same supply-chain network. These two separate branches can be designed for the manufacturer and its consumers. The manufacturer will use a private key encryption for a centralized system, a private blockchain which can be accessed only by the manufacturer. For the consumers, a public blockchain can be designed, and this will be accessed by both of the parties, with a public key encryption. The implementation of blockchain in this sector will have an impact in modifying the linear economy toward circularity and sustainable development (Ajwani-Ramchandani et al. 2021).

10.5.2 Proposed Solutions Toward Green Development

Electronic components (ECs) are a mixture of various other elementary devices such as ICs, RAMs, transistors, semiconductors, diodes, resistors, and capacitors (Debnath et al. 2016b). So as a whole, the volume of ECs is large. This heterogeneous mixture of electronic components is also fabricated with hazardous metals. So, the direct recycling of the metals is difficult and also dangerous toward the environment. Also, as the segregation cost is high, the recycling of these ECs is not always considered to be profitable. Generally, they are dumped somewhere scientifically or transferred to people who are willing to use it in some other way. Now these ECs contain major quantities of critical REMs, and their conservation is very crucial (Watari et al. 2020). The future economic growth of a country definitely depends on the conservation of REMs, and so the EC recycling is important in order to conserve the REMs. The economical demand of the electronic components is high. So, to establish a relatively cheap recovery method of the metals, we need to find some easy and cheap methods. Pyrometallurgy is the most conventional and frequently used method for the separation and the recovery of metals, which includes incineration, fusion in blast furnace, liquefying, and gas-state reactions at high temperatures (Cui and Zhang 2008). However, there are boundaries and challenges to use this method efficiently due to the danger of lethal substance discharge and high dioxin generation and expensive futuristic smelter. Smelting is actually detrimental resulting in the production of residues and additional industrial junks. The ceramics and glass elements in PCBs result in more debris development and elevated depletion of valuable and main metals from PCBs. Eventually, through pyrometallurgy, prolonged time is the issue so as to isolate and recover the expensive metals which are generally not recovered up to the termination of the entire process (Cui and Roven 2011). Now hydrometallurgy is enthusiastically manageable, predictable, more precise, and greener than pyrometallurgy. Hydrometallurgical procedures actually demand a sequence of acid or caustic leaches of e-waste, which is then pursued by separation and purification approaches. The cyanide leaching is currently not in use as a hydrometallurgical recovery method owing to the high toxicity of cyanide. Other hydrometallurgical processes are actually slowly emerging as they are in the early phase of research. The leaching with ligands and bioleaching are less evolved thereby showing a reduced number of economic prospective benefits although thiourea and thiosulfate leaching are the greenest methods than the dangerous acid leaching (Hsu et al. 2019). Although cyanide leaching is effective, it is tremendously harmful toward the environment due to the production of toxic by-products as well as owing to the low selectivity of the process (Jiang et al. 2012). The work on biometallurgical recovery has been going on a huge scale because of its small investment cost, small energy consumption, and small environmental impact (Zhang and Xu 2016). The technique of biometallurgy can be summed up in two major steps: Bioleaching and biosorption (Zhang and Xu 2016). But in a biometallurgical process when the volume is less, the efficiency of the method has been proved to be high. Due to the greater volume issue of these ECs, the efficacy of this method is also less, and it is very much time-consuming. So, the use of biometallurgy is not effective enough to recover the crucial metals from the e-waste. Here, green chemistry has a pivotal role to play. The aforementioned drawbacks can be omitted by using the green approach. Ionic liquids (ILs) are recognized as a novel, exclusive, and thrilling group of nonaqueous solvents which have the potential to substitute volatile organic solvents which are being used in bulk in current times. These ionic liquids (ILs) form polar aprotic solvents which consist of small inorganic or organic anions and large organic cations. ILs have now been treated as new green chemicals which are beneficial for both academic and industrial purposes owing to their excellent characteristics, for instance, insignificant vapor pressure, exceptional thermal stability, fine dissolution properties, and also little flammability (Pu et al. 2007; Keskin et al. 2007; Kubisa 2009; Aslanov 2011). Additionally, as they are a promising agent to bring great benefits toward environment as well as product and process, the industry has been showering these green solvents with a greater and undivided attention. ILs such as salts of N, N-disubstituted imidazolines have attracted significant attention as solvents for cellulose (Kubisa 2009). We can thus try to recover the critical and precious metals using these green solvents. These solvents have lesser impact on the environment than other organic solvents. Currently, the use of green solvents is limited so far with the PCB (Zhu et al. 2012). But we may explore this route to retrieve the REMs from the electronic waste management too. Other than the green chemistry, some certain research works were observed whose fundamental basis is nano-chemistry (Tiwary et al. 2017; Babar et al. 2019). So, we believe that in near future, the use of nanotechnology can also be explored for the retrieval of REMs from the e-waste. Thus, the recovery of these valued and major REMs will certainly help in circular economy.

10.6 CONCLUSION

In this chapter, we have discussed a thorough review of proper dismantling/desoldering of ECs from PCB and focused on suitable treatment and recycling techniques of

ECs to recover the valuable metals. There is a detailed discussion on the retrieval of electronic components from the PCBs and the various dismantling methods to recover metals from the electronic components. We have also studied the classification of the electronic components in a very thorough and detailed fashion. The shape and volume and the chemical composition of the e-waste are reviewed corresponding to the treatment technologies implemented. The numerous methods of recovering the precious REMs including the various metallurgical processes have also been discussed in this chapter. We have also shed some light about the role of the industry on all of these and suggested a greener aspect with the use of green solvents to recover the vital REMs from the electronic waste thereby contributing toward the creation of a better and sustainable world.

REFERENCES

Ajiboye, E. A., E. F. Olasehinde, A. O. Adebayo, O. O. Ajayi, M. K. Ghosh, and S. Basu. "Extraction of Cu, Zn, and Ni from waste silica-rich integrated circuits by sulfation roasting and water leaching." *Chemical Papers* 74, no. 2 (2020): 663–671.

Ajwani-Ramchandani, Raji, Sandra Figueira, Rui Torres de Oliveira, and Shishir Jha. "Enhancing the circular and modified linear economy: The importance of blockchain for developing economies." *Resources, Conservation and Recycling* 168 (2021): 105468.

Alam, Mahabubul, Sreeja Chowdhury, Mark M. Tehranipoor, and Ujjwal Guin. "Robust, low-cost, and accurate detection of recycled ICs using digital signatures." In *2018 IEEE International Symposium on Hardware Oriented Security and Trust (HOST)*, pp. 209–214. IEEE (2018).

Aslanov, L. A. "Ionic liquids: Liquid structure." *Journal of Molecular Liquids* 162, no. 3 (2011): 101–104.

Babar, Santosh, Nana Gavade, Harish Shinde, Anil Gore, Prasad Mahajan, Ki Hwan Lee, Vijaykumar Bhuse, and Kalyanrao Garadkar. "An innovative transformation of waste toner powder into magnetic g-C3N4-Fe2O3 photocatalyst: Sustainable e-waste management." *Journal of Environmental Chemical Engineering* 7, no. 2 (2019): 103041.

Barnwal, Amit, and Nikhil Dhawan. "Physical processing of discarded integrated circuits for recovery of metallic values." *JOM* 72, no. 7 (2020): 2730–2738.

Cecchi, Grazia, Pietro Marescotti, Simone Di Piazza, and Mirca Zotti. "Native fungi as metal remediators: Silver myco-accumulation from metal contaminated waste-rock dumps (Libiola Mine, Italy)." *Journal of Environmental Science and Health, Part B* 52, no. 3 (2017): 191–195.

Ceci, Andrea, Oriana Maggi, Flavia Pinzari, and Anna Maria Persiani. "Growth responses to and accumulation of vanadium in agricultural soil fungi." *Applied Soil Ecology* 58 (2012): 1–11.

Charles, Rhys Gareth, Peter Douglas, Ingrid Liv Hallin, Ian Matthews, and Gareth Liversage. "An investigation of trends in precious metal and copper content of RAM modules in WEEE: Implications for long term recycling potential." *Waste Management* 60 (2017): 505–520.

Chen, Tse-Lun, Tzu-Hao Huang, Ching-Hsiang Hsu, Yi-Hung Chen, Shu-Yuan Pan, and Pen-Chi Chiang. "Removal of fine particles from IC chip carbonization process in a rotating packed bed: Modeling and assessment." *Chemosphere* 238 (2020): 124600.

Cui, Jirang, and Hans Jørgen Roven. "Electronic waste." In *Waste*, pp. 281–296. Academic Press, 2011.

Cui, Jirang, and Lifeng Zhang. "Metallurgical recovery of metals from electronic waste: A review." *Journal of Hazardous Materials* 158, no. 2–3 (2008): 228–256.

Debnath, Biswajit, Ranjana Chowdhury, and Sadhan Kumar Ghosh. "Bio-metallurgical recovery of gold from printed circuit boards (PCB): A review." *Journal of Solid Waste Technology & Management* 42, no. 1 (2016a).

Debnath, Biswajit, Ranjana Chowdhury, and Sadhan Kumar Ghosh. "Sustainability of metal recovery from e-waste." *Frontiers of Environmental Science & Engineering* 12, no. 6 (2018): 1–12.

Debnath, Biswajit, Priyankar Roychowdhury, and Rayan Kundu. "Electronic components (EC) reuse and recycling–A new approach towards WEEE management." *Procedia Environmental Sciences* 35 (2016b): 656–668.

Di Piazza, Simone, Grazia Cecchi, Anna Maria Cardinale, Cristina Carbone, Mauro Giorgio Mariotti, Marco Giovine, and MircaZotti. "Penicillium expansum link strain for a bio-metallurgical method to recover REEs from WEEE." *Waste Management* 60 (2017): 596–600.

Dutta, D., and Goel, S. (2021). Understanding the gap between formal and informal e-waste recycling facilities in India. *Waste Management*, 125, 163–171.

Elkington, Daniel, Nathan Cooling, Warwick Belcher, Paul C. Dastoor, and Xiaojing Zhou. "Organic thin-film transistor (OTFT)-based sensors." Electronics 3, no. 2 (2014): 234–254.

Gadd, Geoffrey M. "Geomycology: Biogeochemical transformations of rocks, minerals, metals and radionuclides by fungi, bioweathering and bioremediation." *Mycological Research* 111, no. 1 (2007): 3–49.

Gahan, Chandra Sekhar, H. A. R. A. G. O. B. I. N. D. A. Srichandan, D. J. Kim, and A. Akcil. "Bio-hydrometallurgy and its applications: A review." In *Advances in Biotechnology*, pp. 71–100. Indian Publisher (2013).

Gibson, R. W., P. D. Goodman, L. Holt, I. M. Dalrymple, and D. J. Fray. U. S. Patent No. 6,641,712. U.S. Patent and Trademark Office (2003).

Guin, Ujjwal, Domenic Forte, and Mark Tehranipoor. "Design of accurate low-cost on-chip structures for protecting integrated circuits against recycling." *IEEE Transactions on Very Large Scale Integration (VLSI) Systems* 24, no. 4 (2015): 1233–1246.

Hall, William J., Norbert Miskolczi, Jude Onwudili, and Paul T. Williams. "Thermal processing of toxic flame-retarded polymers using a waste fluidized catalytic cracker (FCC) catalyst." *Energy & Fuels* 22, no. 3 (2008): 1691–1697.

Hennebel, Tom, Nico Boon, SynthiaMaes, and Markus Lenz. "Biotechnologies for critical raw material recovery from primary and secondary sources: R&D priorities and future perspectives." *New Biotechnology* 32, no. 1 (2015): 121–127.

Hsu, Emily, Katayun Barmak, Alan C. West, and Ah-Hyung A. Park. "Advancements in the treatment and processing of electronic waste with sustainability: A review of metal extraction and recovery technologies." *Green Chemistry* 21, no. 5 (2019): 919–936.

Huang, Kui, Jie Guo, and Zhenming Xu. "Recycling of waste printed circuit boards: A review of current technologies and treatment status in China." *Journal of Hazardous Materials* 164, no. 2–3 (2009): 399–408.

Ilyas, S., Srivastava, R. R., & Kim, H. (2021). O2-enriched microbial activity with pH-sensitive solvo-chemical and electro-chlorination strategy to reclaim critical metals from the hazardous waste printed circuit boards. *Journal of Hazardous Materials*, 416, 125769.

Islam, Md Nazmul, Vinay C. Patii, and Sandip Kundu. "On IC traceability via blockchain." In *2018 International Symposium on VLSI Design, Automation and Test (VLSI-DAT)*, pp. 1–4. IEEE (2018).

Jiang, Ping, Megan Harney, Yuxin Song, Ben Chen, Queenie Chen, Tianniu Chen, Gillian Lazarus, Lawrence H. Dubois, and Michael B. Korzenski. "Improving the end-of-life for electronic materials via sustainable recycling methods." *Procedia Environmental Sciences* 16 (2012): 485–490.

Kaya, Muammer. "Recovery of metals from electronic waste by physical and chemical recycling processes." *International Journal of Chemical, Molecular, Nuclear, Materials and Metallurgical Engineering* 10, no. 2 (2016): 232–243.

Kaya, Muammer, and Ayça Sözeri. "A review of electronic waste (e-waste) recycling technologies "is e-waste an opportunity or treat?"." In *Proceedings 138th Annual Meeting*, pp. 1055–1060. TMS (2009).

Keskin, Seda, DefneKayrak-Talay, Uğur Akman, and Öner Hortaçsu. "A review of ionic liquids towards supercritical fluid applications." *The Journal of Supercritical Fluids* 43, no. 1 (2007): 150–180.

Kubisa, Przemysław. "Ionic liquids as solvents for polymerization processes—Progress and challenges." *Progress in Polymer Science* 34, no. 12 (2009): 1333–1347.

Lee, Weibin, and Chin-Chen Chang. "With traceability property." *Journal of information Science and Engineering* 15 (1999): 669–678.

Lessard, Joseph D., Leonid N. Shekhter, Daniel G. Gribbin, Yuri Blagoveshchensky, and Larry F. McHugh. "A new technology platform for the production of electronic grade tantalum nanopowders from tantalum scrap sources." International Journal of Refractory Metals and Hard Materials 48 (2015): 408–413.

Li, Jia, Zhenming Xu, and Yaohe Zhou. "Application of corona discharge and electrostatic force to separate metals and nonmetals from crushed particles of waste printed circuit boards." *Journal of Electrostatics* 65, no. 4 (2007): 233–238.

Li, Jianzhi, Puneet Shrivastava, Zong Gao, and Hong-Chao Zhang. "Printed circuit board recycling: A state-of-the-art survey." *IEEE Transactions on Electronics Packaging Manufacturing* 27, no. 1 (2004): 33–42.

Liu, Wen-Wu, Chang-Wei Hu, Yu Yang, Dong-Mei Tong, Liang-Fang Zhu, Rui-Nan Zhang, and Bo-Han Zhao. "Study on the effect of metal types in (Me)-Al-MCM-41 on the mesoporous structure and catalytic behavior during the vapor-catalyzed co-pyrolysis of pubescens and LDPE." *Applied Catalysis B: Environmental* 129 (2013): 202–213.

Li-xiang, Zheng, Yang Pei-liang, and Nie Guo-Jian. "The Research of Components Information Platform Construction and Standardized Processing Technology." *Electronics Quality* 6 (2008).

Luda, M. P. (2011). Recycling of printed circuit boards. In Integrated Waste Management-Volume II. IntechOpen.

Mecucci, Andrea, and Keith Scott. "Leaching and electrochemical recovery of copper, lead and tin from scrap printed circuit boards." *Journal of Chemical Technology & Biotechnology: International Research in Process, Environmental & Clean Technology* 77, no. 4 (2002): 449–457.

Mitrasinovic, A., L. Pershin, and J. Mostaghimi. "Electronic waste treatment by high enthalpy plasma jet." In *International Plasma Chemistry Society (IPCS20)*. (2013).

Niu, Bo, Zhenyang Chen, and Zhenming Xu. "An integrated and environmental-friendly technology for recovering valuable materials from waste tantalum capacitors." *Journal of Cleaner Production* 166 (2017a): 512–518.

Niu, Bo, Zhenyang Chen, and Zhenming Xu. "Recovery of tantalum from waste tantalum capacitors by supercritical water treatment." *ACS Sustainable Chemistry & Engineering* 5, no. 5 (2017b): 4421–4428.

Ohajinwa, C. M., Van Bodegom, P. M., Vijver, M. G., & Peijnenburg, W. J. (2017). Health risks awareness of electronic waste workers in the informal sector in Nigeria. *International Journal of Environmental Research and Public Health*, 14(8), 911.

Onofri, S., A. Anastasi, G. Del Frate, S. Di Piazza, N. Garnero, M. Guglielminetti, D. Isola et al. "Biodiversity of rock, beach and water fungi in Italy." *Plant Biosystems-An International Journal Dealing with all Aspects of Plant Biology* 145, no. 4 (2011): 978–987.

Park, Jeongwon, James E. Royer, Corneliu N. Colesniuc, Forest I. Bohrer, Amos Sharoni, SunghoJin, Ivan K. Schuller, William C. Trogler, and Andrew C. Kummel. "Ambient induced degradation and chemically activated recovery in copper phthalocyanine thin film transistors." *Journal of Applied Physics* 106, no. 3 (2009): 034505.

Park, Seungsoo, Seongmin Kim, Yosep Han, and Jaikoo Park. "Apparatus for electronic component disassembly from printed circuit board assembly in e-wastes." *International Journal of Mineral Processing* 144 (2015): 11–15.

Peng, Mou, Wa Layiding, Xiang Dong, Gao Jiangang, and Duan Guanghong. "A physical process for recycling and reusing waste printed circuit boards." In *IEEE International Symposium on Electronics and the Environment, 2004. Conference Record, 2004*, pp. 237–242. IEEE (2004).

Petter, P. M. H., Hugo M. Veit, and Andréa M. Bernardes. "Evaluation of gold and silver leaching from printed circuit board of cellphones." *Waste Management* 34, no. 2 (2014): 475–482.

Pu, Y., N. Jiang, and A. J. Ragauskas. "Ionic liquid as a green solvent for lignin." *Journal of Wood Chemistry and Technology* 27 (2007): 23–33.

Qiu, Ke-Qiang, Qian Wu, and Zhi-hua Zhan. "Vacuum pyrolysis characteristics of waste printed circuit boards epoxy resin and analysis of liquid products." *Journal of Central South University: Science and Technology* 40, no. 5 (2009): 1209–1215.

Qu, Yang, and Bin Lian. "Bioleaching of rare earth and radioactive elements from red mud using Penicillium tricolor RM-10." *Bioresource Technology* 136 (2013): 16–23.

Rath, Swagat S., Pradeep Nayak, Partha Sarathi Mukherjee, G. Roy Chaudhury, and B. K. Mishra. "Treatment of electronic waste to recover metal values using thermal plasma coupled with acid leaching–A response surface modeling approach." *Waste Management* 32, no. 3 (2012): 575–583.

Ruan, Jujun, Jiaxin Huang, Lipeng Dong, and Zhe Huang. "Environmentally friendly technology of recovering nickel resources and producing nano-Al_2O_3 from waste metal film resistors." *ACS Sustainable Chemistry & Engineering* 5, no. 9 (2017a): 8234–8240.

Ruan, Jujun, Jie Zheng, Lipeng Dong, and Rongliang Qiu. "Environment-friendly technology of recovering full resources of waste capacitors." *ACS Sustainable Chemistry & Engineering* 5, no. 1 (2017b): 287–293.

Ruidian, Tan, Wang Tonghua, Tan Suxia, He Xinzhan, and Wu Tao. "Products from microwave heating of waste printed circuit boards." *Environmental Pollution & Control* 8 (2007): 013.

Ruj, B., and J. Chang. "E-waste (cell phone) treatment by thermal plasma technique." In *Proceedings of the 7th International Symposium on Feedstock Recycling of Polymeric Materials (7th ISFR 2013)*. (2013).

Sheng, Peter P., and Thomas H. Etsell. "Recovery of gold from computer circuit board scrap using aqua regia." *Waste Management & Research* 25, no. 4 (2007): 380–383.

Spitczok von Brisinski, Lena, Daniel Goldmann, and Frank Endres. "Recovery of metals from tantalum capacitors with ionic liquids." *ChemieIngenieur Technik* 86, no. 1–2 (2014): 196–199.

Sum, Elaine YL. "The recovery of metals from electronic scrap." *Jom* 43, no. 4 (1991): 53–61.

Sun, Jing, Wenlong Wang, Zhen Liu, Qingluan Ma, Chao Zhao, and Chunyuan Ma. "Kinetic study of the pyrolysis of waste printed circuit boards subject to conventional and microwave heating." *Energies* 5, no. 9 (2012): 3295–3306.

Tiwary, Chandra S., S. Kishore, R. Vasireddi, D. R. Mahapatra, P. M. Ajayan, and K. Chattopadhyay. "Electronic waste recycling via cryo-milling and nanoparticle beneficiation." *Materials Today* 20, no. 2 (2017): 67–73.

Veit, Hugo Marcelo, Andréa Moura Bernardes, Jane Zoppas Ferreira, Jorge Alberto Soares Tenório, and Célia de Fraga Malfatti. "Recovery of copper from printed circuit boards scraps by mechanical processing and electrometallurgy." *Journal of Hazardous Materials* 137, no. 3 (2006): 1704–1709.

Veit, Hugo Marcelo, Taina Rose Diehl, Anderson Paulo Salami, Joel da Silva Rodrigues, Andrea Moura Bernardes, and Jorge Alberto Soares Tenório. "Utilization of magnetic and electrostatic separation in the recycling of printed circuit boards scrap." *Waste Management* 25, no. 1 (2005): 67–74.

Vidyadhar, A., and Avimanyu Das. "Kinetics and efficacy of froth flotation for the recovery of metal values from pulverized printed circuit boards." In *Proceedings of XXVI International Mineral Processing Congress (IMPC) 2012 Proceedings*, pp. 236–243. New Delhi, India, September (2012).

Wang, Jianbo, and Zhenming Xu. "Environmental friendly technology for aluminum electrolytic capacitors recycling from waste printed circuit boards." *Journal of Hazardous Materials* 326 (2017): 1–9.

Watari, Takuma, Keisuke Nansai, and Kenichi Nakajima. "Review of critical metal dynamics to 2050 for 48 elements." *Resources, Conservation and Recycling* 155 (2020): 104669.

Widmer, R., Oswald-Krapf, H., Sinha-Khetriwal, D., Schnellmann, M., & Böni, H. (2005). Global perspectives on e-waste. *Environmental impact assessment review*, 25(5), 436–458.

Xu, Xiaolin, Fahim Rahman, Bicky Shakya, Apostol Vassilev, Domenic Forte, and Mark Tehranipoor. "Electronics supply chain integrity enabled by blockchain." *ACM Transactions on Design Automation of Electronic Systems (TODAES)* 24, no. 3 (2019): 1–25.

Xue, Mianqiang, Guoqing Yan, Jia Li, and Zhenming Xu. "Electrostatic separation for recycling conductors, semiconductors, and nonconductors from electronic waste." *Environmental Science & Technology* 46, no. 19 (2012): 10556–10563.

Zhang, Lingen, and Zhenming Xu. "A review of current progress of recycling technologies for metals from waste electrical and electronic equipment." *Journal of Cleaner Production* 127 (2016): 19–36.

Zhang, Yongliang, Lu Zhan, Bing Xie, Zhenming Xu, and Chao Chen. "Decomposition of packaging materials and recycling GaAs from waste ICs by hydrothermal treatment." *ACS Sustainable Chemistry & Engineering* 7, no. 16 (2019): 14111–14118.

Zhu, P., Y. Chen, L. Y. Wang, and M. Zhou. "Treatment of waste printed circuit board by green solvent using ionic liquid." *Waste Management* 32, no. 10 (2012): 1914–1918.

Zotti, Mirca, Simone Di Piazza, Enrica Roccotiello, Gabriella Lucchetti, Mauro Giorgio Mariotti, and Pietro Marescotti. "Microfungi in highly copper-contaminated soils from an abandoned Fe–Cu sulphide mine: Growth responses, tolerance and bioaccumulation." *Chemosphere* 117 (2014): 471–476.

Part V

Toward Sustainable Circular
Economy in E-waste Sector

Features Approaches that Dictate Sustainable
Circular Economy in the E-waste Sector

11 Approaches to Improving the Circular Economy Paradigm in E-waste Management in India through Informal–Formal Sector Integration

Gautam Mehra, Rachna Arora, Reva Prakash and Abhijit Banerjee

CONTENTS

11.1 INTRODUCTION

Like most traditional societies, India has always had a long-standing repair and reuse culture. Electrical and electronic equipment (EEE) were no exception, and repair mechanics used to be widely available in both formal and informal sectors. However, in recent decades, the business model of the consumer electronics sector has changed globally, with a high emphasis on planned and perceived obsolescence (LeBel 2016; Maitre-Ekern and Dalhammar 2016; Satyro et al. 2018; La Rosa 2020). India in the 21st century, being a liberalised market with a fast-growing and aspirational middle class, has witnessed a boom in all kinds of consumer goods including electronics. As a result, the demand for repair has significantly decreased, while

DOI: 10.1201/9781003301899-16

the generation of e-waste has witnessed a quantum increase (Anantharaman 2014; Bhatt et al. 2016; Chaudhary and Vrat 2019; Down to Earth 2020; Rathore 2020). When e-waste imports are added to the picture, a huge opportunity for recycling and materials recovery has emerged. Taking advantage of these trends, e-waste recycling in India has also expanded dramatically over the last two decades, but is largely dominated by the informal sector, which is characterised by dangerous safety and environmental practices as well as inefficient resource recovery (Awasthi et al. 2016; Gangwar et al. 2019). A formalised recycling sector, as part of a well-organised end-of-life (EoL) ecosystem for e-waste, has the potential to significantly increase resource recovery and contribute to the circular economy goal for the electrical/ electronics goods sector, while reducing environmental and safety risks at the same time. Given that India is highly dependent on the import of critical minerals, promotion of a circular economy is imperative, especially in light of the government's stated priorities in expanding the IT, telecommunications, and manufacturing sectors. Indeed, a circular economy vision is central to "Strategy on Resource Efficiency" published by the NITI Aayog (NITI 2017). Recent government policies to introduce Extended Producer Responsibility (EPR) as a means of improving formal recycling by imposing e-waste recovery targets on manufacturers have produced mixed results so far. While the formal recycling ecosystem has expanded, the majority of e-waste continues to get channelled into the informal sector (NITI and MeitY 2019). The government's approach ignores ground realities that imbue the informal sector with significant advantages vis-à-vis the formal sector. This chapter investigates the obstacles faced by the formal recycling sector through a careful analysis of the entire EoL ecosystem for e-waste and outlines approaches to improve circularity by pursuing judicious integration between the formal and informal sectors.

11.2 RATIONALE FOR PROMOTING CIRCULAR ECONOMY FOR E-WASTE IN INDIA

Ever increasing extraction of primary raw materials contributes to resource depletion and higher prices, countries that are import dependent face higher import bills for raw materials, and serious environmental impacts are associated with the entire value chain from mining/extraction to waste disposal, which include damage to ecosystems and biodiversity loss, air and water pollution, greenhouse gas emissions, and harm to human health. As these challenges have become more acute, many countries are trying to adopt "circular economy" principles (European Commission 2020; Government of Japan 2020; World Bank 2017) that are expected to bring about significant economic benefits in terms of jobs, investment, competitiveness, etc. For the EU, the indisputable world leader in promoting the circular economy, it is estimated that moving towards a circular economy would generate a net economic benefit of EUR 1.8 trillion by 2030 (McKinsey 2015). A similar study on India estimated a half a trillion USD opportunity by 2030 (FICCI and Accenture Strategy 2018).

India is highly import-dependent for a range of critical minerals (IGEP 2013) such as cobalt (100%), molybdenum (100%), nickel (100%), copper (95%), etc., that has serious consequences on its manufacturing prospects as well as trade balance. Moreover, in recent years, the specter of "resource nationalism" has reared its ugly head, putting further pressure on import-dependent countries (Smyth 2020). India is endowed with considerable mineral resources, but, unfortunately, much of identified reserves occur in regions of rich biodiversity and/or tribal land, creating the potential for terrible environmental destruction and social conflict (CSE 2008). For all these reasons, it is imperative that Indian policymakers adopt the circular economy approach as a high priority.

E-waste contains 60 elements from the periodic table which are present in complex forms with materials categorised as rare earths and hazardous and precious metals. According to the Global E-waste Monitor, the estimated value of materials recoverable from e-waste globally in 2019 stood at USD 57 billion, with a documented collection and recycling rate of only 17.4% (Forti et al. 2020). For waste streams such as e-waste that are typically concentrated in urban conglomerations, the concept of "urban mining" has the potential to contribute to the overall circular economy vision by providing a significant and reliable source of secondary raw materials.

The demand for electronic products in India has reached USD 400 billion by 2020 with fantastic growth rates over the last two decades; the Compound Annual Growth Rate (CAGR) jumping dramatically from 9.6% for the 2010–2016 period to 41% for the 2016–2020 period (NITI and MeitY 2019). By 2019, India had become the third largest producer of e-waste in the world after the United States and China, generating 3.2 million metric tonnes in that year, even if per capita generation remains low as compared to high-income countries (Forti et al. 2020). However, for a variety of reasons, resource recovery rates in India remain much lower compared to high performing countries in the EU (UNEP 2016). With the right framework in place, the e-waste sector in India has an estimated potential of approximately USD 1 billion by 2030 (FICCI and Accenture Strategy 2018).

A circular economy approach to e-waste enjoys strong congruence with several key policy priorities of the Government of India. The "Swachh Bharat (Clean India) Mission" focuses on extracting resources from waste, under the rubric of improved waste management, which can then re-enter the productive use cycle. The "Make in India" mission envisions a significant rise in manufacturing as a share of GDP and can benefit from secondary resources recovered from waste. The "Digital India" mission, which emphasises a high digitisation of government services, would trigger a higher demand for electronic products which can be fulfilled through the Make in India mission, which in turn can draw heavily on secondary resources recovered through the Swachh Bharat Mission. Finally, the recycling industry is expected to increase the demand for skilled jobs, thus contributing to the goals of the "Skill India Mission". If fully realised, this inter-linking dynamic (depicted in Figure 11.1) has the potential to push India significantly towards a resource-efficient and circular economy.

In 2019, the NITI Aayog and the Ministry of Electronics and Information Technology (MeitY) released the Strategy on Resource Efficiency in the Electrical and Electronic Equipment Sector (NITI Aayog and MeitY 2019), as part of the Strategy for Resource Efficiency for India developed by NITI Aayog in collaboration with the EU Resource Efficiency Initiative for India. The analysis and recommendations in this chapter are based broadly on the NITI/MeitY Strategy.[1]

---

[1] The authors were key contributors to NITI Aayog and MeitY 2019.

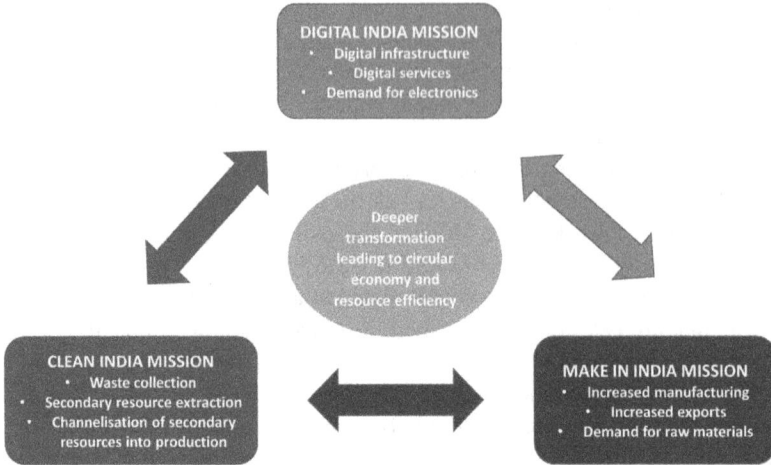

FIGURE 11.1 Congruence of key Government of India priorities with the goal of a resource-efficient and circular economy.

11.3 CURRENT E-WASTE RECYCLING SCENARIO IN INDIA: DOMINANCE OF THE INFORMAL SECTOR

Most recycling in India is dominated by the informal sector, which plays a vital role in its contribution to a circular economy as well as significant employment and income generation for the vast "unskilled" population (Chikarmane et al. 2008). E-waste recycling is also dominated by the informal sector, and this informal economy has expanded significantly in recent decades with the increase in e-waste generation (Dutta and Goel 2021). The informal recycling sector, especially in cities, has thrived because it can provide economic opportunity in dense urban spaces with minimal investment that may include only rudimentary equipment; sometimes, workers' houses double up as "workshops", reducing the need for space rental (Awasthi et al. 2018). Prominent informal clusters handling e-waste include Shastri Park in Delhi and Moradabad in Uttar Pradesh. Like the case of other recycling clusters, such as those for End-of-Life Vehicles, over time, these centres have developed elaborate networks and even a surprising degree of specialisation. For example, the Moradabad cluster has specialised units for recycling printed circuit boards (PCBs) and draws on PCB sourcing from multiple states. An e-waste aggregator on the outskirts of Kolkata may be involved only till dismantling and generates profit by sending the extracted PCBs to Moradabad (NITI Aayog and MeitY 2019).

The informal sector handling e-waste is composed of different types of actors with specialised roles, and it is essential to understand the role of these actors for gaining proper appreciation of the sector. The following actors may be distinguished:

A. Itinerant buyer: Individuals who roam in neighbourhoods and provide doorstep collection of e-waste, in exchange for on-the-spot cash payment. Popularly known as *kabadiwallahs* in north India.

B. Small sorter/aggregator: Individuals or small enterprises engaged in basic sorting of collected e-waste before selling on to large aggregators. Typically operate out of homes with no need for rented space.

C. Large aggregator: Larger enterprises that collect and sort a wide range of e-waste and associated equipment. They typically employ a few people and need some storage space. Large aggregators have well-developed networks both upstream (with small collectors/sorters) and downstream (dismantlers).

D. Dismantler: The next level in the chain that involves dismantling e-waste products into basic components such as cathode ray tubes and printed circuit boards. Dismantling is carried out using rudimentary hand-held tools and may sometimes be specialised, say only for computers. Extracted valuable components such as PCBs are sold on to recyclers while material with no market value is typically burnt or dumped indiscriminately.

E. Recycler: Informal recyclers use crude methods such as acid baths to extract valuable materials from dismantled e-waste components. Extracted materials such as precious metals like gold and silver are then either sold off to informal producers (for example, local producers of jewellery) or secondary raw material aggregators who supply raw material to the formal manufacturing sector. Waste products generated from informal recycling are typically burnt or dumped indiscriminately.

The informal e-waste value chain is depicted diagrammatically in Figure 11.2. It is important to note the following points:

1. Actors A and B are also involved in diverting usable products and components to the second-hand market, which may involve some repairing. This second-hand market is also mostly informal.

2. There is often some degree of overlap among the different actors. For example, *kabadiwallahs* may be engaged in some degree of sorting/aggregating, aggregators may be engaged in some dismantling, and dismantlers may be engaged in some recycling.

11.4   SHORTCOMINGS OF THE INFORMAL SECTOR

Despite its many advantages in terms of doorstep collection and employment generation, the informal sector faces significant shortcomings and poses serious challenges for society.

1. *Workers' health*: Since the informal sector is unable to invest in advanced technology and equipment, crude tools and methods are widely used for dismantling and recycling. This leads to a high incidence of injuries and accidents, exacerbated by the lack of personal protective equipment (PPE). Further, workers are exposed to dangerous levels of pollution at different stages: fine particulates from shredding, toxic fumes from burning material like plastics, toxic fumes from acid and other chemical baths used for materials recovery, etc. Women and children are also exposed to such pollution

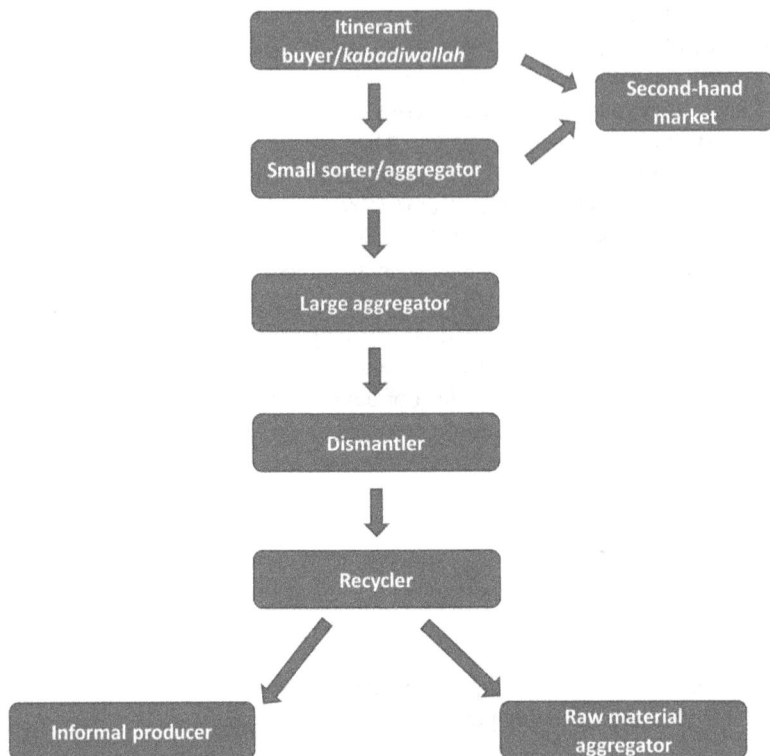

FIGURE 11.2   Actors and steps in the informal e-waste recycling sector.

since these dangerous activities are often carried out in homes and back-yards, and they are even employed as workers in some cases. Damaging effects on workers' health in the informal e-waste recycling sector in India have been extensively documented in the research literature (Basel Action Network 2002; Greenpeace 2005; Awasthi et al. 2016; Gangwar et al. 2019).

2. *Environmental pollution*: Since the direct health impacts on workers are disregarded in the informal e-waste recycling sector, it is not surprising that the wider consequences of environmental pollution would remain unaddressed as well. This is partly due to ignorance and partly a result of lack of environmental enforcement in informal clusters which are often situated inside densely populated settlements. Solid waste generated from the processes, including leftover material with no market value, is typically disposed indiscriminately and is often found in dumps and ditches immediately outside such clusters/settlements. Toxic fumes and smoke from burning plastic or other crude chemical processes spread all over the vicinity and contribute to overall air pollution. Liquid waste from acid/chemical baths is simply poured into nearby drains, which ends up contaminating the soil, groundwater, and nearby surface water bodies. Levels and impacts of toxic pollution generated by informal e-waste recycling have been

extensively documented in the research literature (Basel Action Network 2002; Greenpeace 2005; Awasthi et al. 2016; Gangwar et al. 2019).

3. *Resource recovery*: The informal sector is not only limited by technology and finance, but also by what immediate market value can recovered materials have in their own limited networks. This automatically means that they are only able to recover and recycle a relatively small fraction of recoverable materials in the e-waste stream (Dutta and Goel 2021). Experts estimate that the material recovery efficiency of the Indian informal e-waste recycling sector is in the range of 20% to 30%, mostly restricted to precious metals. In contrast, the best e-waste recycling plants in the world have a material recovery efficiency of more than 95%. Government of India research institutes, such as the Centre for Materials for Electronics Technology (C-MET) and the Central Institute of Petrochemicals Engineering and Technology (CIPET), have developed indigenous technologies that can achieve material recovery efficiencies of 80% to 90% (NITI Aayog and MeitY 2019). From a circular economy perspective, this is the most important argument for upgrading e-waste recycling efforts in India.

11.5 EXTENDED PRODUCER RESPONSIBILITY (EPR): A GAME CHANGER?

Extended Producer Responsibility (EPR) implies that the producer should shoulder some responsibility for products even after the EoL stage, and EPR has become an internationally significant environmental policy tool to promote circular economy (OECD 2014). In India, the E-waste Management and Handling Rules 2012 had introduced the concept of EPR, but there was little initial progress due to the absence of specific guidelines on compliance. The E-waste Management Rules 2016 were promulgated to make effective EPR implementation central to e-waste recycling.

The 2016 Rules introduced targets for each producer on the basis of their annual sales, and each producer is supposed to meet these targets by channelling e-waste to formal recyclers with verifiable documentation. Targets are supposed to be gradually increased annually, with a 70% target expected to be achieved within 7 years. Many options are available to producers to meet their respective targets. They may channelise e-waste on their own or through partnerships. They are also free to utilise economic instruments such as Advance Recycling Fee (ARF) or Deposit Refund Scheme (DRS) that provide incentives to consumers to bring back used equipment to authorised dealers (NITI and MeitY 2019).

The 2016 Rules have brought about certain visible changes. Large electronics manufacturers have established Producer Responsibility Organisations (PROs) for collection and channelling of e-waste to formal recyclers with documentation, determining that such a collective approach would be more efficient and effective than individual efforts. Substantial budgetary allocations have been made by producers to support PROs, in light of the significant penalties arising from non-compliance with recycling targets. Due to the potential opportunity arising out of these developments, close to 150 formal companies have set up facilities in India (Hinchliffe et al. 2019).

However, PROs and formal recyclers have not found their journey smooth sailing. Several years after the 2016 Rules, the majority of e-waste continues to be handled by the informal sector, reflecting its entrenched advantages. Leakage of e-waste to the informal sector reduces the volume available for formal recyclers, potentially jeopardising their investment. Such leakage is also compounded by "cherry picking", when materials with lower recycling value end up eventually reaching the formal channel after more valuable materials have been culled out of the e-waste stream (Hinchliffe et al. 2019). Formal recycling is a capital-intensive business requiring expensive technology, often imported from abroad. Such enterprises' viability depends on revenue generated from selling recovered material to exceed their own material and operating costs. High acquisition cost of e-waste has emerged as another significant problem. Consumers are used to getting an "attractive" price from informal sector collectors and are loathe to deposit e-waste in formal channels organised by PROs. Informal sector collectors are able to offer relatively high prices to consumers since they are often able to extract working components for the refurbished market, and informal recycling does not have to spend on any sort of legal compliance, environmental or otherwise (NITI and MeitY 2019; Hinchliffe et al. 2019). Responsible PROs who invest in developing the entire ecosystem with substantial outreach to a wide range of stakeholders also find themselves at a disadvantage compared to less scrupulous operators who cut corners to maximise profit from valuable materials, even flouting strict documentation mandates (Singhal 2019).

11.6 APPROACHES TO IMPROVE RECYCLING AND RECOVERY: FORMAL–INFORMAL SECTOR INTEGRATION

For the foreseeable future, the informal sector is likely to be a dominant part of the waste management landscape in India. Therefore, any efforts to create a completely formal recycling ecosystem that bypasses the informal sector are unlikely to succeed. Acknowledging this reality is the key to conceptualisation of approaches that attempt to integrate the informal and formal sectors for achieving the best possible outcome under Indian conditions.

The informal sector faces significant barriers that, on the surface, make it difficult to attempt reform. These include the following (Hinchliffe et al. 2019; NITI and MeitY 2019):

- While the bigger actors may regularly undertake significant financial transactions, they are locked out of the formal financial system since they are unregistered enterprises. Financing at the scale necessary for formalisation is only possible through loans from banks or other financial institutions.
- Due to a combination of a lack of awareness, lack of technical skills, and lack of financing, advanced equipment that may ensure high recovery rates with safety and pollution control cannot be purchased and utilised.
- Due to the lack of education and awareness, they are typically uninformed about government rules associated with formal enterprises as well as government programmes that may exist to aid the informal sector.

- Since they operate in a legal void, it is impossible for them to access land necessary for formal enterprise establishment and operation.
- Since they have long existed on the margins of society, engaging in enterprises that have no legal recognition, they are vulnerable to exploitation by unscrupulous state and non-state actors. Unsurprisingly, they have developed an unhealthy suspicion and scepticism about legal mandates, government paperwork, and intervention by "outsiders".

Despite the existence of these serious barriers, experience has shown that through patient, collaborative work with multiple stakeholders, progress is often possible. Various actions, interventions, and initiatives have been undertaken by a variety of actors such as civil society groups (Toxics Link, Chintan), social enterprises (SAAHAS), informal sector associations and unions (SWaCH, HRA, SEWA), manufacturing companies (Microsoft/Nokia), and international agencies like GIZ and EU. These pilot initiatives have demonstrated limited success in legal registration, start-up financing, and establishing upstream–downstream links. However, the role of the informal sector has not been recognised under the 2016 Rules, and such pilot interventions have not been actively supported or scaled up by the government. This has often led to further demotivation in the informal sector (Hinchliffe et al. 2019).

Informed by prior experiences, it is important to conceptualise an alternative approach that is cognisant of Indian realities. The basic idea is simple: Harnessing the collection efficiency of the informal sector and the recycling efficiency of the formal sector simultaneously, while establishing effective linkages between them. The proposed idea is easily understood with the help of a

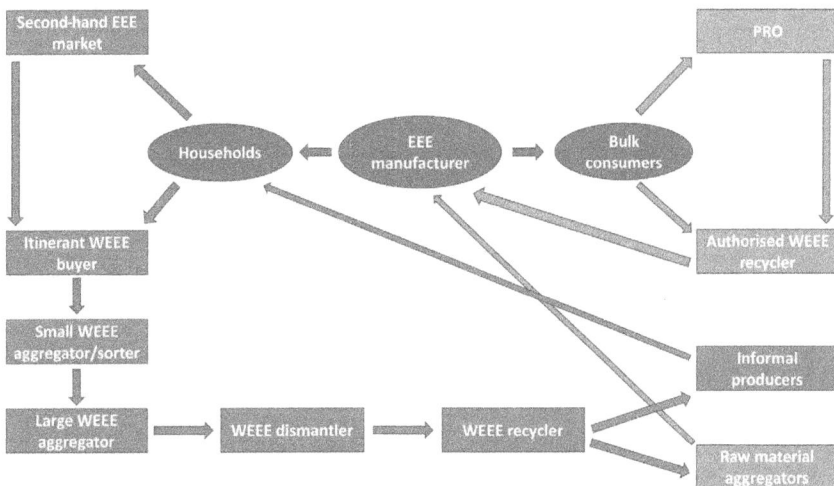

FIGURE 11.3   Current scenario depicting the fate of e-waste in India.

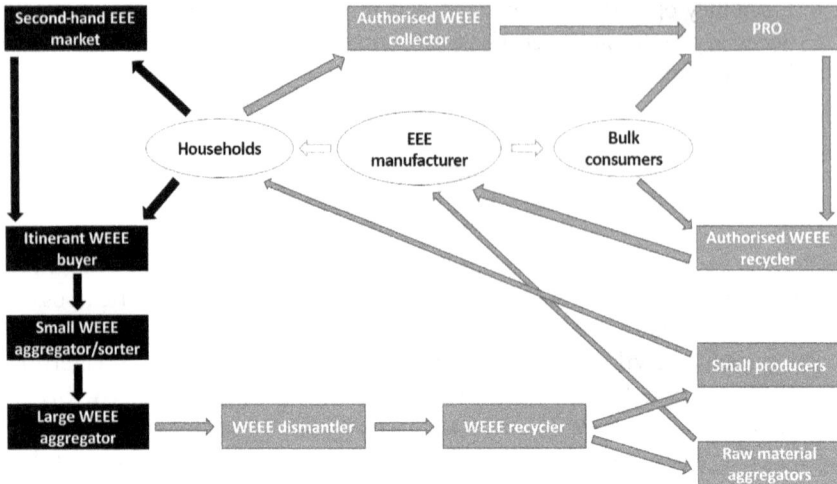

FIGURE 11.4 Proposed alternative scenario for an improved e-waste management system. WEEE, (Waste) Electronic and Electrical Equipment; Dark grey box/arrow, Actor/material flow in the post-consumer informal sector; Light grey box/arrow, Actor/material flow in the post-consumer formal sector; White box/arrow, Actor/material flow in the pre-consumer sector.

schematic diagram. Figure 11.3 depicts the existing situation where the informal and formal actors and value chains exist largely independent of each other. In contrast, Figure 11.4 depicts the proposed alternative approach reflecting partial formalisation of some informal sector actors and judicious integration between the two value chains.

Pertinent points to note about the proposed alternative approach:

- Waste collectors and aggregators do not create environmental hazards and can remain in the informal sector, continuing highly efficient collection and channelisation of e-waste for refurbishment or dismantling.
- Informal dismantlers and recyclers, who are typically larger enterprises, may be enticed to formalise through appropriate capacity development and handholding. Upon legal registration, they will have access to land, finance, and new technology. Even upon formalisation, they will belong to the category of small and medium enterprises (SMEs) and can be clearly distinguished from the large corporate recyclers currently working with PROs. Access to better technology and skills, coupled with formal regulation, will ensure minimisation of environmental and safety issues.
- Several informal enterprises can pool together to set up a single formalised entity, thereby reducing the cost of land, technology, and so on to each actor.

- Adoption of indigenous technology developed by CIPET/C-MET, mentioned in Section 11.4, will reduce investment and operating cost, helping to make their enterprises more viable. (This may be applicable to the large formal recycling companies as well.)
- Using better technology, the newly formalised SMEs can extract a wider range of materials from e-waste. This would increase their revenue and justify their upfront investment, enabling them to make a bigger profit.
- Judicious linkages would be established between various actors in the informal and formal sectors, as depicted in Figure 11.4. Since informal sector dismantling/recycling is often quite specialised (see Section 11.3), it is conceivable that profitable linkages will develop on the basis of geographic concentration, material specialisation, and transport distance optimisation. Further, there will be a limit to the range of materials the SMEs can extract; therefore, linkages with the formal sector will allow the extraction of a wider range of materials, potentially at mutual benefit.
- The large formal recycling companies, working closely with PROs, will need to target bulk generators to ensure their long-term viability, while the informal sector may concentrate on collection from households.
- Through formalisation, the newly created SMEs will gain legal recognition and improved social status, in addition to better revenue and profitability. In addition, they will be able to access the many benefit schemes available under the Ministry of Micro, Small and Medium Enterprises, especially the capital subsidy scheme.

For this alternative vision to be realised, several steps need to be taken:

1. The central role of the informal sector must be recognised by the government and reflected accordingly in policy. Formalisation of dangerous unregulated industries is in the government's interest, since it reduces hazards and generates new tax revenue. Support for small, pilot initiatives will enable scaling up, and demonstration of success will lead to wider emulation in a virtuous cycle. Different government schemes already exist to aid SMEs, including capital subsidy, attractive financing, etc.; a supportive framework is needed to incentivise informal entities to become registered SMEs.
2. Even if the government plays a supportive role, informal entities will need substantial capacity development as well as patient handholding interventions to guide them through many aspects of the formalisation process including registration, permits, applying for loans/subsidies, choice of appropriate technology, access to land, etc. A wide range of stakeholders from civil society organisations to international aid agencies already have experience in such interventions, and scaling them up will only be possible if the government continues to play a supportive role.
3. Access to land is a common problem in urbanised areas of India, and the government will have to allocate suitable land at concessional rates to enable these newly formalised SMEs to take off. Centralised industrial parks may

not always be workable; distances from existing e-waste clusters may make such enterprises unviable. The ideal solution would be the identification and allocation of land close to existing e-waste clusters. Since only SMEs are being conceived, large land parcels will not be necessary, making the goal attainable.

4. A wide-ranging multi-stakeholder dialogue would be necessary that brings the informal sector, manufacturers, PROs, formal recyclers, government agencies, and civil society organisations together. This is of utmost importance to develop workable linkages between the informal and formal sectors that are often in competition. If a supportive framework for SMEs is put in place, through ongoing dialogue, the formal recycling sector will realise its own long-term benefit from formal–informal integration. If informal recyclers become regulated SMEs, then they will have to pay taxes and meet environmental/safety regulations; therefore, they cannot compete "unfairly" with the formal sector. In turn, the new SMEs will realise the potential for higher revenue and profits by channelling materials to the formal sector, especially materials they do not have the technology to extract.

5. As already noted, the formal recyclers, working with PROs, will need to depend heavily on e-waste collected from bulk generators. Large quantities can be collected from institutional actors at a single collection event, making collection/transportation cost-effective. Overall, larger volumes from bulk generators will contribute to the financial viability of the formal recyclers. The 2016 Rules clearly state that bulk generators have to dispose of their e-waste through authorised channels with proper documentation, but awareness and compliance are seriously lacking. The government has to significantly increase outreach efforts to bulk generators, roping in other stakeholders as necessary, coupled with appropriate enforcement of compliance.

11.7 CONCLUSION

In high-income countries, the informal sector does not exist, and the waste management ecosystem is simplified. But for the foreseeable future, the informal sector in India is likely to exist and be an important part of the waste management sector. Therefore, it is incumbent on policymakers to take advantage of the benefits provided by the informal sector in terms of employment and efficient collection, while minimising the harm caused from dangerous recycling practices. Movement towards the formalisation of the large dismantlers and informal recyclers is the only reasonable path forward; this approach has the potential to get the best out of each set of actors—the informal sector's collection efficiency and the formal sector's recycling efficiency.

Of course, improving recycling is only part of the answer in the quest for a resource-efficient and circular economy. Other approaches, outside the scope of this chapter, ought to be pursued in parallel. These include design changes that prolong life and allow repair, tackling the scourge of planned obsolescence, sharing economy, and so on. But many of these other approaches demand a longer time horizon, while improving recycling may be considered "low-hanging fruit" that can provide measurable benefits in the short-to-medium term.

REFERENCES

Anantharaman, M. 2014. Networked ecological citizenship, the new middle classes and the provisioning of sustainable waste management in Bangalore, India. *Journal of Cleaner Production* 63:173–183.

Awasthi, A. K., M. Wang, Z. Wang, M. K. Awasthi, and J. Li. 2018. E-waste management in India: A mini-review. *Waste Management and Research* 36(5):408–414.

Awasthi, A. K., X. Zeng, and J. Li. 2016. Environmental pollution of electronic waste recycling in India: A critical review. *Environmental Pollution* 211:259–270.

Basel Action Network. 2002. *Exporting Harm: The High-Tech Trashing of Asia.* Seattle, WA: Basel Action Network. http://svtc.org/wp-content/uploads/technotrash.pdf

Bhatt, G., M. Khanna, B. Pani, and A. Chaturvedi. 2016. Consumption pattern, behaviour and awareness towards e-waste among mobile users in New Delhi. *International Journal of Environmental Policy and Decision Making* 2(1):28–40.

Centre for Science and Environment [CSE]. 2008. *Rich Lands Poor People: Is Sustainable Mining Possible?* State of India's Environment 6: A Citizen's Report. New Delhi: Centre for Science and Environment.

Chaudhary, K., and P. Vrat. 2019. An investigation into consumer behaviour towards e-waste disposal practices in India. *Industrial Engineering Journal* 12(1):14–29.

Chikarmane, P., L. Narayan, and B. Chaturvedi. 2008. *Recycling Livelihoods: Integration of the Informal Recycling Sector in Solid Waste Management in India.* Report prepared for GTZ by SNDT Women's University and Chintan Environmental Research and Action Group, New Delhi.

Down to Earth. 2020. E-waste up 43% in 3 years. *Down to Earth*, September 23. www.downtoearth.org.in/news/agriculture/as-told-to-parliament-september-23-2020-e-waste-up-43-in-3-years-73512

Dutta, D., and S. Goel. 2021. Understanding the gap between formal and informal e-waste recycling facilities in India. *Waste Management* 125:163–171.

European Commission. 2020. *EU Circular Economy Action Plan.* https://ec.europa.eu/environment/circular-economy/

FICCI and Accenture Strategy. 2018. *Accelerating India's Circular Economy Shift: A Half-Trillion USD Opportunity—Future-Proofing Growth in a Resource-Scarce World.* New Delhi: FICCI and Accenture Strategy. http://ficci.in/spdocument/22977/FICCI-Circular-Economy.pdf

Forti, V., C. P. Baldé, V., R. Kuehr, and G. Bel. 2020. *The Global E-Waste Monitor 2020: Quantities, Flows and the Circular Economy Potential.* Bonn/Geneva/Vienna: United Nations University (UNU), International Telecommunication Union (ITU) & International Solid Waste Association (ISWA). http://ewastemonitor.info/

Gangwar, C., R. Choudhari, A. Chauhan, A. Kumar, A. Singh, and A. Tripathi. 2019. Assessment of air pollution caused by illegal e-waste burning to evaluate the human health risk. *Environment International* 125:191–199.

Government of Japan. 2020. *Circular Economy Vision 2020 Compiled.* Ministry of Economy, Trade and Industry. www.meti.go.jp/english/press/2020/0522_003.html

Greenpeace. 2005. *Recycling of Electronic Wastes in China and India: Workplace and Environmental Contamination.* Exeter, UK: Greenpeace International. www.greenpeace.org/static/planet4-international-stateless/2005/08/ee56bf32-recycling-of-electronic-waste.pdf

Hinchliffe, D., M. Hemkhaus, and R. Arora. 2019. Informal-formal partnerships in the Indian electronics waste sector. *Vikalpa: The Journal for Decision Makers* 44(3):136–138.

Indo-German Environment Programme [IGEP]. 2013. *India's Future Needs for Resources: Dimensions, Challenges and Possible Solutions*. New Delhi: GIZ. www.resource-recovery. net/sites/default/files/6_resourceefficiency_report_final.pdf

La Rosa, E. 2020. Planned obsolescence and criminal law: A problematic relationship? In *Sustainability and Law: General and Specific Aspects*, eds. V. Mauerhofer, D. Rupo, and L. Tarquinio, 221–236. Cham, Switzerland: Springer.

LeBel, S. 2016. Fast machines, slow violence: ICTs, planned obsolescence, and e-waste. *Globalizations* 13(3):300–309.

Maitre-Ekern, E., and C. Dalhammar. 2016. Regulating planned obsolescence: A review of legal approaches to increase product durability and reparability in Europe. *Review of European, Comparative and International Environmental Law* 25(3):378–394.

McKinsey and Co. 2015. *Europe's Circular Economy Opportunity*. McKinsey Center for Business and Environment. www.mckinsey.com/~/media/McKinsey/Business%20Functions/ Sustainability/Our%20Insights/Europes%20circular%20economy%20opportunity/ Europes%20circulareconomy%20opportunity.ashx

NITI. 2017. *Strategy on Resource Efficiency*. New Delhi: Government of India. https://niti.gov. in/writereaddata/files/document_publication/StrategyOnResouceEfficiency_0.pdf

NITI Aayog and MeitY. 2019. *Strategy on Resource Efficiency in Electrical and Electronic Equipment Sector*. New Delhi: Government of India. www.eu-rei.com/pdf/publication/ NA_MeitY_RE%20Strategy%20in%20EEE%20Sector_Jan%202019.pdf

OECD. 2014. *The State of Play on Extended Producer Responsibility (EPR): Opportunities and Challenges*. Issues Paper for the Global Forum on Environment: Promoting Sustainable Materials Management through Extended Producer Responsibility (EPR). Tokyo, Japan: OECD. www.oecd.org/environment/waste/Global%20Forum%20Tokyo%20Issues%20 Paper%2030-5-2014.pdf

Rathore, G. 2020. Circulating waste, circulating bodies? A critical review of e-waste trade. *Geoforum* 110:180–182.

Satyro, W., J. Sacomano, J. Contador, and R. Telles. 2018. Planned obsolescence or planned resource depletion? A sustainable approach. *Journal of Cleaner Production* 195:744–752.

Singhal, P. 2019. Disrupting the status quo via systematic transformation: PROs and e-waste. *Vikalpa: The Journal for Decision Makers* 44(3):151–153.

Smyth, J. 2020. Industry needs a rare earths supply chain outside China. *Financial Times*, July 28. www.ft.com/content/fc368da6-1c86-454b-91ed-cb2727507661

United Nations Environment Programme [UNEP]. 2016. *Global Material Flows and Resource Productivity*. Assessment Report for the UNEP International Resource Panel. www.resource panel.org/reports/global-material-flows-and-resource-productivity-database-link

World Bank. 2017. *China Circular Economy Promotion Law*. Public-Private-Partnership Legal Resource Center of the World Bank Group. https://ppp.worldbank.org/public-private-partnership/library/china-circular-economy-promotion-law

12 Consumer Behaviors in the Circular Economy with Special Focus on E-products

Tetiana Shevchenko and Yuriy Danko

CONTENTS

12.1 INTRODUCTION

It is impossible to imagine modern life without electrical and electronic products (e-products) which are constantly being improved. Over time, e-products become obsolete or reach end-of-life (EoL) and eventually go into the waste—e-waste. According to the Wang and colleagues (2013) study, the developed world produces more than 900 types of electronic products which include more than 1,000 different materials and substances, most of which are toxic, such as mercury, lead cadmium, arsenic, and chromium (Mudgal et al. 2013). Current estimates predict close to 50 Mt of waste electrical and electronic equipment (WEEE) worldwide per year (Wang et al. 2013). It was calculated that in the United States, on average, more than 5 Mt of e-waste is generated annually. Each citizen of the EU generates about 15 kg of e-waste per year for an estimated total of more than 7 Mt annually (Tanskanen 2013). According to the Baldé and colleagues (2017) study, only 20% (8.9 Mt) of e-waste worldwide is documented for recycling, and 80% (35.8 Mt) is undocumented. The United States, United Kingdom, and the European Union are the key exporters

DOI: 10.1201/9781003301899-17

of e-waste to India, Nigeria, and China (Baldé et al. 2017; Nduneseokwu et al. 2017; Chi et al. 2014).

Today, many developed countries in the world use a CE paradigm as a conceivable alternative to intensify the sustainability of the existing e-waste management system. The last momentum of the CE conceptualization has been catalyzed by the 'butterfly figure' presented by the Ellen MacArthur Foundation outlining the circular strategies' hierarchy by priorities: 'recycling', 'remanufacturing', 'reuse', 'refurbishment', 'repair', and 'repurpose' (EMF 2013). To follow the hierarchy, Blomsma and Brennan (2017) highlight that, after all, the circular economy concept implies moving away from evaluating singular strategies and moving toward the evaluation of two or more circular strategies working in parallel or in sequential configuration. However, the implementation of circular strategies remains in the early stages (McDowall et al. 2017), because the transition from a linear business to a circular one brings a variety of practical challenges and creates a range of barriers (Mont et al. 2017; Bocken et al. 2016) including barriers related to consumer behavior. The significance of consumer behavior barriers is explored in multifarious studies on CE-related barriers, e.g. Kirchherr et al. (2018), Camacho-Otero et al. (2018), van Eijk (2015), and Jesus and Mendonça (2018). Kirchherr and colleagues' (2018) study ascertained that cultural barriers, which include behavioral barriers, determine the existence of other types of barriers to a clear extent.

In order to understand the mechanism and tools to mitigate possible behavioral obstacles and challenges, this review attempts to clarify more circular e-products' consumption alternatives and behaviors, in light of the needs of circular strategies to be implemented. Definitively speaking, the consumer acceptance of e-products generated by circular business (circular e-products) and proper EoL or obsolete e-product discarding are vital for companies capitalizing on the residual value of products and materials. These two points will be given due attention in providing a holistic vision of circular-economy-related behaviors in terms of all possible consumer contributions along the lines of the circular strategies. To conduct a literature review, the research articles were selected from scientific peer-reviewed journals in English. Such search engines as Web of Science, Scopus, and Google Scholar were used for the selection of literature with the keywords 'circular economy', 'e-product', 'e-waste' 'consumer behavior', 'circular behavior', 'recycling behavior', 'purchasing behavior', 'product acceptance', 'circular product', and 'circular consumption'. In addition, some papers were searched for references of the initially identified relevant studies.

12.2 RELATED STUDIES

12.2.1 UNDERSTANDING CONSUMER BEHAVIOR IN A CE

The significance of the consumer in the success of pro-circular initiatives is actively discussed by the scientific community in various countries regardless of the progress achieved in moving toward more circular production and consumption alternatives (Mugge et al. 2017; Weelden et al. 2016; Hazen et al. 2017; Wang et al. 2013; Borthakur and Govind 2016; Nduneseokwu et al. 2017). The CE-related behaviors have been examined across different product groups including household electronics (Pérez-Belis et al. 2015; Yla-Mella et al. 2015), clothing and fabric (Laitala and Klepp 2018; Bianchi and Birtwistle 2012; Goworek et al. 2012), furniture (Edbring et al. 2016), vehicles (Matsumoto et al. 2018; Despeisse et al. 2015), etc. For e-products, the most discussed issues today are remanufactured e-products acceptance and e-waste proper discarding or recycling behavior. From a regional perspective, countries in North America and Europe are focused on the issue of remanufactured e-product acceptance, while countries in Asia and Africa are more concerned with the issue of consumer recycling behavior.

There are several comprehensive studies as an initial step for deeper understanding of the consumer role in the transition to circular business models. The study of Wastling and colleagues (2018) focused on defining key consumer behaviors to operationalize circular businesses and outlining how design can contribute to enhancing circular behaviors (Wastling et al. 2018). Edbring and colleagues (2016) outlined consumer attitudes to more circular consumption alternatives, namely collaborative consumption, models for extending the life of product, and access-based consumption. The scholars conclude that consumer attitudes depend on the product groups and vary with the mentioned consumption models (Edbring et al. 2016).

There are attempts to define the essence of consumer behavior contributing to the CE. In this line, frequent terms such as 'circular behavior', 'pro-circular behavior', and 'consumer switching behavior' are used in literature. In particular, Muranko and colleagues (2018) defined a 'pro-circular behavior' as activities that are brought due to the prioritizing of resource-efficiency. However, no contribution to resource-efficiency is due to circular business models. To highlight the point, Bocken and colleagues (2016) argued that narrowing resource flow or resource efficiency is aimed at involving fewer resources per product and has no concern with the multiple usage of materials and products. These scientists concluded that only slowing and closing the loop strategies reflect the essence of material and product circularity. Nevertheless, as highlighted by Shevchenko and Kronenberg (2020), extending the material life-cycle leads to the changing of the entire resource cycle for which the material cycle is a segment. Hence, the resource efficiency can be stipulated by pro-circular activity as the consequence of slowing and closing the loop but not the ultimate goal of circular behavior as determined by Muranko et al. (2018).

To define consumer behaviors and activities related to operationalization of circular business models, Wastling and colleagues (2018) used the term 'circular behavior'. The scholars argued that the essence of 'circular behavior' can be inter preted over an influence to slowing and closing resource loops according to understanding provided by Bocken et al. (2016) and Stahel (2010). According to these authors' interpretations, slowing of resource loops means the highest utilization of products, either through extending product lifetime or via product service systems or sharing schemes. Closing resource loops implies ensuring recycling of materials in a closed-loop fashion at the product EoL. Mugge (2018) provides interpretation of behaviors toward the circular products in a similar manner highlighting the necessity to realize the full potential of the so-called inner loops of the circular economy (Mugge 2018).

12.2.2 Consumer Acceptance of E-products Generated
by Circular Businesses

In the past decade, a huge number of studies have explored consumer willingness to pay across various product categories including e-products that contribute to a CE (Harms and Linton 2016; Michaud and Llerena 2011; Anstine 2000; Hamzaoui-Essoussi and Linton 2010).

Table 12.1 presents recent studies on consumer acceptance of CE-related e-products. As noted before, the most debated issue is remanufactured product acceptance probably due to the trend of remanufacturing in developed countries (Abbey et al. 2017; Khor and Hazen 2017; Matsumoto et al. 2018). For instance, remanufacturing is a significant share of the UK economy. It represents UK-wide savings of 270 kt of raw materials and 800 kt of CO_2, at an estimated value of £5 billion (Parker and Butler 2007). The recycled products' consumer acceptance issue (Sun et al. 2018; Hamzaoui-Essoussi and Linton 2010) has less debate than the first one, despite the fact that recycled products contribute toward waste avoidance today and considering that it's a final and beginning point for cycling a material. The least under discussion question today is consumer choice of reusable and recyclable products as circular alternatives to linear products without any potential for reuse and recycling. Moreover, the issue of these products' acceptance is considered in the literature in the context of green behavior (Biswas and Roy 2015; Böcker and Meelen 2017; Akehurst et al. 2012), thereby revealing that the issue is not gaining due relevance.

Further, the literature review revealed the interest in studying consumer behavior in the context of consumer activities contributing to two or three phases of the consumption stage of a product lifecycle. For instance investigations of Wastling et al. (2018) and Bovea et al. (2018) focused on products covering accepting (purchasing), using, and disposal activities. There are a number of studies aimed at '*pro-circular procurement*' and '*careful usage*' behaviors (Böcker and Meelen 2017; Mugge et al. 2017; Piscicelli et al. 2018; Young et al. 2010) or '*pro-circular procurement and appropriate disposal behaviors*' (Pérez-Belis et al. 2015). Additionally, in light of consumer accepting activity, there are the studies which focus on several products that contribute to different strategies, namely remanufactured and recycled products (Sun et al. 2018; Hamzaoui-Essoussi and Linton 2014), recycled and up-cycled products (Böcker and Meelen 2017; Park and Lin 2018), and remanufactured and shared products (Camacho-Otero et al. 2018). However, literature review brought to light the lack of studies that focused on product category having several characteristics in one, for instance, reusable recycled product or recyclable remanufactured product. These products definitely have other contributions to a CE, i.e., such products contribute to two or more circular strategies working in parallel or in sequence, versus the products with one characteristic.

12.2.3 Central Matters of Discussion in the Issue on
Consumer Acceptance of E-products for the CE

The matters of discussion of studies dealing with the acceptance of e-products that contribute to the CE are as follows: attitude and intention, key determinants, barriers

TABLE 12.1

Recent Studies on Consumer Acceptance of CE-related E-products

Study Affiliation	Recyclable and Reusable Product	Remanufactured Refurbishment Product	Recycled Product	Product for Sharing	Up-cycled Product	Second-hand Product
Bovea et al. 2018						X
Sun et al. 2018		X	X			
Abbey et al. 2017		X				
Böcker and Meelen 2017	X		X	X	X	
Wilson 2016					X	
Mugge et al. 2017		X				
Weelden et al. 2016		X				
Gaur et al. 2015		X				
Park and Lin 2018			X		X	
Goworek et al. 2012						X
Biswas and Roy 2015	X					
Jiménez-Parra et al. 2014		X				
Khor and Hazen 2017		X				
Piscicelli et al. 2018				X		X
Barbarossa and Pastore 2015	X					
Hamzaoui-Essoussi and Linton 2010			X			
Park 2015					X	
Harms and Linton 2016		X				
Pérez-Belis et al. 2015				X		X
Hazen et al. 2017		X				
Edbring et al. 2016				X		X
Young et al. 2010	X		X			X
Wang et al. 2013		X				
Bei and Simpson 1995			X			
Camacho-Otero et al. 2018		X		X		
Bridgens et al. 2018					X	
Neto et al.2016		X				X
Matsumoto et al. 2018		X				
Godelnik 2017				X		
Yu and Lee 2019					X	
Hamzaoui-Essoussi and Linton 2014		X	X			
Wastling et al. 2018				X		
Mokan et al. 2018	X					

of willingness to pay, pricing, perceived quality, marketing tools, motivation and incentives (Table 12.2).

E-product remanufacturing is an EoL product strategy aimed at recovering the residual value of used-before products. This strategy stipulates restoring old or obsolete e-products to resell to a new consumer (Weelden et al. 2016). Although EoL products can be refurbished or remanufactured at a lower cost in comparison with the initial manufacturing cost, consumer value of these products is less than new ones (Hazen et al. 2017; Abbey et al. 2017). Hence, consumer intention to purchase a remanufactured product is lower than for new product (Wang et al. 2013). Understanding of the determinants that influence consumer acceptance of remanufactured products was elucidated in the studies marked in Table 12.2. One of the first attempts to investigate how the determinants underpin consumer purchase intention concerning remanufactured products was made by Wang and colleagues (2013). The authors identified that purchase intention is directly influenced by purchase attitude attended by perceived behavioral control and indirectly influenced by knowledge about product via attitude, perceived risk, and perceived benefit (Wang et al. 2013).

An empirical justification concerning remanufactured product costing and quality assumptions was conducted by Hazen and colleagues (2017). These scientists examined the issue of willingness to pay for remanufactured products in light of the two factors ambiguity tolerance and perceived quality, thereby, trying to explain why consumers perceive remanufactured products as lower quality hence have less willingness to pay them. They concluded that enhancing the willingness to pay for remanufactured products can be accomplished by reducing the level of ambiguity related to their remanufacturing processes (Hazen et al. 2017). In similar vein, an empirical investigation about consumer perception of remanufactured products in terms of price discount and brand equity was made by Abbey and colleagues (2017). The authors indicated that discounting has a linear effect on refurbished/remanufactured product attractiveness whereas the brand was less important to the consumer than remanufactured product quality.

According to Michaud and Llerena's (2011) and Abbey et al. (2017) studies, a refurbished or remanufactured product should be considered, to a certain extent, as a green product because of the environmental benefits of the remanufacturing process. Michaud and Llerena (2011) argued that consumers are more willing to pay for a refurbished product when they know the product is 'green', i.e., consumers tend to value the refurbished product less than the conventional one if they are not informed about its respective environmental impacts. Developing this thought, Abbey and colleagues (2017) introduces the concept of 'negative attribute perceptions', i.e., disgust, which has a detrimental effect on attractiveness of remanufactured products. Searching for the tools restraining the effect of this attribute, these authors highlight that consumers who consider a remanufactured product as a green one typically found such product to be significantly more attractive (Abbey et al. 2017).

The next type of e-products that contribute to the CE by slowing and closing the loop in the future is potentially circular products, namely reusable and recyclable e-products. Such products have various features such as recyclable, reusable, repairable, durable, and dismountable ones. Building on the thoughts about consumer recycling behavior as a part of green behavior (Zhao et al. 2014; D'Souza et al. 2007),

TABLE 12.2
Key Matters of Discussion on Consumer Acceptance of CE-related E-products

	Remanufactured/ Refurbished E-products	Reusable and Recyclable E-products	Recycled E-products	Shared E-products	Used Before/Second-Hand E-products
			E-product Types Contributing to CE		
Drivers	Gaur et al. 2015	Pickett-Baker and Ozaki 2008	Park and Lin 2018	Piscicelli et al. 2018	Bovea et al. 2018
	Wastling et al. 2018	Mokan et al. 2018	Wong and Mo 2013	Cohen and Muñoz 2016	Pérez-Belis et al. 2015
	Harms and Linton 2016	Biswas and Roy 2015		Godelnik 2017	Piscicelli et al. 2018
Determinants	Weelden et al. 2016	Paco et al. 2014	Bei and Simpson 1995	Roos and Hahn 2019	Piscicelli et al. 2018
	Abbey et al. 2017	Trivedi et al. 2015	Anstine 2000	Zhu et al. 2019	Neto et al. 2016
	Michaud and Llerena 2011	Akehurst et al. 2012	Wong and Mo 2013	Davidson et al. 2017	
Attitudes and intention	Khor and Hazen 2017	Mobley et al. 1995	Young et al. 2010	Camacho-Otero et al. 2018	Young et al. 2010
	Wang and Hazen 2016	Young et al. 2010	Mobley et al. 1995	Piscicelli et al. 2015	Edbring et al. 2016
	Wang et al. 2013		Park and Lin 2018	Edbring et al. 2016	
Barriers	Guide and Li 2010	Barbarossa and Pastore 2015	Sun et al. 2018	Bielefeldt et al. 2016	Edbring et al. 2016Neto et al. 2016
	Matsumoto et al. 2018	Biswas and Roy 2015	Park and Lin 2018	Godelnik 2017	
	Sun et al. 2018				
Pricing, quality, marketing, motivation	Neto et al. 2016	Lieder et al. 2017	D'Souza et al. 2007	Piscicelli et al. 2018	Goworek et al. 2012
	Hamzaoui-Essoussi and Linton 2014	Vadde et al. 2007	Hamzaoui-Essoussi and Linton 2014	Park and Armstrong 2017	Edbring et al. 2016
	Hazen et al. 2017	Laroche et al. 2001	Bei and Simpson 1995	Böcker and Meelen 2017	

recyclable and/or reusable product can be also considered as green product, to a certain extent, since its end-of life processing has positive environmental benefits due to closing or slowing of the loop in the future. As for green products, even though consumers are significantly worried about environmental degradation, the existing market share of these products remains quite small (Barbarossa and Pastore 2015) due to a gap between attitude and behavior. According to the study of Young (2010), 30% of consumers report their concerns about environmental issues but struggle to translate these concerns into purchases. The scholar interprets such situations as 'values–action gap' or 'attitude–behavior gap'.

One more type of e-products that contributes to the CE by closing the loop in the past is recycled products. Despite the tremendous growth of recycled product production today, few studies were identified dedicated to the consumer acceptance of these products. Anstine (2000) argued that the product produced from recycled materials is perceived to have caused less damage to the environment by some consumers than a product produced from virgin material only. The Mobley et al.'s (1995) study concludes that the presence of recycled materials in a product is a favorable influence to consumers, regardless of the type of product. At the same time, the positive effects of recycling held only for recognized brands, not for new ones. Determinants of recycled product purchasing behavior are addressed more deeply by Bei and Simpson (1995), Wong and Mo (2013), and so on. Bei and Simpson (1995) investigated the determinants of consumer's purchase probabilities toward 11 recycled products based on 'acquisition-transaction utility' theory. These scientists introduced the concept of psychological benefit related to recycled products' acceptance and demonstrated a positive correlation between the psychological benefit from the purchase of recycled products and the probability of purchasing recycled products (Bei and Simpson 1995). Psychological benefit for the consumer from the purchase of a recycled product was added as part of the purchase utility, thereby providing a viewpoint of consumers to aid in the marketing strategies' development toward promoting recycled products (Bei and Simpson 1995). Further, the impact of perceived product quality and environmental knowledge on recycled products' purchasing intention was examined by Sun (2018). Based on the theory of expectancy value and the theory of reasoned action, this investigation indicates that the perception of recycled product risk has a significant negative impact on the perception of this product quality (Sun 2018).

Shared and second-hand products are two more types of e-products that contribute to the CE by slowing the loop, thereby avoiding wastage today. Sharing, gifting, renting, bartering, swapping, lending, and borrowing are alternative ways of consuming aimed at intensifying the use of idle assets, preventing new product purchases, and promoting possessions' reuse (Piscicelli et al. 2015; Roos and Hahn 2019; Zhu et al. 2019; Davidson et al. 2017). Regarding the applicability and feasibility of product service systems, which enable collaborative consumption, there are three main uncertainties: adoption by companies, acceptance by consumers, and environmental implications (Tukker 2015). One of the most valued attributes for consumers are control over things, artifacts, and life itself, in the opinion of Tukker (2015). Product service systems are often less accessible than the competing product since such systems do not allow consumers as much behavioral freedom (Tukker 2015). The issue of second-hand e-products' acceptance was also actively discussed in academia (Bovea et al. 2018; Piscicelli et al. 2018; Edbring et al. 2016). Pérez-Belis et al. (2015) studied the issue of willingness to pay or rent second-hand toys. The result of the survey showed that about 65% of participants were willing to rent their toys for children and 58% to buy second-hand toys (Pérez-Belis et al. 2015). Next, we proceed to examine e-waste recycling behavior of consumers as a debated issue today.

12.2.4 CONSUMER E-WASTE RECYCLING BEHAVIOR

To achieve a high e-waste collection rate, the key question to be answered is how to engage consumers in an appropriate discarding for this type of waste. In this line, different factors affect consumer e-waste recycling attitude, including habits, convenience, knowledge, motivations, incentives as well as socio-economic aspects. Dixit and Vaish (2015) and Wagner (2013) highlight that 'convenience' is a crucial determinant of a proper returning of the EoL or obsolete e-product among other ones. To define the essence of 'convenience', these scholars proceed from minimizing consumer's transaction costs, i.e., all possible consumer efforts required for proper e-waste discarding. Posselt and Gensler (2002) identified several types of transaction costs, namely time costs, planning costs, inventory costs, transportation costs, information cost, and psychological costs. To justify consumer recycling behavior, Wagner (2013) proposed the categorization of 'convenience' in terms of consumer efforts for EoL product collection. This author identified the following five principal categories of 'convenience': propinquity to collection location (minimum distance), convenience–knowledge requirements (minimum time to be informed about requirements), the simplicity of the process, prospects of dropping off EoL products and peripherals (maximum days), and the draw of the collection site (services offered). In addition, it was highlighted that customers use resources while returning EoL products, e.g., money, time, and effort (Dixit and Vaish 2015). Comparing behavioral costs of various collection schemes, it was concluded that, in the case of EoL e-product, average consumer behavioral costs for a drop-off system are higher than for a curbside recycling scheme (Best and Kneip 2011).

In terms of various determinants that affect consumer e-waste recycling behavior, recent studies noted that some countries have a dominant factor compared to others. In particular, Shevchenko and colleagues (2019) argued that consumer disposal behavior key determinants vary across the countries of America, Europe, Asia, and Africa. Namely, the prevalent factor of European consumer disposal behavior is knowledge and the level of awareness. 'Convenience' is a dominant factor of rational e-waste disposal conduct for an American consumer. Financial attributions significantly determine e-waste recycling attitude for Asian consumers in contrast to European or American ones. The informal collection sector for EoL e-products is gaining momentum in some Asian countries since economic incentives are a prevalent factor for the consumer and due to the lack of legislative norms taken place. An analogous situation is in African countries.

For EU countries, according to the Directive 2012/19/EC, consumers have to actively contribute to the EoL e-products' collection and should be encouraged to return EoL products. For this purpose, convenient facilities are created where private households are able to return their waste at least free of charge. Although EoL e-products' collection program aims at providing the convenience in order to minimize consumer transaction costs, i.e., the efforts related to collection (Wagner 2013; Dixit and Vaish 2015), there are apparently still other determinants of consumer recycling behavior that restrain this process in developed countries. In this line, it was highlighted that the proliferation of e-waste collection schemes depends on longevity since recycling habit requires years (Tanskanen 2013). In European countries, e-waste recycling analytics conclude about the existence of competing habits, namely the tendency to store EoL/obsolete e-products of specific categories at home (Bovea et al. 2018; Yla-Mella et al. 2015).

Today, three main challenges of e-waste programs, particularly in developing nations, take place (Table 12.3), namely disposal with generic waste (Nduneseokwu et al. 2017; Nnorom et al. 2009; Darby and Obara 2005, etc.), storing some categories of EoL products at home (Yin et al. 2014; Borthakur and Govin 2016, etc.), and informal sector development (Orlins and Guan 2016; Tansel 2017; Pandey and Govind 2014, etc.).

Improper disposal with generic waste is the 'easiest' solution for consumers, especially for small EEE. The challenges of storing e-waste at home and informal sector development are mainly due to the lack of economic incentives (Yla-Mella et al. 2015; Manomaivibool and Vassanadumrongdee 2012), especially for expensive and quickly becoming morally obsolete e-products. In developing countries, e-wastes are considered as EoL e-products as having intrinsic value according to Borthakur and Govind (2016). Yla-Mella and colleagues (2015) concluded that in developed countries, some EoL e-products, e.g., ICT equipment, are supposed to have some residual value, hence consumers tend to store these e-wastes at home. It should be noted that, irrespective of developing or developed nation, consumer motivation of proper return e-waste depends on economic incentives due to the high price of e-product, and rapid obsolescence increases consumer perception of its residual value. Furthermore, the need for customized incentives for obsolete electronics is evidenced by the fact that e-products are often discarded for fashion reasons or keeping up with technological advances, not because these products have reached their end of life. To describe consumer understanding of specific category product lifetime, Cox and colleagues (2013) developed a product typology. They proposed three types of products: 'up-to-date', 'workhorse', and 'investment'. Relying on the typology, most electronic items are classified as 'up-to-date' goods. An expected lifetime of most e-products is less than 5 years. The consequence is that these products are discarded before the end of functional lifetime for reasons of personal feelings of success or to follow technological advances (Cox et al. 2013; Yla-Mella et al. 2015).

Based on the literature review on consumption alternatives and consumer behaviors contributing to the CE with a special focus on e-products, key insights from the discussion will be highlighted in the next section.

12.3   KEY INSIGHTS AND THOUGHTS FROM THE DISCUSSION

The significance of consumers for CE operationalization is tough to exaggerate since they determine the demand of *pro-circular products*', such as reusable and

TABLE 12.3

Related Works on Key Challenges of Advanced EoL E-products' Recycling Programs

Consumer's Recycling Behavior Challenges	Bovea et al. 2018	Yla-Mella et al. 2015	Tanskanen 2013	Ongondo and Williams 2011	Bakhiyi et al. 2018	Wagner 2013	Manomaivibool and Vassanadumrongdee 2012	Borthakur and Govin 2016	Yin et al. 2014	Nduneseokwu et al. 2017	Nnorom et al. 2009	Baldé et al. 2017	Dwivedy and Mittal 2013	Pandey and Govind 2014	Veenstra et al. 2010	Lu et al. 2015	Wang et al. 2011	Bai et al. 2018	Darby and Obara 2005	Shevchenko et al. 2019	Dixit and Vaish 2015	Orlins and Guan 2016	Tansel 2017
Informal sector development (improper behavior)		•	•	•	•		•	•	•	•	•	•	•	•	•	•	•	•			•	•	•
Storing e-waste at home	•	•		•		•		•	•									•		•			
Disposal with generic waste											•								•				

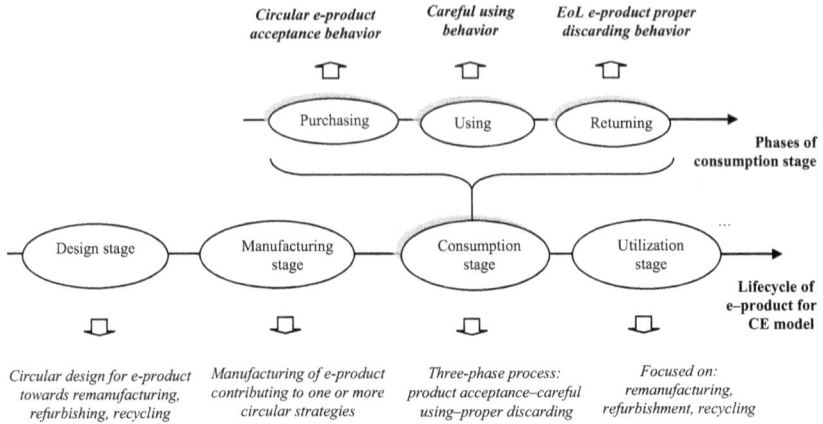

FIGURE 12.1 Lifecycle of e-product in the CE and relevant behaviors.

recyclable ones, refurbished and remanufactured products, recycled products, and repaired and used-before/second-hand products. Furthermore, the consumer returns EoL products to the economic system when they lose their personal value for him, thereby creating preconditions for reuse and recycling. Among all characters in a chain of material and product value maintenance, a consumer is a distinct actor because of the existence of individual factors with underlying behaviors, unlike other actors, namely, the manufacturer, distributor, collector, recycler, and repairer whose activities are more predictable.

Based on the holistic vision of CE-related behaviors via the review provided, it should be noted that the entire pro-circular e-product lifecycle can be considered as a set of stages where the last stage, 'product utilization', gives either new life to the product in the case of refurbishing/remanufacturing, or new life to materials from which the product is formed (Figure 12.1). In the context of consumer behaviors required for circular business, the consumption stage of a product lifecycle can be considered as a three-phase process that includes three types of consumer activities, namely, product purchasing (product acceptance), product using, and EoL product discarding (Shevchenko et al. 2021). According to the phases outlined in Figure 12.1, the following consumer behaviors under consumption stage should be highlighted: circular e-product acceptance behavior, careful using behavior, and EoL e-product proper discarding behavior.

Moreover, in the context of purchasing, using, and recycling activities under the consumption stage of an e-product, the consumer role can be identified as a tri-dimensional one. Consumer activities in terms of three-phase consumption stage, 'to buy—to use—to return', are relevant to specific roles of the product customer, product user, and EoL product holder, respectively. Following on to the study by Zhao and colleagues (2014), we assume that some of these activities can be interdependent. In the context of green behavior, the scholars argued that purchasing behavior is strongly determined by attitudes. An income, age, and perceived consumer effectiveness are the significant predictors of using behavior. In turn, recycling behavior is mainly influenced by using behavior.

Furthermore, purchasing consumer behavior deals with the following types of pro-circular products placed on the market namely, potentially circular product (reusable and recyclable), remanufactured product, refurbished product, recycled product, up-cycled product, second-hand product, and shared product. It should be emphasized that each of these product types contributes to the CE to slowing or closing the loop in the past or future. Further, it is fair to say that recycled products together with recy-clable products are essential since closing the loop is a strategic goal of the CE. In addition, it is noteworthy that careful using and proper e-waste discarding consumer activities contribute to CE each in its own way. Cautious usage of electronic items, their repair, and timely maintenance increase the product lifecycle ensuring advance-ment toward more circular consumption in terms of slogging the loop. Suitable dis-posal of EoL electronic items is a crucial step of consumer activity because of the creation of prerequisites for reuse and recycling. Ultimately recycling behavior is determined by an individual attitude of the consumer to the collection and recycling processes. To change neutral or negative attitudes, tools and techniques of consumer motivation toward as full as possible collection are still under discussion.

Last, collaborative consumption is an alternative model that contributes to circular flows of products and materials. The collaborative consumption implies such activi-ties as renting, sharing, gifting, swapping, bartering, lending, and borrowing that aim at causing a minimization of environmental impacts of production and consumption.

12.4   CONCLUSION

The significance of consumer behavior in the success of circular business models for e-products is actively discussed in academia across different countries of Europe, North and South America, Australia, Asia, Oceania, and Africa. Systematization of studies on CE-related e-products' consumption alternatives and behaviors allows revealing that the most under-discussion issues today are remanufactured e-products' acceptance and e-waste proper discarding or recycling behavior. From a regional perspective, the countries of North America and Europe are more focused on the issue of remanufactured product consumer acceptance, and the countries of Asia and Africa are more concerned by the consumer recycling behavior issue. The recycled products' consumer acceptance issue has less debate despite the fact that these products contribute to avoiding wastage today. The least under-discussion issue today is the consumer choice of reusable and recyclable e-products rather than products without any potential for reuse and recycling. However, the literature review brought to light a lack of studies which focus on a product category that has several characteristics in one, for instance, reusable recycled product or recyclable remanufactured product.

Furthermore, this chapter outlines key insights from the discussion on more cir-cular consumption alternatives and consumer behaviors with the focus on e-products, relying on the literature review. In particular, the consumption stage of the e-product lifecycle can be considered as a three-phase process that includes the fol-lowing types of consumer activities: product purchasing, product using, and EoL product discarding. In the context of purchasing, using, and recycling activities under the consumption stage of an e-product, the consumer role can be identified as a tri-dimensional one. Hence, in terms of three-phase consumption stage 'to buy–to use–to return', consumer activities are relevant to specific consumer roles, namely e-product customer, e-product user, and EoL e-product holder, respectively. In this line, the following consumer behaviors under the consumption stage of an e-product lifecycle should be highlighted: (1) circular e-product acceptance behavior, (2) careful using behavior, and (3) EoL e-product proper discarding behavior. Finally, purchasing consumer behavior deals with the following types of pro-circular products: potentially circular product (reusable and recyclable), remanufactured product, refurbished product, recycled product, up-cycled product, product used before, and shared product. Each of these product types contributes to the CE slowing or closing the loop in the past or future.

We aim to carry out research in this topic and believe that the further investigation should be focused on measuring consumer contribution to circular strategies for managing the process of evolving consumer behavior toward more circular consumption alternatives and behaviors.

ACKNOWLEDGMENTS

This study is co-funded by the Erasmus+ Programme of the European Union within the project "Towards circular economy thinking & ideation in Ukraine according to the EU action plan" (grant number 620966-EPP-1-2020). Also this research was supported by CentraleSupélec of University of Paris-Saclay in France and Sumy National Agrarian University in Ukraine.

REFERENCES

Abbey, J.D., Kleber, R., Souza, G.C., Voigt, G. 2017. The role of perceived quality risk in pricing remanufactured products. *Production and Operations Management* 26:100–115. https://doi.org/10.1111/poms.12628

Akehurst, G., Afonso, C., Gonçalves, H. M. 2012. Re-examining green purchase behaviour and the green consumer profile: New evidences. *Management Decision* 50(5):972–988. https://doi.org/10.1108/00251741211227726

Anstine, J. 2000. Consumers' willingness to pay for recycled content in plastic kitchen garbage bags: A hedonic price approach. *Applied Economics Letters* 7(1):35–39. https://doi.org/10.1080/135048500352068

Bakhiyi, B., Gravel, S., Ceballos, D., Flynn, M. A., Zayed, J. 2018. Has the question of e-waste opened a Pandora's box? An overview of unpredictable issues and challenges. *Environment International* 110:173–192. https://doi.org/10.1016/j.envint.2017.10.021

Baldé, C.P., Forti, V., Gray, V., Kuehr, R., Stegmann, P. 2017. *The Global E-Waste Monitor—2017.* Bonn, Germany: United Nations University; Geneva, Switzerland: International Telecommunication Union; Vienna, Austria: International Solid Waste Association. Available: www.itu.int/en/ITUD/Climate-Change/Documents/GEM%20 2017/Global-E-waste%20Monitor%202017%20.pdf (accessed on 7 January 2019).

Barbarossa, C., Pastore, A. 2015. Why environmentally conscious consumers do not purchase green products: A cognitive mapping approach. *Qualitative Market Research* 18(2):188–209. https://doi.org/10.1108/QMR-06-2012-0030

Bei, L.-T., Simpson, E. M. 1995. The determinants of consumers' purchase decisions for recycled products: An application of acquisition-transaction utility theory, in *NA—Advances in Consumer Research*, vol. 22, eds. Frank R. Kardes, Mita Sujan, Provo, UT: Association for Consumer Research, 257–261.

Best, H., Kneip, T. 2011. The impact of attitudes and behavioral costs on environmental behavior: A natural experiment on household waste recycling. *Social Science Research* 40:917–930. https://doi.org/10.1016/j.ssresearch.2010.12.001

Bianchi, C., Birtwistle, G. 2012. Consumer clothing disposal behaviour: A comparative study. *International Journal of Consumer Studies* 36(3):335–341. https://doi.org/10.1111/j.1470-6431.2011.01011.x

Bielefeldt, J., Poelzl, J., Herbst, U. 2016. What's mine isn't yours—Barriers to participation in the sharing economy. *Die Unternehmung Seite* 4–25.

Biswas, A., Roy, M. 2015. Leveraging factors for sustained green consumption behavior based on consumption value perceptions: Testing the structural model. *Journal of Cleaner Production* 95:332–340. https://doi.org/10.1016/j.jclepro.2015.02.042

Blomsma, F., Brennan, G. 2017. The emergence of circular economy a new framing around prolonging resource productivity. A special issue on the circular economy. *Journal of Industrial Ecology* 21(3):603–614. https://doi.org/10.1111/jiec.12603

Bocken, N., Pauw, I., Bakker, C., Grinten, B. 2016. Product design and business model strategies for a circular economy. *Journal of Industrial and Production Engineering* 33:308–320. https://doi.org/10.1080/21681015.2016.1172124

Böcker, L., Meelen, T. 2017. Sharing for people, planet or profit? Analysing motivations for intended sharing economy participation. *Environmental Innovation and Societal Transitions* 23:28–39.

Borthakur, A., Govind, M. 2016. Emerging trends in consumers' e-waste disposal behaviour and awareness: A worldwide overview with special focus on India. *Resources, Conservation and Recycling*, 117:102–113. https://doi.org/10.1016/j.resconrec.2016.11.011

Bovea, M., Ibáñez-Forés, V., Pérez-Belis, V., Juan, P. 2018. A survey on consumers' attitude towards storing and end of life strategies of small information and communication technology devices in Spain. *Waste Management* 71:589–602. https://doi.org/10.1016/j.wasman.2017.10.040

Bridgens, B., Powell, M., Farmer, G., Walsh, C., Reed, E., Royapoor, M., Gosling, P., Hall, J., Heidrich, O. 2018. Creative upcycling: Reconnecting people, materials and place through making. *Journal of Cleaner Production* 189:145–154. https://doi.org/10.1016/j.jclepro.2018.03.317

Camacho-Otero, J., Boks, C., Pettersen, I. 2018. Consumption in the circular economy: A literature review. *Sustainability* 10:2758. https://doi.org/10.3390/su10082758

Chi, X., Wang, M., Reuter, M. 2014. E-waste collection channels and household recycling behaviors in Taizhou of China. *Journal of Cleaner Production* 80:87–95. https://doi.org/10.1016/j.jclepro.2014.05.056

Cohen, B., Muñoz, P. 2016. Sharing cities and sustainable consumption and production: Towards an integrated framework. *Journal of Cleaner Production* 134:87–97. https://doi.org/10.1016/j.jclepro.2015.07.133

Cox, J., Griffith, S., Giorgi, S., King, G. 2013. Consumer understanding of product lifetimes. *Resource Conservation and Recycling* 79:21–29. https://doi.org/10.1016/j.resconrec.2013.05.003

Darby, L., Obara, L. 2005. Household recycling behaviour and attitudes towards the disposal of small electrical and electronic equipment. *Resources, Conservation and Recycling* 44(1):17–35. https://doi.org/10.1016/j.resconrec.2004.09.002

Davidson, A., Habibi, M., Laroche, M. 2017. Materialism and the sharing economy: A cross-cultural study of American and Indian consumers. *Journal of Business Resources* 82:364–372. https://doi.org/10.1016/j.jbusres.2015.07.045

Despeisse, M., Yusuke, K., Masaru, N., Barwood, M. 2015. *Towards a Circular Economy for End-of-Life Vehicles: A Comparative Study UK—Japan*. The 22nd CIRP conference on Life Cycle Engineering. Procedia CIRP 29:668–673.

Dixit, S., Vaish, A. 2015. Perceived barriers, collection models, incentives and consumer preferences: An exploratory study for effective implementation of reverse logistics. *International Journal of Logistics Systems and Management* 21(3):304–318. https://doi.org/10.1504/IJLSM.2015.069729

D'Souza, C., Taghian, M., Khosla, R. 2007. Examination of environmental beliefs and its impact on the influence of price, quality and demographic characteristics with respect to green purchase intention. *Journal of Targeting, Measurement and Analysis for Marketing* 15(2):69–78.

Dwivedy, M., Mittal, R. K. 2013. Willingness of residents to participate in e-waste recycling in India. *Environmental Development* 6:48–68. https://doi.org/10.1016/j.envdev.2013.03.001

Edbring, E. G., Lehner, M., Mont, O. 2016. Exploring consumer attitudes to alternative models of consumption: motivations and barriers. *Journal of Cleaner Production* 33:5–15. https://doi.org/10.1016/j.jclepro.2015.10.107

Ellen MacArthur Foundation. 2013. *Towards the Circular Economy. Economic and Business Rationale for an Accelerated Transition*, vol. 1.

Gaur, J., Amini, M., Banerjee, P., Gupta, R. 2015. Drivers of consumer purchase intentions for remanufactured products. *Qualitative Market Research* 18:30–47. https://doi.org/10.1108/QMR-01-2014-0001

Godelnik, R. 2017. Millennials and the sharing economy: Lessons from a 'buy nothing new, share everything month' project. *Environmental Innovation and Societal Transitions* 23:40–52. https://doi.org/10.1016/j.eist.2017.02.002

Goworek, H., Fisher, T., Cooper, T., Woodward, S., Hiller, A. 2012. The sustainable clothing market: An evaluation of potential strategies for UK retailers. *International Journal of Retail & Distribution Management* 40(12):935–955. https://doi.org/10.1108/09590551211274937

Guide, V. D., Li, J. J. 2010. The potential for cannibalization of new products sales by remanufactured products. 41(3):547–572. https://doi.org/10.1111/j.1540-5915.2010.00280.x

Hamzaoui-Essoussi, L., Linton, J. D. 2010. New or recycled products: How much are consumers willing to pay? *Journal of Consumer Marketing* 27(5):458–468. https://doi.org/10.1108/07363761011063358

Hamzaoui-Essoussi, L., Linton, J. D. 2014. Offering branded remanufactured /recycled products: At what price? *Journal of Remanufacturing* 9:1–15. https://doi.org/10.1186/s13243-014-0009-9

Harms, R., Linton, J. D. 2016. Willingness to pay for eco-certified refurbished products: The effects of environmental attitudes and knowledge. *Journal of Industrial Ecology* 20:893–904. https://doi.org/10.1111/jiec.12301

Hazen, B. T., Boone, C. A., Wang, Y., Khor, K. S. 2017. Perceived quality of remanufactured products: Construct and measure development. *Journal of Cleaner Production* 142:716–726. https://doi.org/10.1016/j.jclepro.2016.05.099

Jesus, A., Mendonça, S. 2018. Lost in transition? Drivers and barriers in the eco innovation road to the circular economy. *Ecological Economics* 145:75–89. https://doi.org/10.1016/J.ECOLECON.2017.08.001

Jiménez-Parra, B., Rubio, S., Vicente-Molina, M. 2014. Key drivers in the behavior of potential consumers of remanufactured products: A study on laptops in Spain. *Journal of Cleaner Production* 85:488–496. https://doi.org/10.1016/j.jclepro.2014.05.047

Khor, K. S., Hazen, B. T. 2017. Remanufactured products purchase intentions and behaviour: Evidence from Malaysia. *International Journal of Production Research* 55:2149–2162. https://doi.org/10.1080/00207543.2016.1194534

Kirchherr, J., Piscicelli, L., Bour, R., Kostense-Smit, E., Muller, J., Huibrechtse-Truijens, A., Hekkert, M. 2018. Barriers to the circular economy: Evidence from the European Union (EU). *Ecological Economics* 150:264–272. https://doi.org/10.1016/j.ecolecon.2018.04.028

Laitala, K., Klepp, I.G. 2018. Care and production of clothing in Norwegian homes: Environmental implications of mending and making practices. *Sustainability* 10:2899. https://doi.org/10.3390/su10082899

Laroche, M., Bergeron, J., Barbaro-Forleo, G. 2001. Targeting consumers who are willing to pay more for environmentally friendly products. *Journal of Consumer Marketing* 18(6):503–520. https://doi.org/10.1108/EUM0000000006155

Lieder, M., Asif, F.M.A., Rashid, A. 2017. Towards circular economy implementation: An agent-based simulation approach for business model changes. *Autonomous Agents and Multi-Agent Systems* 31(6):1377–1402. https://doi.org/10.1007/s10458-017-9365-9

Lu, C.Y., Zhang, L., Zhong, Y.G., Ren, W.X., Tobias, M., Mu, Z.L., Ma, Z.X., Geng, Y., Xue, B. 2015. An overview of e-waste management in China. *Journal of Material Cycles and Waste Management* 17(1):1–12. https://doi.org/10.1007/s10163-014-0256-8

Manomaivibool, P., Vassanadumrongdee, S. 2012. Buying back household waste electrical and electronic equipment: Assessing Thailand's proposed policy in light of past disposal behavior and future preferences. *Resources, Conservation and Recycling* 68:117–125. https://doi.org/10.1016/j.resconrec.2012.08.014

Matsumoto, M., Chinen, K., Endo, H. 2018. Remanufactured auto parts market in Japan: Historical review and factors affecting green purchasing behavior. *Journal of Cleaner Production* 172:4494–4505. https://doi.org/10.1016/j.jclepro.2017.10.266

McDowall, W. et al. 2017. Circular economy policies in China and Europe. *Journal of Industrial Ecology* 21(3):651–661. https://doi.org/10.1111/jiec.12597

Michaud, C., Llerena, D. 2011. Green consumer behaviour: An experimental analysis of willingness to pay for remanufactured products. *Business Strategy and the Environment* 20(6):408–420. https://doi.org/10.1002/bse.703

Mobley, A.S., Painter, T.S., Untch, E.M., Unnava, H.R. 1995. Consumer evaluation of recycled products. *Psychology and Marketing* 12(3):165–176. https://doi.org/10.1002/mar.4220120302

Mokan, K.V., Lee T.C., Bhoyar, M.R. 2018. The intention of green products purchasing among Malaysian consumers: A case study of Batu Pahat, Johor. *Indian Journal of Public Health Research and Development* 9(10):996–1001. https://doi.org/10.5958/0976-5506.2018.01300.1

Mont, O., Plepys, A., Whalen, K., Nußholz, J. 2017. *Business Model Innovation for a Circular Economy Drivers and Barriers for the Swedish Industry—The Voice of REES Companies.* Available: http://lup.lub.lu.se/search/ws/files/33914256/MISTRA_REES_Drivers_and_Barriers_Lund

Mudgal, S., Sales, K., Guilcher, S., Lockwood, S., Morgan, V. 2013. *Equivalent Conditions for Waste Electrical and Electronic Equipment (WEEE) Recycling Operations Taking Place Outside the European Union.* Final Report. European Commission–DG Environment. Available: http://ec.europa.eu/environment/waste/weee/pdf/Final%20report_E%20C%20S.pdf (accessed on 27 February 2019).

Mugge, R. 2018. Product design and consumer behaviour in a circular economy. *Sustainability* 10(10): 3704. https://doi.org/10.3390/su10103704

Mugge, R., Jockin, B., Bocken, N. 2017. How to sell refurbished smartphones? An investigation of different customer groups and appropriate incentives. *Journal of Cleaner Production* 147:284–296. https://doi.org/10.1016/j.jclepro.2017.01.111

Muranko, Z., Andrews, D., Newton, E., Chaer, I., Proudman, P. 2018. The pro-circular change model (P-CCM): Proposing a framework facilitating behavioural change towards a circular economy. *Resources, Conservation and Recycling* 135:132–140. https://doi.org/10.1016/j.resconrec.2017.12.017

Nduneseokwu, C. K., Qu, Y., Appolloni, A. 2017. Factors influencing consumers' intentions to participate in a formal e-waste collection system: A case study of Onitsha, Nigeria. *Sustainability* 9(6):1–17. https://doi.org/10.3390/su9060881

Neto, J. Q. F., Bloemhof, J., Corbett, C. 2016. Market prices of remanufactured, used and new items: Evidence from eBay. *International Journal of Production Economics* 171:371–380. https://doi.org/10.1016/j.ijpe.2015.02.006

Nnorom, I. C., Ohakwe, J., Osibanjo, O. 2009. Survey of willingness of residents to participate in electronic waste recycling in Nigeria—A case study of mobile phone recycling. *Journal of Cleaner Production* 17:1629–1637. https://doi.org/10.1016/j.jclepro.2009.08.009

Ongondo, F. O., Williams, I. D. 2011. Mobile phone collection, reuse and recycling in the UK. *Waste Management* 31:1307–1315. https://doi.org/10.1016/j.wasman.2011.01.032

Orlins, S., Guan, D. 2016. China's toxic informal e-waste recycling: Local approaches to a global environmental problem. *Journal of Cleaner Production* 114:71–80. https://doi.org/10.1016/j.jclepro.2015.05.090

Paco, A., Alves, H., Shiel, C., Filho, W. L. 2014. An analysis of the measurement of the construct "buying behaviour" in green marketing. *Journal of Integrative Environmental Sciences* 11(1):55–69. https://doi.org/10.1080/1943815X.2014.894082

Pandey, P., Govind, M. 2014. Social repercussions of e-waste management in India: A study of three informal recycling sites in Delhi. *International Journal of Environmental Studies* 71:241–260. https://doi.org/10.1080/00207233.2014.926160

Park, H., Armstrong, C. M. J. 2017. Collaborative apparel consumption in the digital sharing economy: An agenda for academic inquiry. *International Journal of Consumer Studies* 41:465–474. https://doi.org/10.1111/ijcs.12354

Park, H. H. 2015. The influence of LOHAS consumption tendency and perceived consumer effectiveness on trust and purchase intention regarding upcycling fashion goods. *International Journal of Human Ecology* 16:37–47.

Park, H. J., Lin, L. M. 2018. Exploring attitude–behaviour gap in sustainable consumption: Comparison of recycled and upcycled fashion products. *Journal of Business Research* 117:623–628. https://doi.org/10.1016/j.jbusres.2018.08.025

Parker, D., Butler, P. 2007. *Introduction to Remanufacturing*. Centre for Remanufacturing and Reuse, 16. www.remanufacturing.org.uk/pdf/story/1p76.pdf

Pérez-Belis, V., Bovea, M. D., Simy, A. 2015. Consumer behaviour and environmental education in the field of waste electrical and electronic toys: A Spanish case study. *Waste Management* 36:277–288. https://doi.org/10.1016/j.wasman.2014.10.022

Pickett-Baker, J., Ozaki, R. 2008. Pro-environmental products: marketing influence on consumer purchase decision. *Journal of Consumer Marketing* 25(5):281–293. https://doi.org/10.1108/07363760810890516

Piscicelli, L., Cooper, T., Fisher, T. 2015. The role of values in collaborative consumption: Insights from a product-service system for lending and borrowing in the UK. *Journal of Cleaner Production* 97:21–29. https://doi.org/10.1016/j.jclepro.2014.07.032

Piscicelli, L., Ludden, G., Cooper, T. 2018. What makes a sustainable business model successful? An empirical comparison of two peer-to-peer goods-sharing platforms. *Journal of Cleaner Production* 172:4580–4591. https://doi.org/10.1016/j.jclepro.2017.08.170

Posselt, T., Gensler, S. 2002. Ein transaktions kost en orientierter Ansatz zur Erklärung von Handelsbetriebstypen (A transaction cost based framework to explain retail company types). *Die Betriebswirtschaft* 60:182–198.

Roos, D., Hahn, R. 2019. Understanding collaborative consumption: An extension of the theory of planned behavior with value-based personal norms. *Journal of Business Ethics* 158:679–697.

Shevchenko, T., Kronenberg, J. 2020. Management of material and product circularity potential as an approach to operationalise circular economy. *Progress in Industrial Ecology, An International Journal* 14(1):30–57. https://doi.org/ 10.1504/PIE.2020.105193

Shevchenko, T., Laitala, K., Danko, Y. 2019. Understanding consumer e-waste recycling behavior: Introducing a new economic incentive to increase the collection rates. *Sustainability* 11:2656. https://doi.org/10.3390/su11092656

Shevchenko, T., Vavrek, R., Hubanova, O., Danko, Yu., Chovancova, J., Mykhailova, L. 2021. Clarifying a circularity phenomenon in a circular economy under the notion of "potential": New dimension and categorization. *Problemy Ekorozwoju* 16(1):79–89. https://ekorozwoj.pollub.pl/index.php/number-1612021/clarifying-a-circularity-phenomenon-in-a-circular-economy-under-the-notion-of-potential/

Stahel, W. R. 2010. *The Performance Economy*. Hampshire, UK: Palgrave Macmillan.

Sun, H., Teh, P.-L., Linton, J.D. 2018. Impact of environmental knowledge and product quality on student attitude toward products with recycled/remanufactured content: Implications for environmental education and green manufacturing. *Business Strategy and the Environment* 27(7):935–945. https://doi.org/10.1002/bse.2043

Tansel, B. 2017. From electronic consumer products to e-wastes: Global outlook, waste quantities, recycling challenges. *Environment International* 98:35–45. https://doi.org/10.1016/j.envint.2016.10.002

Tanskanen, P. 2013. Management and recycling of electronic waste. *Acta Materialia* 61:1001–1011. https://doi.org/10.1016/j.actamat.2012.11.005

Trivedi, R. H., Patel, J.D., Savalia, J.R. 2015. Pro-environmental behaviour, locus of control and willingness to pay for environmental friendly products. *Marketing Intelligence & Planning* 33(1):67–89. https://doi.org/10.1108/MIP-03-2012-0028

Tukker, A. 2015. Product services for a resource-efficient and circular economy–A review. *Journal of Cleaner Production* 97:76–91. https://doi.org/10.1016/j.jclepro.2013.11.049

Vadde, S., Kamarthi, S.V., Gupta, S.M. 2007. Optimal pricing of reusable and recyclable components under alternative product acquisition mechanisms. *International Journal of Production Research* 45(18–19):4621–4652. https://doi.org/10.1080/00207540701449973

Van Eijk, F. 2015. *Barriers & Drivers Towards a Circular Economy*. Available: http://www.circulairondernemen.nl/uploads/e00e8643951aef8adde612123e824493.pdf

Veenstra, A., Wang, C., Fan, W.J., Ru, Y.H. 2010. An analysis of e-waste flows in China. *International Journal of Advanced Manufacturing Technology* 47:449–459. https://doi.org/10.1007/s00170-009-2356-5

Wagner, T.P. 2013. Examining the concept of convenient collection: An application to extended producer responsibility and product stewardship frameworks. *Waste Management* 33:499–507. https://doi.org/10.1016/j.wasman.2012.06.015

Wang, Y., Hazen, B.T. 2016. Consumer product knowledge and intention to purchase remanufactured products. *International Journal of Production Economics* 181:460–469. https://doi.org/10.1016/j.ijpe.2015.08.031

Wang, Y., Wiegerinck, V., Krikke, H., Zhang, H. 2013. Understanding the purchase intention towards remanufactured product in closed-loop supply chains: An empirical study in China. *International Journal of Physical Distribution & Logistics Management* 43(10):866–888. https://doi.org/10.1108/IJPDLM-01-2013-0011

Wang, Z., Zhang, B., Yin, J., Zhang, X. 2011. Willingness and behavior towards e-waste recycling for residents in Beijing city, China. *Journal of Cleaner Production* 19:977–984. https://doi.org/10.1016/j.jclepro.2010.09.016

Wastling, T., Charnley, F., Moreno, M. 2018. Design for circular behaviour: considering users in a circular economy. *Sustainability* 10:1743. https://doi.org/10.3390/su10061743

Weelden, E., Mugge, R., Bakker, C. 2016. Paving the way towards circular consumption: Exploring consumer acceptance of refurbished mobile phones in the Dutch market. *Journal of Cleaner Production* 113:743–754. https://doi.org/10.1016/j.jclepro.2015.11.065

Wilson, M. 2016. When creative consumers go green: Understanding consumer upcycling. *Journal of Product & Brand Management* 25(4):394–399. https://doi.org/10.1108/JPBM-09-2015-0972

Wong, M. W., Mo, H. F. 2013. Purchase behaviour related to recycled product in China. *Review of Business Research* 13(2):125–131.

Yin, J., Gao, Y., He, X. 2014. Survey and analysis of consumers' behaviour of waste mobile phone recycling in China. *Journal of Cleaner Production* 65:517–525. https://doi.org/10.1016/j.jclepro.2013.10.006

Yla-Mella, J., Keiski, R. L., Pongracz, E. 2015. Electronic waste recovery in Finland: Consumers' perceptions towards recycling and re-use of mobile phones. *Waste Management* 45:374–384. https://doi.org/10.1016/j.wasman.2015.02.031

Young, W., Hwang, K., McDonald, S., Oates, C. 2010. Sustainable consumption: Green consumer behaviour when purchasing products. *Sustainable Development* 18(1):20–31. https://doi.org/10.1002/sd.394

Yu, S., Lee, J. 2019. The effects of consumers' perceived values on intention to purchase upcycled products. *Sustainability* 11(4):1034. https://doi.org/10.3390/su11041034

Zhao, H., Gao, Q., Wu, Y., Wang, Y., Zhu, X. 2014. What affects green consumer behaviour in China? A case study from Qingdao. *Journal of Cleaner Production* 63:143–151. https://doi.org/10.1016/j.jclepro.2013.05.021

Zhu, Y., Ma, L., Jiang, R. 2019. A cross-cultural study of English and Chinese online platform reviews: A genre-based view. *Discourse and Communication* 13(3):342–365. https://doi.org/10.1177/1750481319835642

13 Towards Circular Economy via Sorbent Prepared from E-waste
A Chemical Engineering Perspective

Suparna Bhattacharyya and Tathagata Mukherjee

CONTENTS

13.1 INTRODUCTION

The rising necessity and use of electronic and electrical appliances in our daily life-style and its end-of-life (EoL) waste have shaped their way of an extra environmental and well-being challenge (Chandrasekaran et al., 2018). Nevertheless, with suitable recycling procedure, these wastes have the potential to preserve the natural resources and also, in a way, put a stop to water and air pollution (Dutta et al., 2015). Electronic waste is generally termed as e-waste. It originates from electrical and electronic

DOI: 10.1201/9781003301899-18

industry or from household waste (Bhattacharyya et al., 2019a) which have been rejected for use as a result of malfunctioning or having quality flaws at the time of production or in use (Weidenhamer et al., 2010). These comprise fragments of electronic or electrical hardware like videotape recorders, mobile phones, electronic scanners, laser printers, fax machines, electronic tablets, microwave oven, DVDs, x-ray machineries, kitchen appliances, and few laboratory or scientific equipment. Enormous quantities of e-waste are formed as a consequence of constant consumption and/or continual scientific progression and replacement, specifically in the computer and mobile phone industries (Noon et al., 2011). Near-about four hundred and forty-seven lakh tons of e-waste were produced in 2016 and are estimated to touch more than 550 lakh tons by 2021 with a yearly progress rate of 3% to 4% in calculation. Additionally, the generation of e-waste from outdated computers in countries like South Africa and China is likely to increase by a peak of 200% to 400% (Noon et al., 2011; Chandrasekaran et al., 2018). Universally, the collective e-waste production has increasingly progressed, and it also threatens the well-being of humans and other living organisms as well as the environment because of its hazardous and exceedingly toxic elements (Weidenhamer et al., 2010). On the basis of the number of cell phones and personal computers (PCs) which are manufactured, it is predictable that 17 million PCs and 100 million cell phones will be disposed on a yearly basis. The production of these increases every year, thereby increasing e-waste at an equal rate (Chandrasekaran et al., 2018). Constructing a money-making as well as eco-friendly recycling technique encompasses classifying and computing the valued resources and risky constituents. This will permit us to realize the physical characteristics of e-waste and upsurge the metal recovery chance and thus reserve the natural resources (Arukwe et al., 2012; Weidenhamer et al., 2010). This will aid in delivering an ecological resolution of e-waste management. Figuring out a way to use the EOL by-products (e-wastes) as a forerunner of raw materials for the production of innovative and advanced high-end merchantable products could kindle the recycling process and create economic revenues (Noon et al., 2011).

Two widely used heavy metals like Cobalt and Nickel are utilized in innumerable manufacturing units like refining factory and paint and textile and metal finishing industries (Weidenhamer et al., 2010). Thus, the challenge is primarily linked to the treatment of wastewater, which is discharged consisting of nickel and cobalt from these production units and is vital owing to the threat on human health and its ecological effect (Chandrasekaran et al., 2018; Noon et al., 2011) along with the use of nickel composites for different mechanical and chemical treatment processes (Bhattacharyya et al., 2023). There are several approaches for eliminating the heavy metal ions from wastewater along with precipitation of chemicals, electrocoagulation technique, filtration methods, adsorption, and reverse osmosis (Bhattacharyya et al., 2018); among these procedures, adsorption has proved to be a capable technique not possessing the shortcomings like low effectiveness and high cost (Noon et al., 2011; Bhattacharyya et al., 2019b; (Bhattacharyya, 2019). Reviewing and optimizing different thermochemical technologies for resource recovery from sewage sludge are also stated in brief (Bhattacharyya and Ghosh, 2019; Bhattacharyya et al., 2016).

Currently, cheap adsorbents comprising carbon-containing waste materials such as peat, bamboo, bagasse tyre, and bone char serve a good purpose of wastewater pollutant exclusion. Nevertheless, two most important problems exist with these adsorbents

(Noon et al., 2011)—their quality of low yielding and extremely high processing temperatures which are essential for their activation process. Zeolites is another class of adsorbent which is required for industrial use. Their molecular structure provokes them to be precisely attractive to adsorption because of the highly ordered fashion of the nano-structured framework (Bhattacharyya, 2019; Bhattacharyya et al., 2019b, not only in sol but also as an aqueous solution (Bhattacharyya et al., 2021). The principal cost for zeolitic adsorbent is more than activated carbons. Several researchers examined the use of zeolite-structured waste products for heavy metal adsorption from ions of effluents (Chandrasekaran et al., 2018; Noon et al., 2011; Malav et al., 2020). In this chapter, a detailed study has been conducted in the preparation of sorbents from e-waste. It summarizes the different methods that have been employed to synthesize sorbents from e-wastes.

13.2 METHODOLOGY

The procedure assumed is elaborated in this chapter. A detailed literature research was performed to obtain knowledge regarding sorbent materials. Various searching browsers were surveyed using various keywords like 'sorbent'; 'sorbent + circular economy'; 'sorbent + e-waste'; 'PCB + DVD'; 'sorbent + WEEE'; and 'CD'. Then the knowledge acquired from the literature review was sorted from the rest of the papers which explicitly dealt on sorbent materials obtained from waste products. Subsequently, the research papers were categorized into dealing with sorbents which emerged from electronic waste. An isolated inspection was conducted for detecting numerous areas of application. Based on the literature review on principles of circular economy, a draft framework was established. Then the idea of was enhanced with summing up of innovative and new technologies such as blockchain, Big data, and IoT.

13.3 TECHNIQUES OF SORBENTS' PREPARATION FROM DIFFERENT E-WASTE

13.3.1 PCB

The fastest growing subject of concern in the field of waste management is electronic waste (e-waste) (Mundada et al., 2004). Its alarming rate of increase is causing different unavoidable environmental problems which can only be solved with immediate and necessary measures against it (Li et al., 2009). The huge production of electronic and electrical products due to increasing urban population, increasing revenue, and wasteful lifestyle of modern generation is considered as one of the highlighting reasons for e-waste generation (Li et al., 2009). Printed Circuit Boards (PCB) is considered as a source of e-waste generation that causes toxicity in landfills and agricultural areas (Hadi et al., 2013a). The contamination is not only restricted in the land areas but also gets spread in the soil and groundwater in the adjacent regions causing various life risk diseases and incurable health issues. Additionally, heavy metals, furans, and toxic dioxins get liberated when these PCBs get burnt out and mix with the atmosphere causing fatal health disorder (Bi et al., 2010; Owens et al., 2007). So, recycling and reuse are the best substitutes to protect the environment from these kind of toxic waste products. Along with them, the process of recovery of precious materials can also be applicable from the reuse and recycling of PCBs. Different useful metallic and non-metallic elements are extracted from these wastes which are further used in various other processing techniques like treatment of wastewater, solid waste, and processing bio-fuels. (Zhou et al., 2010; Flandinet et al., 2012; Zhou et al., 2010; Zhou and Qiu, 2010). PCB acts as a major recycling material that recycles the toxic landfills. Both its metallic and non-metallic constituents are widely used for various major ecological problem breakthrough. The metallic component of PCB is widely used for economic benefits (Bi et al., 2010; Hadi et al., 2013b). The non-metallic part is used as adsorbent and sorption material. The PCBs are generally activated and chemically treated for further use. The treated PCBs act as a very efficient and economic adsorbent as compared to the normal commercial adsorbing and absorbing materials (Hadi et al., 2013a).

13.3.2 CD

Compact Discs (CDs) are the major source of e-waste materials that remain in high demand for the preparation of carbon black (CB). CB is a novel and most demanding adsorbent for adsorbing dye from industrial and textile wastewater. Among the different physico-chemical methods like photo catalytic degradation, coagulation, flocculation, membrane filtration, precipitation, ion exchange, solvent extraction, ozonation, irradiation, and adsorption, it proved to be the best of the lot which is widely accepted among the recent researchers because of its benefits such as easy design techniques, good efficacy, toxic materials insensitivity, and capability for industrial scale applications (Noorimotlagh et al., 2016; Landi et al., 2010; Ashiq et al., 2012). About, 3% to 5% of total electrical and electronic wastes per year is produced globally as a phenolic waste which is vastly used as landfills and source of the phenolic compounds. These phenolic materials act as core pollutants for groundwater and wastewater. These pollutants can effectively be treated by the adsorbents prepared from CDs and electronic wastes of PCs. Moreover, collection and recycling of these materials decrease the health risk and environmental hazards that can be caused by these kinds of materials. The non-degradable compounds in landfills also get reduced due to the recycling and proper use of e-wastes (Cardoso et al., 2012). E-waste generated from CDs is an inexpensive alternative for carbon sources and is very readily recycled. This not only reduces the environmental risk but also utilizes materials in a better and constructive manner. Hence, it can be finally concluded that the electronic waste recycling is a major step for the production of value-added materials. It is an interesting alternative for landfills and incineration, which results in a proper solid waste management outcome (Prola et al., 2013).

13.3.3 DVD

Digital versatile disc (DVD) and Blue Ray discs on the other hand are also a great alternative to the e-waste of the contemporary day. These electronic wastes can be transformed into value-added products using recent technologies. Some of the cost-effective methods for utilizing e-waste are carbonizing and incinerating these wastes into carbon powder and activated carbon which can readily be used as adsorbing materials. Many other countries produce activated carbon from different natural wastes like coconut shell and nutshells to reduce ecological risks and surrounding pollution (Jorfi et al., 2017). Thus, the process of transforming e-waste into an adsorbent or treated product for adsorption is a great help in reducing environmental risks. DVDs are the great source of forming carbonaceous materials. Using these carbon materials as adsorbing agents or as filtering products is a novel technological use of recent times. Not only the carbon compound is produced directly from the incinerated waste, but the treated powder is also capable of being used in various ways (Ferreira et al., 2017). Thus, the PC parts, CDs, and DVDs are the most usable cheapest alternatives of activated carbon production and an efficient saving mechanism from ecological risks and upcoming pollutions (Jamshidi et al., 2016).

13.4 DISCUSSION AND ANALYSIS FROM THE CHEMICAL ENGINEERING PERSPECTIVE

E-waste can be degraded by various thermochemical processes. Pyrolysis is one of the important techniques that leads to the disintegration of the polymeric groups and is quite significant in e-waste treatment. The end product of pyrolysis of e-wastes varies depending on the type of wastes, type of reactor, and the catalyst that maybe present, and so on (Mukherjee et al., 2019). The products such as pyro-oil, char, gases, and ash are obtained with varying percentages on the basis of using different catalysts. Examples of a few catalysts are zeolites, limestone, minerals, metal-based catalysts, silica-alumina, oyster shells, meso-structured catalysts, and FCC catalysts. Vacuum pyrolysis is more beneficial as it works under low temperature and pressure. An efficient and green thermochemical processing method is co-pyrolysis of e-waste and biomass. Gasification is an important arena which is yet to be discovered vividly (Weidenhamer et al., 2010; Mukherjee et al., 2019). In most cases, the polymeric part of e-waste undergoes gasification. In chemical plants, one of the gaseous byproducts produced known as 'thin gas' is highly rich in oxygen and can be used as a raw material. The impact of mixed molten carbonate configuration on the preparation of hydrogen through the method of steam gasification of activated carbon which has been recovered from plastic waste has been reported. Chemical kinetics plays an important role in the waste decomposition reactions. Proper study helps one to understand the process and carry out the experiments along with the equipment. Literature review states many works in the field of e-waste kinetics. The kinetics should contain both the primary and secondary reactions though literature mostly reports of primary kinetic reactions. There are other parameters that need to be included along with the reactions such as temperature and the production of raw materials.

13.4.1 KINETICS AND ISOTHERM OF SORPTION PROCESS

For designing and selecting the optimum working conditions for a life-sized lot method, the adsorption rate estimation proposes vital data. Kinetics, isotherm, and thermodynamics data of the adsorption process provides critical information about the reaction's mechanism and its paths as detailed in Table 13.1 for e-waste precursors (Mitra et al., 2020).

13.4.1.1 The Langmuir Isotherm

The Langmuir isotherm assumes a monolayer adsorption of an adsorptive surface containing an inadequate number of adsorption sites. The Langmuir isotherm (Eq. 13.1) also assumes that there is no lateral interaction within the dye molecules as well as zero migration of the adsorbate. Furthermore, the model takes on only a secure quantity of energetically inconsistent active sites. (Arukwe et al., 2012; Mukherjee et al., 2019; Debnath et al., 2021):

$$\frac{c_e}{q_e} = \frac{1}{Q_0 b} + \frac{c_e}{Q_0}$$

(13.1)

where q_e is equilibrium adsorbent-phase concentration of adsorbate (mg L^{-1}); c_e is adsorbate equilibrium concentration (mg L^{-1}); Q_0 is the monolayer adsorption capacity (mg g^{-1}); and b is the Langmuir constants for free energy.

13.4.1.2 Freundlich Isotherm

The Freundlich Isotherm (Eq. 13.2) assumes diverse surface energy scattering of the active sites which relies on the surface coverage as well. Freundlich model equation is shown in its linear form as here (Weidenhamer et al., 2010; Mukherjee et al., 2019):

$$log\, q_e = log\, K_d + \frac{1}{n} x\, log\, C_e$$

(13.2)

where K_d is the Freundlich Constant and $1/n$ is an empirical constant. The value of $1/n$ varies with the system, and it is used to interpret the adsorption intensity.

13.4.1.3 The Pseudo-First-Order Kinetic Model

The pseudo-first-order model is grounded on the supposition that the adsorption degree is correlated with the existing adsorption site number, and it is stated in the equation given here (Bhattacharyya et al., 2018; Ghosh et al., 2019):

$$\frac{dq_t}{dt} = K_1 \left(q_e - q_t \right)$$

(13.3)

where K_1 (1/min) is the rate constant, q_t and q_e (mg/g) are the quantities of adsorbate per gram of adsorbent at time 't' and at equilibrium, respectively. Integrating Eq. (13.3) with the boundary conditions (at $t = 0$, $q_t = 0$ and at $t = t$, $q_t = q_t$), we obtain Eq. (13.4).

$$ln\left(q_e - q_t \right) = ln\, q_e - K_1 t$$

(13.4)

13.4.1.4 The Pseudo-Second-Order Kinetic Model

The pseudo-second-order kinetic model (Eq. 13.5) was originated based on the theory that the adsorption rate is equivalent to the square of the number of vacant adsorption sites.

$$\frac{dq_t}{dt} = K_2 \left(q_e - q_t \right)^2 \tag{13.5}$$

where K_2 (g/mg min) represents the adsorption rate. Integrating Eq. (13.5) with the boundary conditions (at $t = 0$, $q_t = 0$ and at $t = t$, $q_t = q_t$), we obtain Eq. (13.6).

$$\frac{1}{\left(q_e - q_t \right)} = \frac{1}{q_e} + K_2 t \tag{13.6}$$

Eq. (13.6) is mostly stated as given next:

$$\frac{t}{q_t} = \frac{1}{K_2 q_e^2} + \frac{t}{q_e} \tag{13.7}$$

K_2 and q_e^2 are considered as the introductory adsorption rate ($t \rightarrow 0$).

TABLE 13.1

Kinetics and Efficiency of Adsorbents from e-waste precursor in existing literature

Sl. No.	Precursor	Adsorption/Kinetic Model study	Removal Efficiency	Reference
1	Printed Circuit Board	The modified precursor enhances the adsorption capacity thus increasing the removal efficiency of the material.	Hugely enhanced	Hadi et al., 2013a
2	E-waste materials	Langmuir for cobalt and Freundlich model for Ni showing the best fit curve	3.6 mmol/gram	Hadi et al., 2013b
3	Printed Circuit Board	Langmuir fits the best	-	Hadi et al., 2014
4	CDs and DVDs	Freundlich given the best plot following pseudo-second-order kinetics	R2 = 0.9996	
5	CDs	Effectively increasing in comparison to the untreated material.	Adsorption capacity increases drastically	Mallakpour et al., 2020

13.4.2 THERMODYNAMIC STUDIES

The Van't Hoff equation (Eq. 13.8) is used to determine the thermodynamic parameters, i.e. change in enthalpy ($\Delta H°$) and change in entropy ($\Delta S°$). Whereas change in Gibb's free energy ($\Delta G°$) is determined using Eq. (13.9) (Mitra et al., 2020):

$$\ln K = \frac{\Delta S^0}{R} - \frac{\Delta H^0}{RT} \tag{13.8}$$

$$\Delta G^0 = -RT \ln K \tag{13.9}$$

where R is the universal gas constant (8.314 J/mol K), T is the absolute temperature, and K is the equilibrium constant. Chroma et al. produced sorbents from CDs and DVSs. The $\Delta G°$ value ranged from 39.41 to 58.94 J/kmol, while the $\Delta H°$ and $\Delta S°$ values were 8.59 kJ/mol and −6.5 respectively (Choma et al., 2015).

13.4.3 SCOPES FOR REACTOR DESIGN AND UPSCALING

The reactor geometry is based on many aspects, out of which specific importance is given to the thermodynamic and kinetics of chemical reactions which are carried out. The category of reactors which are used for specific procedures has a great effect on the overall productivity of the process. Adsorbent production from e-waste involves heat treatment like pyrolysis. E-waste and plastic pyrolysis, fluidized bed reactor, tube reactor, and fixed bed reactor are utilized (Soler et al., 2018; Miskolczi et al., 2008; Areeprasert and Khaobang, 2018; Jadhao et al., 2020). Debnath and Ghosh (2015) employed a furnace reactor for the preparation of adsorbent from printed circuit boards.

From the perspective of chemical engineering, it is said that fluidized bed aids to perk up the degradation of polymer, acting as a positive catalyst for promoting mass and heat transfer, thus dissolving the melted polymer in thin layers (Debnath et al., 2019). Nevertheless, research works have been carried out on gasification of printed circuit boards used semi-batch reactors as an alternative of traditional updraft or downdraft gasifiers in gasification purposes (Zhang et al., 2013). Probably due to complex materials of PCB materials and several other materials exists which does not react or breaks down throughout the gasification process. In the case of e-waste and biomass co-pyrolysis, promising results have been attained in a fixed bed reactor or FBD (Liu et al., 2013). Other reactor configurations like entrained bed reactor (Liang et al., 2019), horizontal bed reactor, spouted bed reactor (Yang et al., 2021), and auger reactor need to be explored.

13.4.4 PROPOSED CIRCULAR ECONOMY MODEL

The intention in circular economy is to reduce waste and pollution. Parts, products, and finished goods are repaired, taken care of, used, recycled, and reused as much as probable. With the intention of preferred substitute to the dominant economic development by "take, make, and dispose". The intention is to close the economic

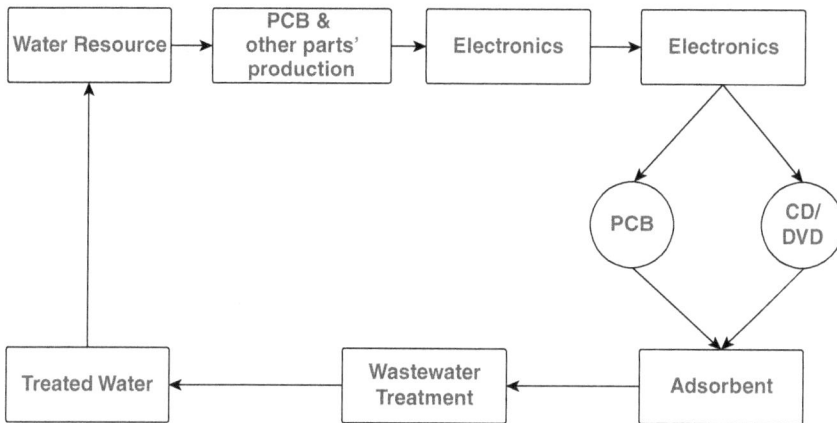

FIGURE 13.1 Implementation of circular economy concept in sorbent production from e-waste.

loop and exercise circular economy. This begins from advanced product design and their utilization, via product and their part's life extension, to beneficial applications (Mitra et al., 2020; Mukherjee et al., 2019; Debnath et al., 2021). Thus far, circular economy is well known to everyone in the form of 3R principles: Recycle, Reuse, and Reduction. A debate is going on, regarding the attractiveness of circular economy for various shareholders. Benefits of circular economy are not just associated with environment and health, but they also have the potential for encouraging economic growth and employment. Figure 13.1 shows the implementation of circular economy in e-waste recycling to sorbents for wastewater purification.

13.5 CONCLUSION

An evolution of circular economy generates prospects for many new forms of creation and new services, both of which are within sectors. Although at the time when circular economic models are implemented, the intersections and potential tensions among diverse sectors and dissimilar governing frameworks are presented to be more superficial. Encounters like this are needed to be submissive, in all sectors, only if the circular economy model is to be comprehended truly. By observation and a close study of a systematic view of technological intervention and policy frameworks and inspecting governing needs in all steps of potential value chains (in place of treating distinct steps in seclusion), it is easily identifiable, and steps can be taken to resolve multiple challenges.

Sorbents arranged from several waste materials as predecessor is a fascinating theme and is the most important field of research contributory in the prime literature direction of resource valorization. Sorbents have a multipurpose efficiency with an extensive usage in effluent handling, gas purification, oil spillage, removal of dye and heavy metals, and so on. The entire assurance of researchers from numerous fields like material science, chemical engineering, green chemistry, and nano technology has provided us few novel adsorbents possessing enhanced features related to marketable counterparts. This chapter is a collection of such aids precisely with waste materials as predecessor materials. The removal efficiency of pollutants has been discussed in great details focusing on the sorbents formed from e-wastes. A section is dedicated specifically which specifies the multipurpose use of sorbent materials.

ACKNOWLEDGEMENT

The authors would like to acknowledge the support received from Jadavpur University, India, and Tirthankar Mukherjee, Griffith University, Australia, for arranging and conceptualizing the manuscript.

REFERENCES

Areeprasert, Chinnathan, & Chanoknunt Khaobang. Pyrolysis and catalytic reforming of ABS/PC and PCB using biochar and e-waste char as alternative green catalysts for oil and metal recovery. *Fuel Processing Technol.* 182 (2018) 26–36.

Ashiq, M.N., Najam-Ul-Haq, M., Amanat, T., Saba, A., Qureshi, A.M., & Nadeem, M. Removal of methylene blue from aqueous solution using acid/base treated rice husk as anadsorbent. *Desalin. Water Treat.* 49 (2012) 376–383.

Bhattacharyya, S. Removal of ranitidine from pharmaceutical waste water using graphene oxide (GO). In *Waste Management and Resource Efficiency* (pp. 1253–1262). Springer, Singapore (2019).

Bhattacharyya, S., Banerjee, P., Bhattacharya, S., Rathour, R.K.S., Majumder, S.K., Das, P., & Datta, S. Comparative assessment on the removal of ranitidine and prednisolone present in solution using graphene oxide (GO) nanoplatelets. *Desalin Water Treat.* 132 (2018) 287–296.

Bhattacharyya, S., Das, P., & Datta, S. Removal of ranitidine from pharmaceutical waste water using activated carbon (AC) prepared from waste lemon peel. In *Waste Water Recycling and Management* (pp. 123–141). Springer, Singapore (2019a).

Bhattacharyya, S., & Dutta, M. *Optimisation of Adsorption Parameter for Removal of Acetaminophen from Pharmaceutical Waste Water Using Box-Behnken Design.* (2016).

Bhattacharyya, S., Mohanty, D., Kumar, P., Das, S. K., Sahoo, P., Pal, S. K., & Chakraborty, S. (2023). A corrosion and tribo-failure analysis of Ni-P-Cu coated mild steel (AISI-1040) at varied copper concentration. *Engineering Failure Analysis*, 107063.

Bhattacharyya, S., & Ghosh, S.K. A review on various thermochemical technologies for resource recovery from sewage sludge. *Waste Valorisation and Recycling.* (2019) 1–17.

Bhattacharyya, S., Manna, S., & Medda, S.K. ZrO2 incorporated TiO2 based solar reflective nanocomposite coatings on glass to be used as energy saving building components. *SN Applied Sciences.* 1(11) (2019b) 1–8.

Bhattacharyya, S., Medda, S.K., & Naskar, M.K. A preparative approach of TiO2-ZrO2 coating using aquo-based TiO2 precursor useful for light reflective application. *Transactions of the Indian Ceramic Society.* 80(4) (2021) 227–233.

Cardoso, N.F., Lima, E.C., Royer, B., Bach, M.V., Dotto, G.L., Pinto, L.A.A., & Calvete, T. Comparison of Spirulina platensis microalgae and commercial activated carbon as adsorbents for the removal of reactive red 120 dye from aqueous effluents. *J. Hazard. Mater.* 241 (2012) 146–153.

Chandrasekaran, S.R., Avasarala, S., Murali, D., Rajagopalan, N., & Sharma, B.K. Materials and energy recovery from e-waste plastics. *ACS Sustain. Chem. Eng.* 6 (2018) 4594–4602.

Choma, J., Marszewski, M., Osuchowski, L., Jagiello, J., Dziura, A., & Jaroniec, M. Adsorption properties of activated carbons prepared from waste CDs and DVDs. *ACS Sustain. Chem. Eng.* 3 (2015) 733–742.

Debnath, Biswajit, Ranjana Chowdhury, & Sadhan Kumar Ghosh. An analysis of e-waste recycling technologies from the chemical engineering perspective. In *Waste Management and Resource Efficiency* (pp. 879–888). Springer, Singapore (2019).

Debnath, Biswajit, & Sadhan K. Ghosh. Preparation and characterization of adsorbent from waste printed circuit boards. *J. Solid Waste Technol. Manag.* 41(4) (2015).

Debnath, Biswajit, Indrashis Saha, Tirthankar Mukherjee, Shweta Mitra, Ankita Das, & Abhijit Das. Sorbents from waste materials: a circular economic approach. In *Sorbents Materials for Controlling Environmental Pollution* (pp. 285–322). Elsevier (2021).

Dutta, M., Das, U., Mondal, S., Bhattachriya, S., Khatun, R., & Bagal, R. Adsorption of acetaminophen by using tea waste derived activated carbon. *Int. J. Environ. Sci.* 6(2) (2015) 270.

Ferreira, G.M.D., Ferreira, G.M.D., Hespanhol, M.C., de Paula Rezende, J., dos Santos Pires, A.C., Gurgel, L.V.A., & da Silva, L.H.M. Adsorption of red azo dyes on multiwalled carbon nanotubes and activated carbon: a thermodynamic study. *Colloids Surf. A Physicochem. Eng. Asp.* 529 (2017) 531–540.

Ghosh, Sadhan Kumar, & Tirthankar Mukherjee. Circular economy through treatment and management of industrial wastewater. In *Waste Water Recycling and Management* (pp. 1–13). Springer, Singapore (2019).

Hadi, Pejman, John Barford, & Gordon McKay. Synergistic effect in the simultaneous removal of binary cobalt–nickel heavy metals from effluents by a novel e-waste-derived material. *Chem. Eng. J.* 228 (2013a) 140–146.

Hadi, Pejman, Ping Gao, John P. Barford, & Gordon McKay. Novel application of the nonmetallic fraction of the recycled printed circuit boards as a toxic heavy metal adsorbent. *J. Hazardous Mat.* 252 (2013b) 166–170.

Hadi, Pejman, Chao Ning, Weiyi Ouyang, Carol Sze Ki Lin, Chi-Wai Hui, & Gordon McKay. Conversion of an aluminosilicate-based waste material to high-value efficient adsorbent. *Chem. Eng. J.* 256 (2014) 415–420.

Jadhao, Prashant Ram, Ejaz Ahmad, K.K. Pant, & K.D.P. Nigam. Environmentally friendly approach for the recovery of metallic fraction from waste printed circuit boards using pyrolysis and ultrasonication. *Waste Manag.* 118 (2020) 150–160.

Jamshidi, M., Ghaedi, M., Dashtian, K., Ghaedi, A.M., Hajati, S., Goudarzi, A., & Alipanahpour, E. Highly efficient simultaneous ultrasonic assisted adsorption of brilliant green and eosin B onto ZnS nanoparticles loaded activated carbon: artificial neural network modeling and central composite design optimization. *Spectrochim. Acta A Mol. Biomol. Spectrosc.* 153 (2016) 257–267.

Jorfi, S., Soltani, R.D.C., Ahmadi, M., Khataee, A., & Safari, M. Sono-assisted adsorption of a textile dye on milkvetch-derived charcoal supported by silica nano powder. *J. Environ. Manage.* 187 (2017) 111–121.

Landi, M., Naddeo, V., & Belgiorno, V. Influence of ultrasoundon phenol removal by adsorption on granular activated carbon. *Desalin. Water Treat.* 23 (2010) 181–186.

Li, J., Lu, H., Liu, S., & Xu, Z. Optimizing the operating parameters of corona electrostatic separation for recycling waste scraped printed circuit boards by computer simulation of electric field. *J. Hazard. Mater.* 153 (2008) 269–275.

Liang, Dingcheng, Qiang Xie, Zheng Wei, Chaoran Wan, Guangsheng Li, & Junya Cao. Transformation of alkali and alkaline earth metals in Zhundong coal during pyrolysis in an entrained flow bed reactor. *J. Analyt. Appl. Pyrolysis.* 142 (2019) 104661.

Liu, Wen-Wu, Chang-Wei Hu, Yu Yang, Dong-Mei Tong, Liang-Fang Zhu, Rui-Nan Zhang, & Bo-Han Zhao. Study on the effect of metal types in (Me)-Al-MCM-41 on the mesoporous structure and catalytic behavior during the vapor-catalyzed co-pyrolysis of pubescens and LDPE. *Appl. Catalysis B: Environ.* 129 (2013) 202–213.

Malav, L.C., Yadav, K.K., Gupta, N., Kumar, S., Sharma, G.K., Krishnan, S., . . . & Bach, Q.V. A review on municipal solid waste as a renewable source for waste-to-energy project in India: current practices, challenges, and future opportunities. *J. Cleaner Product.* 277 (2020) 123227.

Mallakpour, Shadpour, & Vajiheh Behranvand. Modification of polyurethane sponge with waste compact disc-derived activated carbon and its application in organic solvents/oil sorption. *New J. Chem.* 44(36) (2020) 15609–15616.

Miskolczi, N., Hall, W.J., Angyal, A., Bartha, L., & Williams, P.T. Production of oil with low organobromine content from the pyrolysis of flame retarded HIPS and ABS plastics. *J. Analyt. Appl. Pyrolysis.* 83(1) (2008) 115–123.

Mitra, Shweta, Tirthankar Mukherjee, & Prasad Kaparaju. Prediction of methyl orange removal by iron decorated activated carbon using an artificial neural network. *Environ. Technol.* (2020) 1–16.

Mukherjee, T., M. Rahaman, A. Ghosh, & S. Bose. Optimization of adsorbent derived from non-biodegradable waste employing response surface methodology toward the removal of dye solutions. *Int. J. Environ. Sci. Technol.* 16(12) (2019) 8671–8678.

Mundada, M.N., Kumar, S., & Shekdar, A.V. E-waste: a new challenge for waste management in India. *Int. J. Environ. Stud.* 61 (2004) 265–279.

Noorimotlagh, Z., Darvishi Cheshmeh Soltani, R., Shams Khorramabadi, G., Godini, H., & Almasian, M. Performance of wastewater sludge modified with zinc oxidenanoparticles in the removal of methylene blue from aqueous solutions. *Desalin. Water Treat.* 57 (2016) 1684–1692.

Pant, D., Joshi, D., Upreti, M.K., & Kotnala, R.K. Chemical and biological extraction of metals present in e waste: a hybrid technology. *Waste Manage.* 32 (2012) 979–990.

Prola, L.D.T., Machado, F.M., Bergmann, C.P., de Souza, F.E., Gally, C.R., Lima, E.C., Adebayo, M.A., Dias, S.L.P., & Calvete, T. Adsorption of direct blue 53 dye from aqueous solutions bymulti-walled carbon nanotubes and activated carbon. *J. Environ. Manage.* 130 (2013) 166–175.

Soler, Aurora, Juan A. Conesa, María E. Iñiguez, & Nuria Ortuño. Pollutant formation in the pyrolysis and combustion of materials combining biomass and e-waste. *Sci. Total Environ.* 622 (2018) 1258–1264.

Weidenhamer, J.D., Newman, B.E., & Clever, A. Assessment of leaching potential of highly leaded jewelry. *J. Hazard. Mater.* 177 (2010) 1150–1152.

Yang, Shiliang, Ruihan Dong, Yanxiang Du, Shuai Wang, & Hua Wang. Numerical study of the biomass pyrolysis process in a spouted bed reactor through computational fluid dynamics. *Energy.* 214 (2021) 118839.

Zhang, Shangzhong, Kunio Yoshikawa, Hideki Nakagome, & Tohru Kamo. Kinetics of the steam gasification of a phenolic circuit board in the presence of carbonates. *Appl. Energy.* 101 (2013) 815–821.

Index